Deirdre Madden

FOOD AND NUTRITION

Gill and Macmillan

Published in Ireland by
Gill and Macmillan Ltd
Goldenbridge
Dublin 8
with associated companies throughout the world
© Deirdre Madden, 1980
© Artwork, Gill and Macmillan Ltd, 1980
7171 0882 1
Cover picture reproduced by courtesy of An Foras Talúntais
(The Agricultural Institute)
Print origination in Ireland by
Joe Healy Typesetting, Dublin

All rights reserved.
No part of this publication may be copied,
reproduced or transmitted in any form or by
any means, without permission of the publishers.
Photocopying any part of this book is illegal.

Photocopying
prohibited
by law

Contents

Handwritten note: + Electricity. kitchens. kitchen equipment. Nutrients (that go tog). Food labels.

1 **The Science of Nutrition** 1
   proteins
   carbohydrates
   lipids
   vitamins
   mineral elements
   water
   energy value of food

2 **The Diet** 33
   planning a balanced diet
   diets for:
      babies and children
      adolescents
      adults
      pregnant and nursing mothers
      old people
      invalids
   other diets
   meal planning
   entertaining
   budgeting

3 **Protein Foods** 51
   meat
   poultry
   fish
   eggs
   milk and milk products
   cheese

4 **Energy Foods** 82
   fats and oils
   cereals
   sugar
   vegetables
   fruits

5 **Food Preparation** 100
   basic cooking methods:
      by dry heat
      by fat or oil
      by moist heat
      pressure cooking
      fuel saving
   homemade soups and sauces
   home baking
   convenience foods
   contaminants and additives

6 **Microbiology and Food Spoilage** 119
   fungi
   yeasts
   bacteria
   viruses
   food spoilage
   infectious diseases
   food poisoning
   refuse disposal

7 **Food Preservation** 136
   causes of decay
   methods of preservation
   home preserving:
      jam making
      bottling
      freezing

8 **Human Physiology** 152
   cells and tissues
   the heart
   the blood
   the lymphatic system
   breathing
   the digestive system
   excretion
   the nervous system
   the endocrine system
   reproduction

Appendix A.   **Basic Science** 196

Appendix B.   **Cookery Terms** 204

Appendix C.   **Food Composition Tables** 210

Index 215

Eat more fibre-rich foods.
at leas sugar, fat, less alcoholic drinks.

1. The Science of Nutrition

An unbalanced diet: rice is the staple food of these Vietnamese children (W.H.O.)

A good diet is the most basic human need. All living things require food in order to maintain the normal processes of living. Without sufficient food, we would die.

The science of *nutrition* deals with the nature and composition of foods; the amounts required by the body and the physical and chemical changes brought about by the intake of food.

Food science embraces also such subjects as growth, preservation and manufacture of food, microbiology and the study of food additives.

Food is Eaten:
1. To promote growth and to make good the wear and tear on body cells.
2. To produce heat and energy.
3. To regulate the health of the body, maintain normal body processes and prevent disease.

Food can therefore be defined as any substance, solid or liquid, which performs one or more of these functions. A food which is considered *nutritious,* such as milk, will contain several nutrients while other foods may have only one nutrient present. A *balanced diet* is a habit of eating which supplies all the essential nutrients in the correct proportion for the needs of the body at a particular time — for example, a diet during pregnancy should contain a greater than normal proportion of iron and calcium. The normal diet should also contain sufficient roughage to stimulate the intestine and prevent constipation. *A varied diet containing plenty of raw fruit, vegetables and cereals together with some meat, fish, eggs and milk and/or cheese should provide the correct balance of nutrients.*

Nutrients are classified under the following headings:
proteins;   vitamins;
carbohydrates;   mineral elements;
lipids or fats;   water.

Oxygen is necessary for the utilisation of most nutrients.

Nearer home, many foods make up a typical meal (Sally and Richard Greenhill)

Malnutrition may be caused by eating too much, as well as eating too little (W.H.O.)

Malnutrition

This can be defined as long-term dietary imbalance. It is brought about when the total intake of one or more nutrients is out of proportion to the needs of the individual. This may occur when the body suffers a deficiency disease through *lack* of a nutrient, e.g. beri-beri, a disease brought about by an inadequate supply of thiamine (vitamin B_1).

In some cases malnutrition may be caused by eating *too much* of a nutrient, e.g. overindulgence in foods with a high energy value may cause obesity.

Malnutrition is often confused with undernourishment or starvation.

Undernutrition

This occurs when the intake of nutrients is less than that required by the body. In most cases the total energy intake is inadequate and the diet is deficient in many essential nutrients, resulting, in severe cases, in starvation.

In many parts of the world, millions of people die from starvation each year. A change in agricultural techniques would reduce this number considerably. Farmers in developing countries might be trained in the use of fertilisers, pesticides, disease resistant strains of crops. Irrigation and the intensive rearing of animals could also help to increase the food supply.

Diseases which frequently occur in underdeveloped countries as a result of undernutrition include:

marasmus – owing to a total shortage of nutrients causes wizened, shrunken appearance;
zerophthalmia – caused by lack of vitamin A; results in blindness;
kwashiorkor – occurs as a result of an inadequate supply of protein during infancy; the pot-bellied infants with spindle limbs are familiar to us all from photos of famine-stricken lands;
anaemia – due to a deficiency of iron.
goitre – caused by a shortage of iodine.

More than HALF the population of the world – that is 2,000,000,000 people – are hungry at this moment.

Undernutrition: drought has made this child's diet deficient in essential nutrients (W.H.O.)

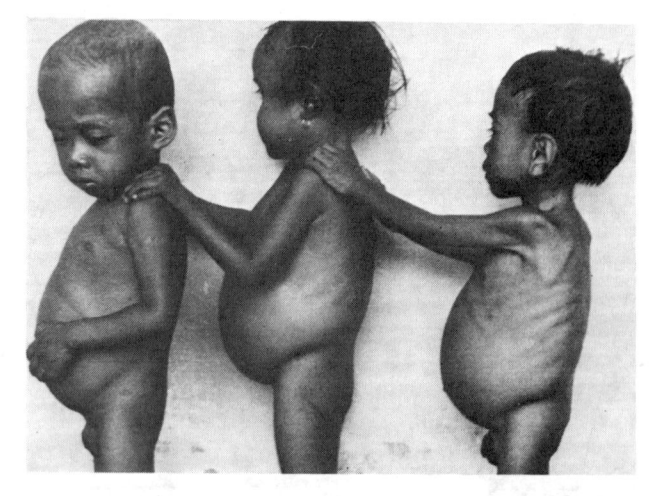

Indonesian three-year-olds suffering from marasmus. Only the first child is able to stand upright without assistance (W.H.O.)

Metabolism

This is the series of chemical reactions within the body which utilise the nutrients in food for building, repair or energy production. Metabolism includes:

Anabolism, the process whereby large molecules are formed from smaller ones. This occurs in the manufacture of cells and involves the *utilisation of energy.*

Catabolism is the breakdown of digested foods such as sugars into their simplest components, e.g. H_2O, CO_2 – *releasing energy.* Growth is the result of anabolism exceeding catabolism.

PROTEINS

Proteins are essential constituents of all cells, both plant and animal. The nucleus and protoplasm of every cell is largely composed of protein.

Protein is absolutely essential in the human diet because it is the only nutrient which supplies nitrogen – and this is vital for cell formation, and therefore, growth. The body is unable to synthesise nitrogen from the plentiful supplies which abound in nature – in the air and in the soil, for instance – so it must be provided with nitrogen in a suitable organic form.

Many plants have the ability to synthesise protein from the nitrates in the soil. We, in turn, obtain our protein from plants or from the animals which have eaten plants and manufactured from them proteins similar to those in the human body.

Elemental Composition

All proteins are composed of the elements carbon, hydrogen, oxygen and nitrogen. They may contain sulphur and phosphorus in small amounts.

Energy Value

1 gram of protein, when oxidised, produces 4 kilocalories or 17 kilojoules.

Chemical Composition

The protein molecule has a large and very complicated structure. It is made up of smaller units called *amino acids* and during digestion is broken down into these simpler substances.

Amino acids

1. These are sub-units of protein. They are relatively small molecules which are generally water-soluble and capable of diffusing through cell walls.
2. There are approximately twenty naturally occuring amino acids, and these can be arranged in any sequence to form proteins. Just as all the words in the English language are formed from twenty-six letters, so the number of proteins which can be made from the twenty or more amino acids is almost infinite.
3. All amino acids have the same basic structure:

$$R - \underset{\underset{NH_2}{|}}{\overset{\overset{H}{|}}{C}} - COOH$$

   a. an amino group NH_2 which is weakly alkaline;
   b. a carboxyl group COOH which is weakly acidic.
4. R varies between one amino acid and the next.

   Examples Glycine R = H
   Alanine R = CH_3
   Methionine R = CH_3-S-CH_2

5. Amino acids have the ability to link together to form protein chains. This occurs when the basic amino group of one amino acid reacts with the acidic carboxyl group of the next with the elimination of water. The link which occurs between the amino acid molecules is called a peptide link (CONH). The long chains which form are known as polypeptide chains.

Formation of a Protein Chain

$$NH_2-\underset{\underset{H}{|}}{\overset{\overset{R}{|}}{C}}-\underset{OH}{\overset{\overset{O}{\parallel}}{C}} \quad \underset{H}{\overset{H}{N}}-\underset{\underset{H}{|}}{\overset{\overset{R}{|}}{C}}-\underset{OH}{\overset{\overset{O}{\parallel}}{C}}$$

condensation (water removed) ↓ ↑ hydrolysis (water added)

The reverse reaction, called *hydrolysis,* occurs when a molecule of water is added, causing the protein to break into its component amino acids. This happens during digestion.

A typical protein molecule consists of several adjacent chains which may be folded, coiled and crosslinked at intervals. Crosslinks can be H bonds, salt links, or most important of all, disulphide links which give great toughness and steady shape. Gluten in flour is a protein containing disulphide links. When ascorbic acid, an oxidising agent, is added to dough it strengthens the disulphide links, making the dough stronger and more elastic. Malt, which contains a reducing enzyme, softens the flour by breaking the links.

Some simple proteins, such as insulin (a pancreatic hormone), may contain as few as fifty amino acid units. Other complicated proteins such as haemoglobin may be made up of over five hundred amino acid units, giving a molecular mass of many hundreds of thousands.

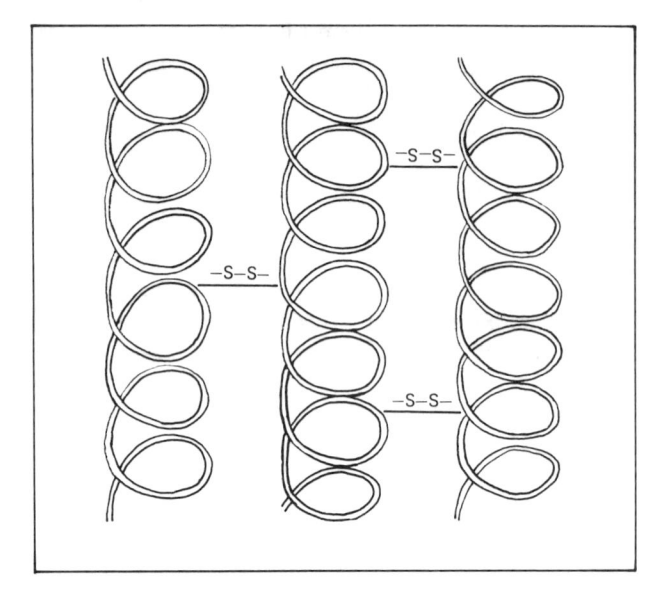

Fig. 1.1 Coiled protein with disulphide cross-links

Types of Protein Structure

a. Globular proteins. Many proteins consist of chains which are coiled into a compact spherical molecule. These are globular proteins. They are water-soluble and easily denatured (see below). *Examples:* albumen and globulin which occur in milk and egg white.

b. Fibrous proteins. A different type of protein is made of polypeptide chains which are twisted into spring-like coils and zig-zags. The spring-like structure gives an elastic property: this type of protein is found in *elastin* – the protein of arteries and tendons – and in *gluten* in flour. The zig-zag structure produces a tough, non-elastic protein such as *collagen* which is found in animal connective tissue.

c. Conjugated proteins. These are proteins which are chemically combined with a non-protein such as a pigment, e.g. pigment + protein = haemoglobin. They are classified according to the non-protein group. *Examples: phosphoproteins* contain phosphate groups such as casein in milk; *lipoproteins* contain lipid and protein; *nucleoproteins* combine nucleic acid and protein (e.g. deoxyribonucleic acid [DNA], the molecule which carries the hereditary factors of an organism).

Essential Amino Acids

Although the human body can manufacture some amino acids, there are eight which cannot be synthesised satisfactorily by the body. These are known as the essential amino acids. A further two are required by growing children, making ten in all.

Essential Amino Acids	*Non-Essential Amino Acids*
Valine	Glycine
Leucine	Alanine
Isoleucine	Norleucine
Phenylalanine	Tryosine
Threonine	Serine
Methionine	Cysteine
Tryptophan	Cystine
Lysine	Ornithine
Arginine } children	Aspartic Acid
Histidine }	Glutamic Acid

Denaturation

This can be described as the untwisting of protein molecules, causing loss of structure and function. It is generally irreversible. It is caused by

a. Heat. This is the most frequent cause of denaturation. The application of strong heat to most proteins, especially the albumen in eggs, causes the linking bonds to break. The protein loses its water binding power and there is considerable shrinkage. The protein coagulates as the protein molecules are packed more tightly together. This causes the skin to form on milk, the custard to curdle or the egg white to harden. If the heat is applied slowly and not raised above 60° (the average coagulation temperature) curdling should not occur.

b. Whipping or shaking a mixture also breaks cross linkages and denatures substances such as egg white.

c. Some chemicals, e.g. alcohol, acids or alkalis, may cause denaturation unless great care is taken to add them slowly and sparingly. *Example:* lemon juice added to milk. *Note:* Alcohol and vinegar have a preservative effect on food as they denature the autolytic enzymes which cause destruction of food.

Biological Value of Protein

This is simply a measure of how nourishing a protein is, that is, how much of the protein will eventually be made available to the body for protein synthesis.

Proteins which contain all the essential amino acids in the proportion required by the human body are said to have a 'high' biological value. Most of their nitrogen is

Important Proteins

Animal	Source	Vegetable	Source
myosin	lean meat	gluten	wheat
caseinogen	milk, cheese	†glycinin	soya bean
lactalbumin	milk	legumin	pulse vegetables
ovalbumin	eggs	oryzenin	rice
*gelatin	bones and connective tissue	†excelsin	nuts
haemoglobin	blood	tuberin	potatoes
vitellin	egg yolk		
collagen	connective tissue		

*Gelatin has no biological value; its main use is as setting agent.
†Glycinin and excelsin are two vegetable proteins with a relatively high number of amino acids.

available for growth and repair. Eggs have the highest biological value.

Biological value of some protein foods

Eggs	100%	Rice	67%
Milk	95%	Cereals (e.g. wheat)	53%
Meat/Fish	80%–90%	Maize	40%
Soya Bean	74%	Gelatine	0%

Proteins of a high biological value are also known as 'complete' proteins or first class proteins and are mostly from *animal* sources.

Proteins which are deficient in one or more of the essential amino acids are known as 'low' biological value proteins. These are also called incomplete or second class proteins, and are usually from *vegetable* sources.

Supplementary value. It is worth noting that it is the *combination* of essential amino acids which is important. Often it is possible to eat a meal which contains some foods deficient in one essential amino acid and others containing large amounts of this amino acid. The ability of one protein to make good the deficiency of another is known as its *supplementary value*. Examples of supplementary (or complementary) proteins are gelatine and bread. Bread is deficient in lysine whereas gelatine is relatively rich in lysine but deficient in tryptophan. Together they provide amino acids in the correct proportion. *Other examples:* fish and rice; maize and beans. A good mixture of vegetable proteins will provide all the essential amino acids required for protein synthesis. Vegans who eat no animal protein whatsoever can stay fit and healthy on this type of diet.

Classification of Protein

Protein can be classified according to its *structure* (globular proteins, fibrous proteins or conjugated proteins) or according to its *biological value* (complete [animal] proteins and incomplete [vegetable] proteins).

Sources of Protein

1st class or animal protein:

Meat	including poultry and processed meats
Fish	including shellfish and frozen and canned fish
Eggs	including preserved eggs
Milk	including skimmed, dried and condensed milk
Cheese	especially hard cheeses

2nd class or vegetable protein

Nuts	peanuts and peanut butter are a good source
Beans	both fresh and processed (e.g. baked beans, frozen beans, dried beans)
Peas	both fresh and processed

Whole cereals and bread
Potatoes and some other vegetables contain very small amounts.

Foods entirely deficient in protein: sugar, most oils, alcohol, tea, coffee.

Biological Functions

1. Essential for growth. It is particularly important in the diet of babies, children, expectant and nursing mothers.
2. Essential for repair of body cells.
3. It is a source of energy (see below).
4. Protein forms enzymes, hormones and antibodies.

Protein as a source of energy (deamination). When protein has been broken down by digestive enzymes into amino acids, most of it is utilised for rebuilding into body proteins. Proteins not used in this way are broken down by the liver into two parts: **a.** the nitrogen-containing amino group NH_2 which is converted by the liver into urea to be removed later by the kidneys, and **b.** the carboxyl group (COOH) which is oxidised to provide heat and energy. This process is called *'deamination'* that is, removing the amino group ($-NH_2$) from the amino acid molecule.

Protein Requirements

The recommended intake of protein each day is 1 gram for each kilogram of body weight; for example, if a person weighs 70 kg (11 stone) he will require 70 g protein each day. More is required by children, teenagers and pregnant and breastfeeding women.

Average requirements: Child 30–50 g
Teenager 60–75 g
Adult 60–70 g
Pregnant woman 90 g

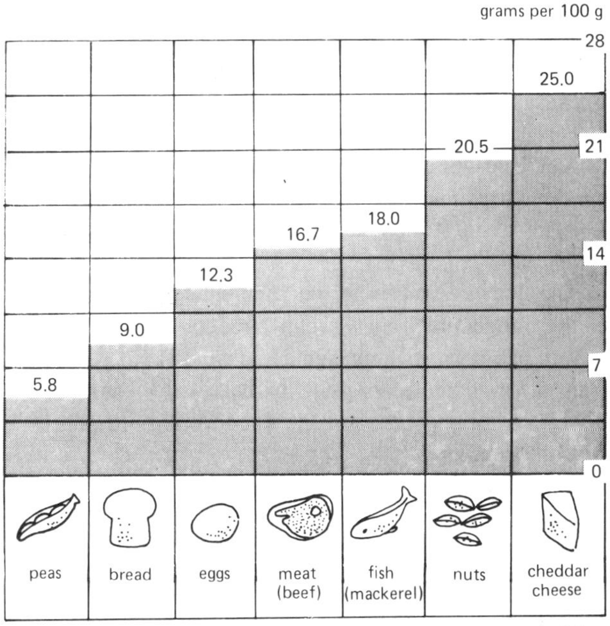

Approximate protein content (grams per 100 g):

peas	5.8
bread	9.0
eggs	12.3
meat (beef)	16.7
fish (mackerel)	18.0
nuts	20.5
cheddar cheese	25.0

Raw meat: protein content of meat increases considerably when cooked due to evaporation of water.

Fig. 1.2 Approximate protein content of some foods

5

Extra protein is also required during illness, convalescence and after surgery. Roughly *one sixth* of the daily intake of food should consist of protein, approximately half from animal and half from vegetable sources.

Effects of Deficiency
1. Degeneration of the body as worn-out cells are not replaced.
2. Retardation of growth in children.
3. Malfunction of various organs due to hormone and enzyme deficiency.
4. Lack of anti-bodies, which makes the body more susceptible to disease.
5. Possible obesity, due to overindulgence in high-calorie foods.
6. Severe deficiency (e.g. in underdeveloped countries) results in weight loss and disease.

Two protein deficiency diseases usually only found in Third World countries are marasmus, caused by a lack of protein and calories, and kwashiorkor, a protein deficiency disease which affects children. In Western Europe the diet of the majority of people contains sufficient protein, but this is not to say that malnutrition does not occur. Underprivileged families, pensioners, itinerants and others on low incomes exist, of necessity, on cheap carbohydrate foods. As meat and other protein foods become more expensive, a greater number of people find it difficult to maintain a nutritionally sound diet. Synthetic meat which has been developed from the soya bean may help to alleviate this problem (see p. 61).

Digestion of Protein
1. The food is masticated in the mouth and passes into the stomach.
2. Here it is reduced to a viscous liquid called chyme by the churning action of the stomach.
3. Pepsin, an enzyme in the gastric juice, reacts with protein foods, breaking them down into simpler substances called peptones.
4. After about four hours the food passes into the duodenum.
5. The enzyme trypsin in the pancreatic juice pours into the duodenum from the pancreas, converting peptones into peptides and amino acids.
6. Another enzyme, erepsin, in the small intestine converts these peptones and any unconverted proteins into amino acids.
7. These are absorbed into the blood stream, through the villi, and pass on via the portal vein to the liver.
8. The nitrogen from excess protein is converted into urea in the liver and remaining protein elements (COOH) are utilised for energy production.

Note: Vitamin B complex is required for protein metabolism.

The Effect of Heat on Protein
1. It coagulates and usually shrinks. It may harden or a skin may form, as in the case of milk.
2. Some amino acids are destroyed.
3. In meat, connective tissue changes to gelatine and extractives are released, making it more digestible.
4. If overcooked, most proteins become tough and indigestible.
5. Some proteins such as meat change colour.
6. It may become denatured, e.g. curdle (see p. 4).

Other Properties of Proteins
1. Most proteins *will not dissolve* in water. Exceptions are egg white, some fish protein and connective tissue.
2. *Maillard reaction:* owing to the effect of heat on amino acids and carbohydrates some foods brown, e.g. toast, fried potatoes.
3. Enzymes can cause *coagulation*, e.g. rennin causes milk to clot in the stomach.
4. *Elasticity:* some proteins such as those in meat muscle, cheese and bread dough have elastic properties because the molecule of these fibrous proteins resembles a coiled spring.

Tests for Protein
1. Millons Test. Put 2 cm of Millons reagent into a test tube. Add some beaten egg white. Boil over a bunsen burner. Result: A pink to dark red colour.

2. Biuret Test. Put some egg white solution into a test tube. Add a few drops of potassium hydroxide. Mix

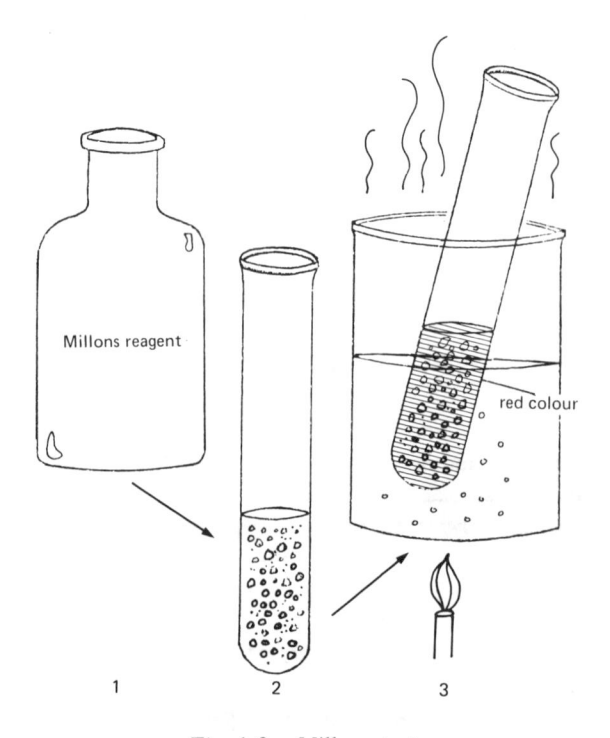

Fig. 1.3   Millons test

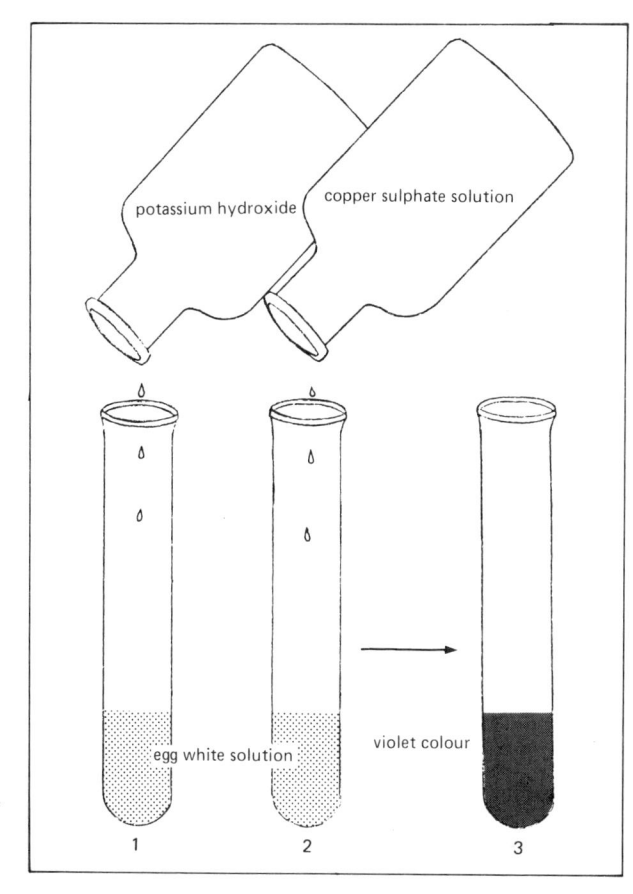

Fig. 1.4 Biuret test

well. Add a few drops of 1 per cent copper sulphate solution. Result: A violet blue colour. *Note:* If the result is rose pink or paler, it indicates the presence of simple proteins i.e. amino acids broken down by digestion.

3. Burn some protein food in a crucible. Note the characteristic smell of burning feathers.

4. To test the coagulation of protein, put a little undiluted egg white in a test tube. Stand the test tube in a beaker of cold water, which should also contain a thermometer. Gradually heat the beaker over a bunsen burner. Note the temperature at which it coagulates (60°C). When heated further, it will become tough (100°C).

5. To test the effect of pepsin on protein, put some water in a beaker and heat to approximately 40°C. Add a little pepsin and mix well. *Very carefully* add a few drops of concentrated hydrochloric acid using a glass rod. Put egg white in a test tube. Add some of this solution. Leave the test tube in beaker of warm water (40°C) for one hour. Test the solution using the Biuret test. Result: a rose pink colour indicating the presence of a simple protein, i.e. peptone or amino acid.

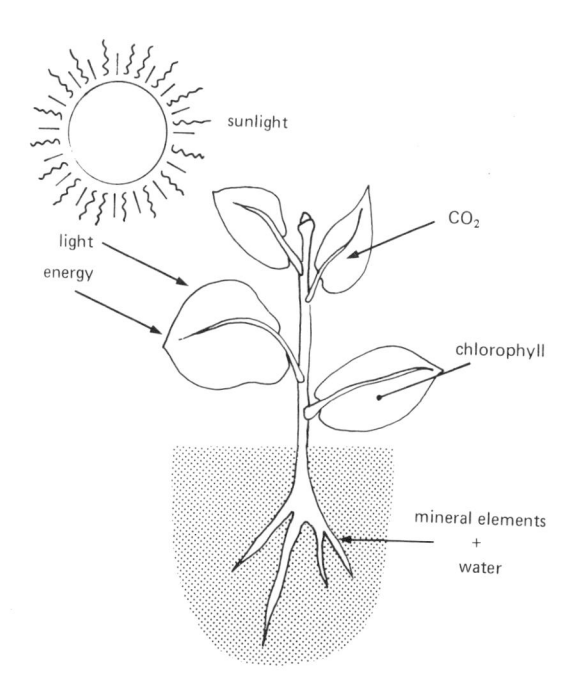

Fig. 1.5 Photosynthesis

CARBOHYDRATES

Carbohydrates are the cheapest and most abundant nutrients in the world. Their primary function is the production of heat and energy. Animals are unable to synthesise carbohydrates, but most plants by a process known as *photosynthesis* can manufacture carbohydrates from carbon dioxide and water. The energy required for this synthesis is obtained from sunlight, which is absorbed by chlorophyll, the pigment in green plants.

The process can be shown by this equation:

$$6CO_2 + 6H_2O \xrightarrow[\text{solar energy}]{\text{chlorophyll absorbs}} C_6H_{12}O_6 + 6O_2$$

Oxygen is liberated during the photosynthesis which, for obvious reasons, can only take place in daylight. Carbon dioxide is absorbed by the leaves from the atmosphere and water is taken in from the soil. The food produced is stored in the root system, stem or seeds. This stored carbohydrate can be oxidised to produce energy either for the plant itself or, when eaten, for the human body.

Elemental Composition

Carbohydrates, as their name implies, are compounds consisting of carbon, hydrogen and oxygen in the ratio 1 : 2 : 1. The hydrogen and oxygen are in the same proportion as in water (H_2O).

Energy Value

1 gram of carbohydrate (starch or sugar) when oxidised produces 3.75–4 kilocalories or 16–17 kilojoules.

Classification

Carbohydrates are classified according to their structure:

1. Monosaccharides ⎫
2. Disaccharides   ⎬ sugars
3. Polysaccharides starch, cellulose

Chemical Structure

Carbohydrates are based on a monosaccharide unit. The number of units making up a carbohydrate may vary from one to several thousands. The molecules may be chemically bonded in pairs, in groups or, as in the case of starch, in large complicated structures made up of thousands of sugar units.

1. Monosaccharides

The most important monoasaccharides have six carbon atoms in the molecule with a general formula $C_6H_{12}O_6$.

a. Glucose. Glucose (also called dextrose or grape sugar) is the unit of which the starch molecule is composed.

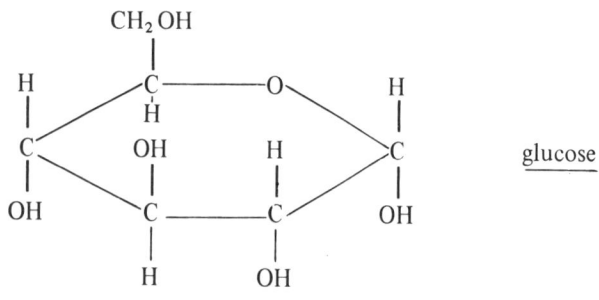

glucose

b. Fructose. Like glucose, fructose is composed of $C_6H_{12}O_6$, but differs owing to the arrangement of the atoms in the molecule. It is the sweetest sugar and is found in sweet fruits and honey.

c. Galactose is produced on hydrolysis of lactose – the milk sugar.

2. Disaccharides

Also known as double sugars, these are formed by the union of two monosaccharide molecules with the loss of a water molecule.

Sucrose is present in some fruits as well as sugar cane and sugar beet. Normal table sugar consists almost entirely of sucrose.

Other important disaccharides are:
Lactose – present in milk. It consists of one molecule of glucose and one molecule of galactose.
Maltose – produced by germinating barley. It is made up of two glucose units.

3. Polysaccharides

The prefix 'poly', meaning many, indicates the presence of many sugar units in the polysaccharide molecule. It occurs when many monosaccharides are bonded together, each with the loss of one H_2O molecule. It can be written as $(C_6H_{10}O_5)_n$ – the n varying according to the starch in question.

Like disaccharides, polysaccharides can be broken down by hydrolysis into smaller components, usually maltose and eventually into glucose. All polysaccharides have a chain-like structure similar to the one above.
Examples: starch, glycogen, pectin and cellulose.

Starch is a form of carbohydrate used by plants to store food. It consists of long chains, often branched, of linked glucose units. Starch is plentiful in cereals such as rice, wheat, oats, and is present in potatoes, pulse vegetables and some root vegetables. The microscopic structure of starch varies from one source to another. Potatoes, for example, consist of large granules whereas rice granules are much smaller. In all cases there is a thick covering on the outside of each granule, thought to be either cellulose or very closely packed starch cells. These make starch in-

Formation of a Disaccharide (sucrose)

fructose + glucose ⇌ (condensation / hydrolysis) sucrose + water
$C_6H_{12}O_6$ + $C_6H_{12}O_6$ → $C_{12}H_{22}O_{11}$ + H_2O

GLUCOSE–o–GLUCOSE–o–GLUCOSE–o–GLUCOSE–o–GLUCOSE
starch

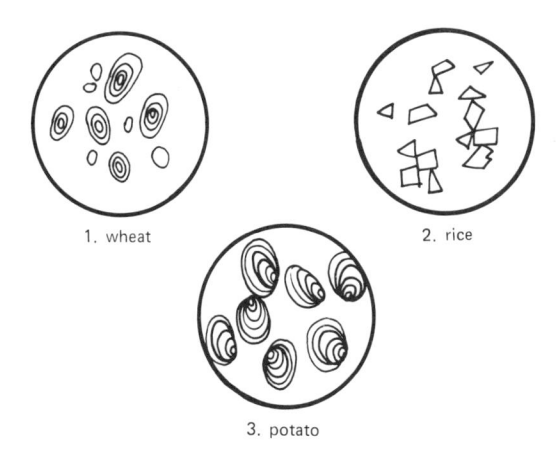

Fig. 1.6 Starch grains

soluble in cold water and difficult to digest. Cooking penetrates or breaks this outer covering, gelatinising the starch granules. Glucose can be obtained from starch by boiling with a strong acid such as hydrochloric acid. This is done commercially to produce powdered and liquid glucose.

Glycogen (animal starch). Just as plants store energy in the form of starch, animals have the ability to convert glucose into a substance similar to starch in order to store it as an energy reserve. This substance, called glycogen, is stored in the liver and is present to a lesser extent in muscle.

Dextrin. When starchy foods such as bread are heated, the large starch molecules nearest the heat break down into simpler, but still fairly large, molecules called dextrins. These brown more readily, resulting in the crusty top on bread or the colour of toast.

Pectin. This is a polysaccharide in plentiful supply in many fruits and some vegetables. Unlike other polysaccharides, its units are not made up of simple sugars but of very complicated molecules. Pectin has no nutritional significance but is important in the setting of jams (see p. 142).

Cellulose. This consists of complicated arrangements of glucose units which form the structural framework of the plant. It is present in all plant foods, generally forming the skin of vegetables and fruit and the outer bran layer of cereals. As there is no enzyme in the human digestive tract which can break down this molecule, it remains undigested. It does, however, have an important function in that it stimulates the muscles of the gut and helps to prevent constipation.

Sources of Carbohydrate
Sugar. Sugar of all types: syrup; treacle; cakes; sweets; biscuits; jam; soft drinks; ice cream. More nourishing sources are milk, fruit, honey and vegetables such as onions and sweetcorn.

Starch. Wheat; flour; bread and cakes made from wheat flour; all other cereals and their products including rice, maize, pasta, potatoes and other root vegetables; breakfast cereals.

Pectin. Unripe fruit.

Cellulose. Fruit, vegetables, cereals, especially in the outer skins and husks.

Biological Functions of Carbohydrate
1. Carbohydrates are oxidised in the body to provide heat and energy. A constant supply is necessary for the functioning of the muscles.
2. Excess carbohydrate is converted into fat, most of which is stored as adipose tissue beneath the skin. This has the advantage of reducing heat loss, but too much causes obesity.
3. Cellulose stimulates the peristaltic movements of the intestine.

Carbohydrate Requirements
About two-thirds of the average diet consists of a mixture of carbohydrates. In poor countries over 80 per cent of the diet may be composed of carbohydrate foods, whereas in wealthy industrial countries carbohydrates form less than 50 per cent of the diet.

Effects of carbohydrate deficiency. Except in cases of actual starvation, it is rare to find insufficient carbohydrate in the diet, as it is the cheapest food available.

Over-processing of carbohydrate foods. Many carbohydrate foods have become over-processed owing to a demand for this type of food: white bread and white sugar are typical examples. The more nourishing parts of the food are removed in processing, resulting in the tasteless, soft foods to which we have become accustomed and which, unfortunately, contain no roughage. In western countries there are increasing incidences of diverticular disease (a bowel disorder) and cancer of the bowel, whereas in more primitive countries where processed foods are seldom eaten these diseases are rare.

Digestion of Carbohydrates
1. In the mouth cooked starch is changed to maltose by the action of ptyalin, an enzyme present in saliva. It passes down the oesophagus.
2. The hydrochloric acid in the stomach neutralises the alkaline ptyalin and no further carbohydrate digestion takes place.

3. When the food reaches the small intestine, amylase, an enzyme present in the pancreatic juice, converts starch to maltose.
4. Also in the small intestine, three enzymes manufactured by the intestinal glands complete carbohydrate digestion:
   a. Maltase acts on maltose, converting it to glucose.
   b. Lactase acts on lactose, converting it to glucose.
   c. Invertase or sucrase acts on sucrose, converting it to glucose.
5. The glucose passes through the villi into the bloodstream and travels via the portal vein to the liver where it
   a. may be released for the production of heat and energy;
   b. may be stored as glycogen until needed for energy production and
   c. the remainder, if not oxidised or stored in the liver, is converted into fat and stored as adipose tissue.

Note: Starch is indigestible and usually needs to be cooked before eating. All forms of sugar are readily digested.

The Effects of Heat on Carbohydrates
1. Starch
   a. Moist heat makes the starch grains swell and as the temperature rises the cell walls break, releasing the starch cells which thicken the liquid. *Example:* white sauce.
   b. Moist heat causes the cell walls of vegetables to soften and become more digestible. If overcooked, vegetables will usually disintegrate. *Example:* boiled potatoes.
   c. Dry heat causes grains to burst. *Example:* popcorn. This happens to a lesser extent when baking bread and pastry.
   d. Dry heat causes the starch grains on the outside to darken and change to dextrin. *Example:* the crust forming on bread.
   e. Overheating causes starch to carbonise or burn.

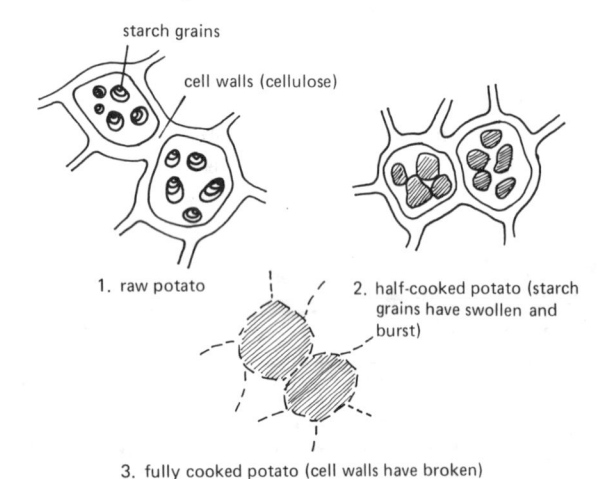

1. raw potato
2. half-cooked potato (starch grains have swollen and burst)
3. fully cooked potato (cell walls have broken)

Fig. 1.7 Effect of cooking on a potato

2. Sugar
   a. Moist heat causes sugar to dissolve.
   b. Further heating will, in turn, change it to syrup, darken it, cause it to caramelise and eventually at 160°C to burn or carbonise.
   c. When exposed to dry heat sugar will quickly melt, brown and then burn.

Other Properties of Carbohydrates
1. Starch
   a. Starch is insoluble in cold water but will dissolve in and thicken hot water. Cellulose is particularly insoluble.
   b. When mixed with water, starch forms a sticky paste. Starchy foods need to be carefully blended with liquid before cooking and stirred during cooking as the starch grains have a tendency to stick together and form lumps.
   c. Uncooked starch is indigestible as the digestive enzymes are unable to penetrate the outer cell walls. Cellulose is particularly indigestible in humans although most animals can digest it.
   d. It is hygroscopic, that is it absorbs water or water vapour. Starch is often mixed with the breadsoda/cream of tartar mixture in baking powder to absorb any moisture in the air so that the acid and alkali will not react prematurely upon each other.

2. Sugar
   a. All sugars are sweet crystalline solids. Fructose is the sweetest and lactose the least sweet.
   b. All sugars are soluble in water, more easily in warm water than in cold.
   c. If too much sugar is added to water it becomes oversaturated and deposits crystals of sugar at the edge of the container. *Example:* in jam making, if too much sugar is used.
   d. Reducing sugars. Most sugars are capable of removing oxygen from another substance; this is called the power of 'reduction'. *Example:* see Fehlings Test at end of chapter. Reducing sugars: glucose, fructose, maltose. Non-reducing carbohydrates: sucrose, starches. These should not be used for the Fehlings test.
   e. Hydrolysis. This is the ability of a compound to react with water. Sugars are capable of hydrolysis, particularly when an acid is added to speed up the reaction. During hydrolysis one water molecule combines with the carbohydrate molecule. Through this process complex sugars and starches can be converted to monosaccharides. This reaction takes place during digestion of carbohydrates and takes place more readily if heat is applied. It also takes place during the manufacture of sweets and in jam making, where the sucrose is converted into 'invert sugar'

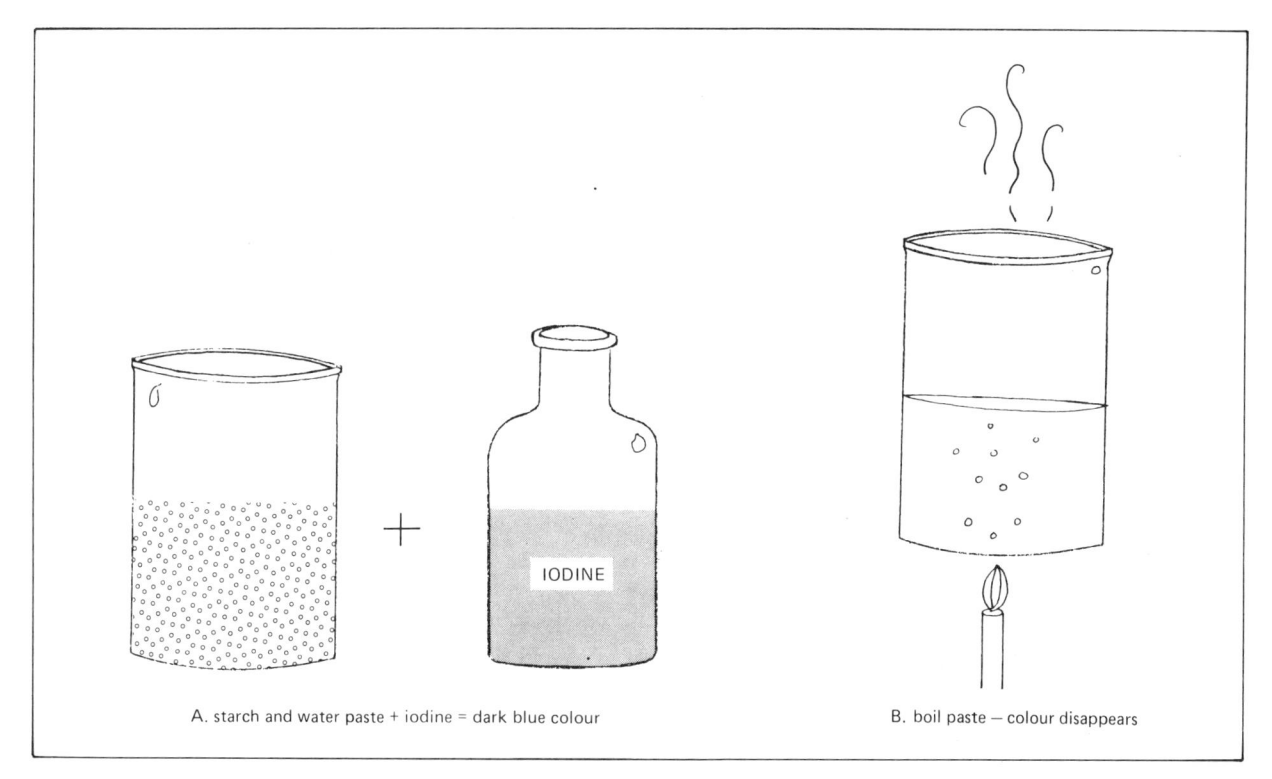

Fig. 1.8 Tests for starch

A. starch and water paste + iodine = dark blue colour

B. boil paste – colour disappears

i.e. a mixture of glucose and fructose, both of which are reducing sugars and are particularly sweet.

f. *Pectin.* This is capable of forming a gel by absorbing water into its molecules. The addition of acid (pH 3.3) is essential for a good result. Pectin will not gel successfully below this point.

Manufacture of Sugar – see p. 92.

Tests for Carbohydrates

1. Presence of starch. Mix some starch and water to a paste in a clean beaker. Add dilute iodine solution. The result, if starch is present, will be a blue inky colour. Bring to the boil and the blue colour will disappear owing to hydrolysis. Test this result with Fehlings solution (see below). The result is a brick-red precipitate proving that the starch has changed to sugar.

2. Solubility of starch. Put some water in a beaker, sprinkle on powdered starch and stir. The starch will settle at the bottom, proving that starch is insoluble in cold water. Now heat the mixture, stirring all the time. The water is taken up by the starch grains and at about 60°C they begin to swell and thicken the liquid. By 85°C a colloidal dispersion (not a true solution) has been formed and by the time boiling point is reached (100°C) the starch grains have burst, the liquid becomes transparent and the mixture gels. Repeat this test using glucose. The glucose readily dissolves especially if the water is heated slightly and upon boiling forms a syrup.

3. Digestion of starch – see p. 171.

4. Fehlings test. This is to establish the presence of glucose. Mix together equal quantities of Fehlings solution A and B. This mixture has a clear blue colour. Make a solution of glucose in a test tube, add a little Fehlings solution and stand in a beaker of hot water. Heat further. The result is a change of colour to a brick red precipitate, proving that the glucose has 'reduced' the Fehlings solution. This experiment will not work with sucrose or starch but other simple sugars can be used. What has happened? Fehlings A and B make a solution of cupric oxide which is readily changed by 'reducing agents' to form cuprous oxide – a cloudy, brick-coloured liquid. One molecule of oxygen is removed from the cupric oxide molecules.

$$\begin{matrix} CuO \\ CuO \end{matrix} + \text{reducing agent} = \begin{matrix} Cu \\ Cu \end{matrix} \!\!> \!\! O + O \text{ (removed)}$$

LIPIDS

A lipid is the biochemical name for the group of greasy substances found in animals and many plants which are hydrophobic (water-hating) and, therefore, insoluble in water. However, they readily dissolve in organic liquids (solvents) such as ether or benzene. Waxes and mineral oils, although lipids, cannot be utilised by the body and are not classified as nutrients.

Lipids are the most concentrated fuel food available to man. They have a high energy value and because they are

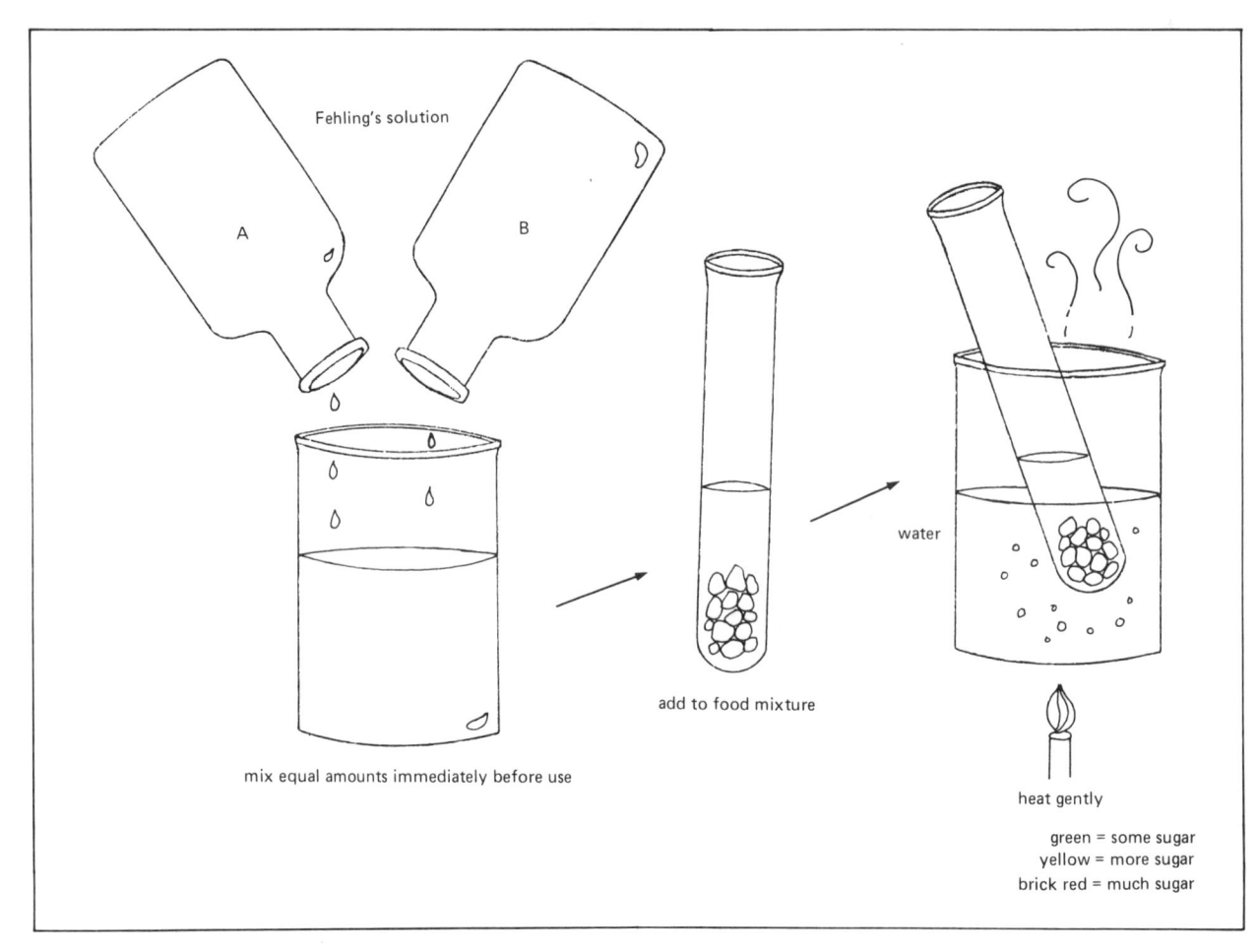

Fig. 1.9 Test for sugar

digested slowly, they delay the feeling of hunger for considerably longer than does a carbohydrate meal. This makes them filling and satisfying.

Lipids are called *oils* when they are liquid at room temperature and *fats* if they are solid at room temperature.

Elemental Composition

Lipids are composed of carbon, hydrogen and oxygen in a different proportion to that of carbohydrates. They contain less oxygen but more carbon and it is this factor which makes them such a powerful source of energy.

Energy Value

1 gram of lipid produces over 9 kilocalories (37 kilojoules) upon oxidation, compared with the average 4 kcalories (16 kjoules) released by carbohydrates or proteins.

Chemical Structure

Fats are formed when fatty acids (alkanoic acids) combine with glycerol (glycerine) with the elimination of water. This combination of alcohol (glycerol) with an organic acid, such as the fatty acids, produces a substance called an *ester*.

Glycerol is a trihydric alcohol, that is it contains three hydroxyl (OH) groups, each of which combines with a fatty acid molecule. Glycerol can be shown simply as an E-shaped structure.

$$\begin{array}{l} \text{—OH} \\ \text{—OH} \\ \text{—OH} \end{array} \quad \text{or more correctly} \quad \begin{array}{l} CH_2OH \\ CHOH \\ CH_2OH \end{array}$$

glycerol

Formation of a lipid

$$\begin{array}{l} \text{—OH} \\ \text{—OH} \\ \text{—OH} \end{array} + 3 \text{ fatty acids} \xrightarrow{\text{condensation}} \text{triglyceride} + 3H_2O$$

The reverse reaction (hydrolysis) occurs during digestion when the enzyme lipase splits lipids into fatty acids and glycerol.

triglyceride + water ⟶ glycerol + fatty acids

The physical properties of a particular lipid depend on the fatty acid joined to the glycerol, but the glycerol molecule always remains the same. In other words, the consistency (whether solid, waxy or liquid at room temperature), colour, texture and flavour of a lipid varies according to the proportion and type of fatty acid present.

Fatty Acids (Alkanoic Acids)

There are about twenty-five fatty acids distributed between plant and animal lipids.

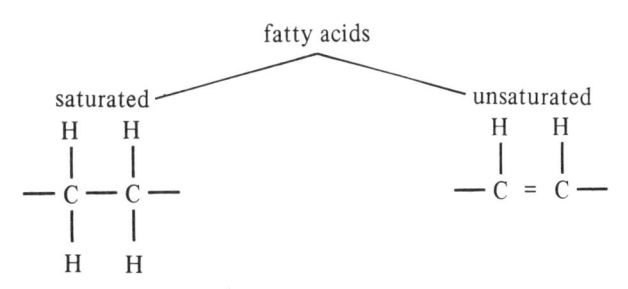

```
            fatty acids
           /          \
      saturated      unsaturated
       H  H            H  H
       |  |            |  |
     — C— C —        — C = C —
       |  |
       H  H
```

Saturated fatty acids have a general formula $R(CH_2)n$. Stearic acid, for example, has the formula

$$CH_3(CH_2)_{16}COOH$$

The complete molecule could be shown like this:

```
   O  H H H H H H H H H H H H H H H H H
   ‖  | | | | | | | | | | | | | | | | |
HO—C—C—C—C—C—C—C—C—C—C—C—C—C—C—C—C—C—H
      | | | | | | | | | | | | | | | | |
      H H H H H H H H H H H H H H H H H
```

carboxyl group   hydrocarbon chain   methyl group

Other saturated fatty acids have an identical structure except that the number of carbon atoms varies: butyric—4; myristic—14; palmitic—16.

Saturated fatty acids are plentiful in hard fats — both animal fats such as suet and lard, and those vegetable fats such as margarine which have been artificially hardened by hydrogenation (see p. 16).

Unsaturated fatty acids have a similar structure but, unlike saturated fatty acids, not all the carbon atoms in the chain are 'saturated' by having two hydrogen atoms linked to them. Instead they have one or more *double bonds*.

```
      H    H
      |    |
    — C  = C —
```

Proportion of Principal Fatty Acids in Some Lipids

Saturated Fatty acids	Cod Liver Oil	Corn Oil	Olive Oil	Egg Yolk	Lard	Mutton Fat	Butter	Coconut Oil
Myristic (C_{14})	6				3	5	12	17
Palmitic (C_{16})	8	13	16	32	24	25	29	9
Stearic (C_{18})	1	4	2	4	18	30	11	2
Unsaturated Fatty Acids								
Oleic (C_{18})	20	29	65	43	42	36	25	6
Linoleic (C_{18})	29	54	15	8	9	4	2	3
Linolenic (C_{18})	25							
Arachidonic (C_{20})	10							
% Unsaturated Fatty Acids	84	83	80	51	51	40	27	9*

*Coconut oil contains 48 per cent lauric acid — another saturated fatty acid. It has a high percentage of saturated fatty acids and a low level of unsaturated fatty acids — unusual in a plant lipid.

Monounsaturated fatty acids: oleic acid (one double bond)
Polyunsaturated fatty acids:
                linoleic acid (two double bonds)
                linolenic acid (three double bonds)
                arachidonic acid (four double bonds)

Unsaturated fatty acids tend to react with other molecules, their double bonds joining with *hydrogen* to form saturated fatty acids or with *oxygen,* causing rancidity (see p. 16).

Essential fatty acids. Polyunsaturated fatty acids are also called the essential fatty acids because while they cannot be synthesised by the body, they are required in small amounts. It is interesting that these essential fatty acids are at present being recommended in the diet for an entirely different reason – to counteract the hardening effect of cholesterol on the coronary arteries, thereby reducing the incidence of heart attacks. Although most fats and oils contain *both* saturated and unsaturated fatty acids, oils and soft fats, usually of marine or vegetable origin, contain a relatively large proportion of unsaturated fatty acids. Marine (fish) oils contain particularly large proportions of the highly unsaturated linolenic and arachidonic fatty acids.

Melting point. The higher the degree of saturation, the higher the melting point of the lipid: hard fats have a high melting point whereas unsaturated oils have a low melting point.

Other Lipid Compounds

Phospholipids. These are composed of one unit of glycerol to which two fatty acid units are attached. The remaining link is taken up by a phosphate radical. Phospholipids are connected with the absorption and utilisation of digested fat in the body. They are useful emulsifiers; the best known is lecithin in egg yolk.

Steroids or sterols. These are basically composed of 4 carbon rings.

They have a high molecular weight and form hard fats. The best known is cholesterol, which causes hardening and blockage of the coronary arteries. It also causes gallstones. Other substances such as vitamin D (cholcalciferol) are derived from a sterol. The sex hormones progesterone and testosterone are sterols which have no connection with the diet.

Sources of Lipids

Both animals and plants tend to store lipids as an energy reserve. A wide range of foods contains fats and oils. For further details see pp. 82–6.

Animal sources
Meat, suet, lard dripping
Oily fish and fish liver oils
Butter, cream, cheese, milk, egg yolk

Plant sources
Nuts and nut oils (e.g. groundnut oil, coconut oil)
Vegetables (e.g. avocados) and vegetable oils (e.g. olive oil)
Margarine and cooking fats (e.g. 'Cookeen')
Whole cereals (e.g. oatmeal), wheat germ

Foods deficient in fat. Most fruit and vegetables; white fish; white flour; sugar.

Biological Functions of Fats

1. Lipids supply large amounts of heat and energy.
2. Fat is stored as a fuel reserve in the adipose tissue beneath the skin (subcutaneous layer). This helps to insulate and protect the body.
3. Fat is stored around certain delicate organs such as the nerves and kidneys to protect them from damage.
4. Lipids act as a vehicle for the fat-soluble vitamins A, D, E and K.
5. As the body cannot synthesise certain fatty acids, these essential fatty acids are considered necessary for general health.
6. Lipids slow down the action of the stomach, delaying the feeling of hunger.

Note: The use of lipids is an essential part of many cooking processes, introducing variety and flavour into the diet.

Lipid Requirements

It is accepted that approximately 20 per cent of our intake of energy foods should be in the form of lipids, at least half of this coming from polyunsaturated fats. People with a high energy output may require a larger proportion and during cold weather or in a cold climate, a greater than average consumption of lipids is normal.

Effects of deficiency

1. Fat deficiency alone has no particular symptoms, but if the total intake of energy foods is too little, there will be weight loss progressing eventually to starvation.
2. A deficiency of essential fatty acids is unlikely as they are widely distributed in food. If it does occur it may

cause general debility and a build-up of cholesterol in the blood.
3. Symptoms of fat-soluble vitamin deficiency may occur when there is a restricted intake of fat.

Dietary problems
a. Overindulgence in fatty foods or foods cooked in fat may lead to obesity.
b. Animal fats are often difficult to digest; exceptions are milk and eggs where the fat is dispersed in a fine emulsion.
c. Saturated fats are thought to make cholesterol accumulate in the coronary arteries, causing arteriosclerosis (see p. 40).

Digestion of Lipids
1. Lipids pass through the mouth and oesophagus chemically unchanged.
2. In the stomach, fats are melted and the food is physically broken down to a liquid called chyme.
3. In the small intestine lipids are emulsified by the combined action of bile from the liver and lipase, the lipid-splitting enzyme present in the pancreatic juice. These break down lipids into their components – fatty acids and glycerol.
4. In this form they pass through the intestinal wall into the lacteals, are converted back into fat and are conveyed by the lymph system to the general circulation of the blood at the neck.
5. Most fat is oxidised to supply heat and energy. The remainder is stored as adipose tissue.

The Effects of Heat on Lipids
Smoke point. Solid fats melt when heated. Normal frying temperatures vary between $175°C$ and $195°C$. If the fat is heated above these temperatures, a blue haze is given off, indicating that the fat is starting to decompose and the glycerol is separating from the fatty acids. This is called the *smoke point*. Fats reach smoke point at a lower temperature than oils. Impurities also lower the smoke point so that fat or oils which have been used many times without straining will have a lower than normal smoke point. Decomposition occurs between $200°C$ (some solid fats) and $250°C$ (cooking oils). Frequent overheating causes the partial hydrogenation of cooking oils: some unsaturated fatty acids link up with hydrogen, thus thickening the oil. Overheating also causes dehydration when the glycerol further decomposes into acrolein, a substance with a characteristic burnt fat smell.
Example
$$CH_2OHCHOHCH_2OH \xrightarrow{heat} CH_2CHCHO + 2H_2O$$
glycerol   acrolein + water

Flash point. When oils and fats reach extremely high temperatures, a vapour rises from the fat, which is highly inflammable. When the temperature is raised so high that the vapour spontaneously ignites, the *flash point* has been reached. Flash point of oil is $320°C-330°C$; that of fat $10°-15°$ lower.

Additives are frequently used in cooking fats and oils in order to lower the smoking temperature and thus prevent food from burning. Antispattering agents such as the emulsifier lecithin may also be used.

Other Properties of Lipids
1. Solubility. Lipids are insoluble in water but will dissolve in organic solvents such as ether and benzene.
2. Emulsions. Although lipids are insoluble in water they have the ability to form a colloidal solution, which is called an *emulsion*. This is formed when two liquids which are insoluble in each other are forced to disperse one within the other. When oil is added to water it does not dissolve in it but forms a distinct layer on top of the water. If the two liquids are shaken vigorously, a *temporary emulsion* is formed with tiny droplets of oil dispersed in the water. This emulsion will revert back to two separate layers before long because it is unstable.

When the oil is dispersed in the water or an aqueous medium such as vinegar, an *oil in water – o/w* emulsion is formed. If the greater proportion of the emulsion is a lipid such as butter or margarine, a *water in oil – w/o* emulsion is formed. In either case the disperse phase (the liquid which is dispersed) forms minute droplets in the continuous phase (the larger amount of liquid).

The reason that oil will not dissolve in water is because of the surface tension of the water. If *emulsifiers* are added they reduce the surface tension and enable a permanent emulsion to be formed. This is how they work. The molecules of emulsifiers contain a hydrophilic (water-loving) and a hydrophobic (water-hating) group. The hydrophilic group attaches itself to the water molecules and the hydrophobic group to the lipid molecules. At the point where the oil and water touch, called the oil/water interface, the emulsifier holds them together.

Emulsifiers occuring naturally in food include the phospholipid lecithin in egg yolk and the proteins gluten in flour and casein in milk. Starch also has emulsifying properties and this factor is utilised for thickening sauces. Many emulsifiers are used commercially in food. *Examples:* lecithin; gums; alginates; pectins and the extensively used glyceryl monosterate (GMS).

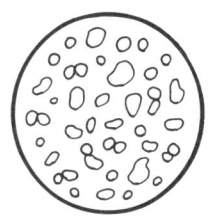

Fig. 1.10 Microscopic appearance of fat globules in an emulsion

Examples of food emulsions
Milk and cream. o/w emulsion, emulsified by caseinogen
Butter. w/o emulsion, emulsified by caseinogen
Mayonnaise. o/w emulsion, emulsified by lecithin or GMS
Margarine. w/o emulsion, emulsified by caseinogen, lecithin and GMS

Stabilisers are substances such as pectin, starch and gelatin which help to *maintain* an emulsion once it is formed. They do this by increasing its viscosity so that the droplets within the emulsion are unable to come in contact with one another and coalesce — that is, revert back to their separate state. Many emulsifiers have a stabilising effect on the emulsion also.

3. Plasticity. So far hard fats and oils have been discussed. Many lipids such as butter and margarine have a definite shape and structure yet they are soft, spreadable and respond to pressure and friction (e.g. creaming or rubbing in with the fingers). These fats do not melt at a fixed temperature but over a range of temperatures, because they contain a mixture of triglycerides, some of which are solid at room temperature and others of which are liquid at the same temperature. The solid lipids form a lattice structure of minute crystals which are surrounded by liquid triglycerides. This enables the crystals to move about slightly, giving them a plastic or pliable texture. As the fats are heated, the solid structure gradually breaks down, the fat becomes softer and more plastic until eventually all the triglycerides have melted and the fat flows. These properties have great importance in cake and pastry making, enabling a single fat to have good creaming, spreading and shortening qualities.

4. Hydrolysis. This occurs when triglycerides react with water to form glycerol and fatty acids (see p. 12). This process is used commercially for rendering down animal fats by heating them under pressure.

5. Saponification. When a fat molecule is hydrolysed with an alkali, salts of the fatty acid (soaps) and glycerol are produced. This is known as saponification or soapmaking. The reaction is hastened by boiling.

6. Rancidity. Oils and fats do not keep well. Spoilage of a lipid is known as rancidity and produces unpleasant odours and flavours. Rancidity may occur in two ways:
a. *Hydrolytic* rancidity occurs when the moisture present in some fats causes hydrolysis of the triglyceride molecules which makes the glycerol separate from the fatty acids. This type of rancidity occurs more rapidly in the presence of certain enzymes and micro-organisms.
b. *Oxidative* rancidity is the most usual type of lipid spoilage. It occurs when the unsaturated fatty acids in the triglyceride molecule react with oxygen rather than hydrogen.

$$-\overset{H}{\underset{}{C}} = \overset{H}{\underset{}{C}}- + O_2 = -\overset{H}{\underset{O}{C}} - \overset{H}{\underset{O}{C}}-$$

unsaturated oil + oxygen = oxyns

The resulting oxyns produce the characteristic unpleasant taste and smell of stale fats. As heat and light speed up this reaction, storage of fats in a cool, dark place such as a refrigerator is recommended. Anti-oxidants prevent oxidative rancidity as they react with any available oxygen in the food during storage. Vegetable oils are the lipids which are most resistant to spoilage, followed by animal fats such as butter and suet. Marine oils decay particularly quickly.

Off flavours may develop owing to the tendency of fats and oils to absorb flavours and odours — for example, butter stored unwrapped in a refrigerator would absorb the flavour of fish or onions stored nearby.

7. Hydrogenation. This is a chemical process whereby unsaturated oils are converted into solid fats by forcing hydrogen gas through them. The unsaturated carbon atoms in the fatty acid molecules take up the hydrogen, becoming 'saturated' in the process.

$$-\overset{H}{\underset{}{C}} = \overset{H}{\underset{}{C}}- + H_2 \longrightarrow -\overset{H}{\underset{H}{C}} - \overset{H}{\underset{H}{C}}-$$

unsaturated molecule + hydrogen → saturated molecule

The reaction is speeded up by the use of nickel which acts as a catalyst. This process is used in the manufacture of margarines (see p. 84). Oils can be fully or partly hydrogenated, depending on the consistency of fat required.

Refining
Oils and fats in their raw state are often strong smelling, with an unpleasant taste and appearance. Before use they must be extracted and refined. Vegetable oils are refined by:
a. neutralising with caustic soda to remove acid impurities;
b. bleaching or decolourising with Fullers earth or charcoal;
c. filtering to remove solid impurities;
d. deodorising by steam distillation to remove strong smells.

Tests for Lipids
1. Put a small amount of grated cheese into a test tube. Add ether and warm it slightly. Pour a little on to a fil-

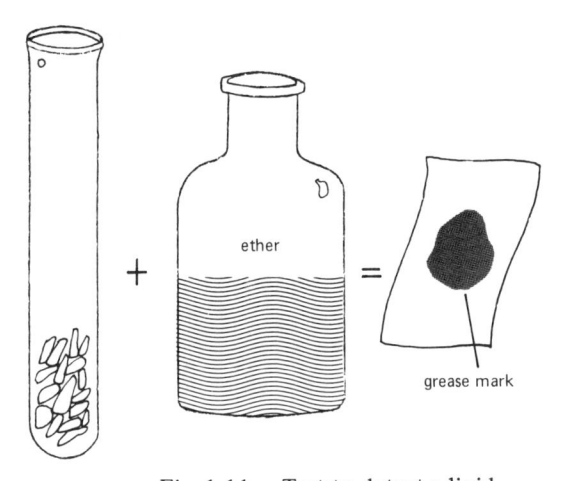

Fig. 1.11   Test to detect a lipid

ter or blotting paper. Result: a grease mark on the paper.

2. *To distinguish saturated/unsaturated lipids* put 15 g chopped suet into a test tube. Add 10 ml ether. Put 4 drops of olive oil in another test tube. Add 10 ml ether. Make up a 1 per cent solution of iodine in alcohol. Start adding the iodine drop by drop, counting the number of drops needed to obtain a good yellow colour. The more unsaturated the fat, the more iodine is needed.

Note: Never use a naked flame when using ether.

3. *Simple test to detect a lipid in food.* Place food on a piece of brown blotting paper for about 30 minutes. If there is a lipid present it will leave a grease mark.

VITAMINS

Vitamins are a relatively new aspect of nutrition. They were discovered in the early part of the twentieth century, although some were not isolated until the 1930s and 1940s. As they are the subject of constant research, new discoveries are being made about them all the time. Vitamins were originally called 'accessory food factors'. In the study of nutrition they are usually identified by a chemical name as well as a letter.

Definition

Vitamins are complex organic substances which are usually obtained by the body from food. They do not produce energy and therefore have no calorific value. Apart from small amounts of vitamins B and K, most vitamins cannot be synthesised by the body. Each vitamin is completely different from any other, has specific functions and cannot take the place of any other.

Vitamins are required by the body in very small amounts but if these are not included in the diet various deficiency diseases will occur. A well-balanced diet usually supplies the required amount of all vitamins and, thanks to improved general knowledge of nutrition, it is unusual to find deficiency diseases in the western world.

It is rarely necessary to supplement the diet with vitamin tonics and pills unless these are prescribed by a doctor. There is no proof that extra vitamins improve health — in fact, where vitamins A and D are concerned, the opposite could be the case.

Note: As vitamins are required in such minute amounts, they are usually measured in milligrammes (mg) — 1/1,000 of a gram — or in microgrammes (µg) — 1/1,000,000 of a gram.

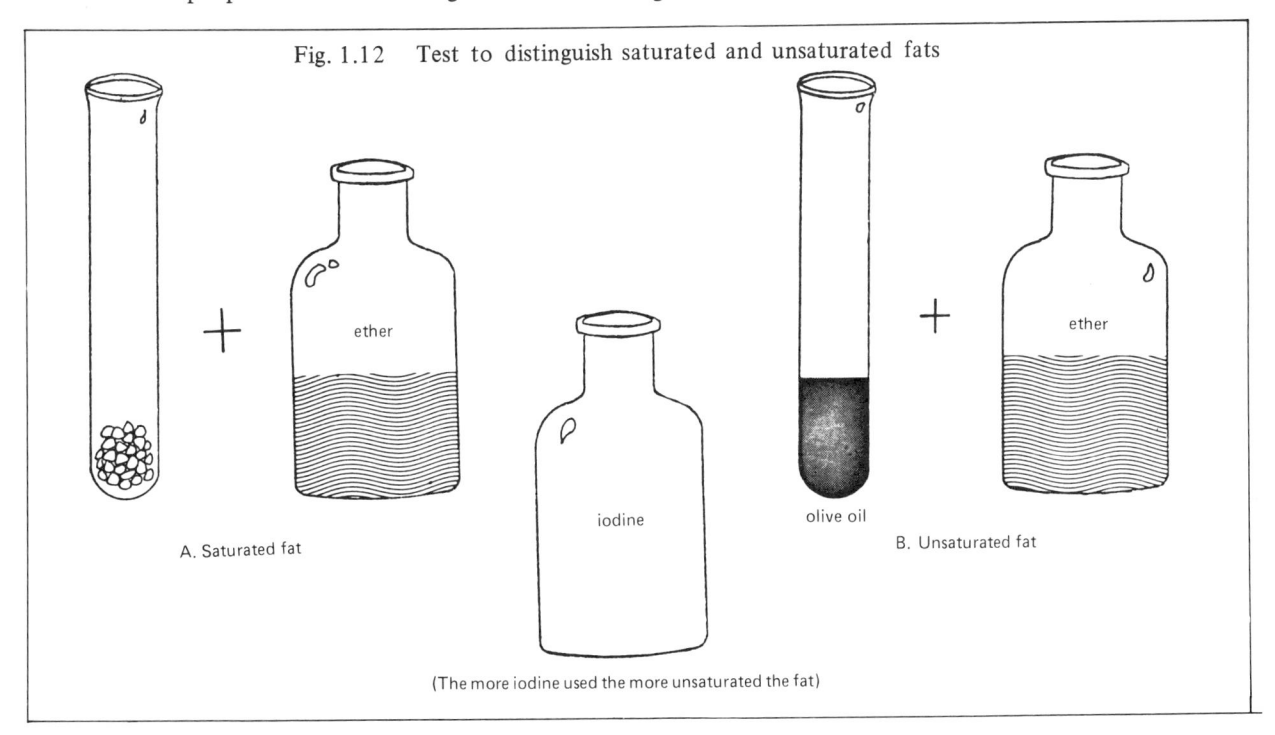

Fig. 1.12   Test to distinguish saturated and unsaturated fats

A. Saturated fat

B. Unsaturated fat

(The more iodine used the more unsaturated the fat)

Classification

Vitamins are classified simply as fat-soluble and water-soluble. This classification can be misleading, however, as there are exceptions among the synthetic vitamins.

Fat-soluble vitamins   A, D, E, K
Water-soluble vitamins   B group, C

Digestion and Absorption

As all vitamins are absorbed in their natural state there is no need for them to undergo any digestive process. Some vitamin B and C is absorbed into the bloodstream through the walls of the stomach. Fat-soluble vitamins A and D and any remaining vitamins B and C are absorbed in the small intestine and pass through the portal vein to the liver where they are stored or pass into the bloodstream for utilisation. Vitamins A and D can be stored for many months and for this reason may become toxic if taken in great amounts. Conversely, water-soluble vitamins B group and C cannot be stored and need to be taken regularly, ideally every day.

Fat-Soluble Vitamins

VITAMIN A (Retinol)

This is a yellow, fat-soluble alcohol. It is sometimes known as the antixeropthalmic vitamin, as deficiency may cause xeropthalmia, a disease which makes the lining membranes of the body — e.g. the nose, throat and eye — dry up, resulting eventually in blindness.

A secondary source of vitamin A is the substance carotene ($C_{40}H_{50}$). This is sometimes called the precursor of vitamin A, or provitamin A. The main type of carotene is known as beta-carotene and is a yellow pigment, very similar in structure to vitamin A. It is converted to vitamin A in the small intestine. Most carotene comes from vegetable sources.

Note: Only one third of the carotene we eat is converted into vitamin A in the body. It is not as easily absorbed as retinol — true vitamin A.

Functions of Vitamin A

1. It is necessary for healthy skin, glands and the moist lining membranes of the body, including the bronchial tubes and the cornea of the eye.
2. It regulates growth, especially that of children.
3. It is necessary for the general health of the eyes, but especially the ability of the eyes to adapt to dim light. Retinol contributes to the formation of the pigment rhodopsin in the retina of the eye, which is sensitive to dim light.

Effects of Deficiency

1. Dryness of skin and lining membranes, leading eventually to xeropthalmia. This disease is very common in the Third World where the diet consists mainly of cereals.
2. Retarded growth in children.
3. Night blindness or inability to see in dim light. This could be dangerous in the case of airline pilots or long-distance lorry drivers.
4. Lowered resistance to infection.

Hypervitaminosis A occurs when too much Vitamin A is taken. As vitamin A can be stored for long periods in the body, it is possible, for example, that if a diet containing adequate vitamin A is supplemented with cod liver oil *taken in excess* of the recommended dosage, poisoning could result. Symptoms of the disease are pains in the bones, nausea, loss of hair and enlargement of the liver. It is very rare.

Vitamin A Sources	Carotene Sources
halibut liver oil	carrots
cod liver oil	spinach
liver	watercress
butter	dried apricots
margarine	tomatoes
cheese	prunes
eggs	cabbage
herrings	peas
milk and cream in summer	lettuce

The foods at the top of the lists contain the largest amount of the vitamin. Most yellow and red fruit and vegetables are rich in carotene. The darker the colour, the more carotene.

Note: It must be remembered that the food with the largest amount of vitamin A is not necessarily the best source of it. For example, a person might be more likely to drink a large amount of milk regularly than to eat liver or cod liver oil regularly. Similarly watercress, a good source of vitamin A, is usually used only in small amounts as a garnish and even then it is often discarded, so it rarely constitutes a major source of the vitamin in our diet.

Recommended daily allowance. The World Health Organisation recommends

children 300–750 µg (rising according to age)
adults 750 µg
pregnant and nursing mothers 1200 µg

It is important that children and pregnant and nursing mothers take sufficient vitamin A to meet the requirements for growth.

Properties and Stability

This vitamin is heat stable so cooking and various forms of preservation by heat have little effect on it. There is some loss during dehydration, especially when old fashioned methods, such as sun drying, are used. As vitamin A is almost insoluble in water, moist cooking methods have little effect on it. There may be slight loss of carotene due to oxidation if a vegetable is grated or chopped. Air should be excluded where possible, so saucepans should be covered during cooking.

Test for Vitamin A

Dissolve some drops of cod liver oil in five times its volume of chloroform. Take a drop of this solution and add 2–3 drops of saturated antimony trichloride ($SbCl_3$) in chloroform. Result: a brilliant blue colour indicates the presence of vitamin A.

VITAMIN D (Calciferols)

Vitamin D can be obtained by the body in two ways: either directly from food or by the action of sunlight on the skin. It is a white crystalline steroid (fatty substance) and can occur as cholecalciferol or ergocalciferol. *Ergocalciferol* is formed by subjecting oils to radiation, and is used in the manufacture of synthetic vitamin D. *Cholecalciferol* is formed in the skin when the ultra-violet rays of the sun are absorbed by the provitamin present in the skin which is called 7 dehydroxy cholesterol.

Functions of Vitamin D
1. Vitamin D, together with calcium and phosphorus, is essential in the formation of bones and teeth. It is of vital importance during pregnancy and early childhood as it controls the calcification or laying down of calcium in children's bones.
2. It assists the absorption and distribution of calcium in the body.

Effects of Deficiency
1. Rickets, a disease which causes malformation of the bones in children. It is directly due to lack of vitamin D, which is therefore sometimes called the anti-rachitic vitamin. Rickets is rare in developed countries.
2. Osteomalacia, a disease similar to rickets which occurs in adults. The bones gradually lose calcium and the calcium is not replaced. It is sometimes found in elderly people and in women after successive pregnancies.
3. Dental decay.

Note: In tropical countries vitamin D deficiency is rare because of constant sunshine. In industrial cities in Victorian and pre-war England, rickets was a common disease because of the foggy, polluted air, through which the sun could not penetrate. However, two laws, one forbidding the use of smoke-producing fuels in built-up areas and another providing free cod liver oil to pregnant women and young children, have virtually eliminated rickets from England.

Hypervitaminosis D. Commercial vitamin preparations taken in excess in conjunction with a diet plentiful in vitamin D may result in large deposits of calcium in the blood and heart. This is especially dangerous in elderly people, but is fortunately very rare.

Vitamin D Sources
1. *Sunlight* reacting on the skin is the main source.
2. *Dietary sources:*
   egg yolk
   margarine
   oily fish
   fish liver oils
   dairy produce in summer

Properties and Stability

One of the most stable of vitamins, vitamin D is not affected by heat, oxidation, water, acids or alkalis, so loss in cooking and preservation is negligible.

VITAMIN E (Tocopherols)

Although not always regarded as a true vitamin, vitamin E is known to be concerned with metabolism. It belongs to a group of oily substances called tocopherols.

Functions of Vitamin E
1. Thought to be essential for normal metabolism.
2. A very useful antioxidant as tocopherols themselves oxidise first, delaying normal oxidation and rancidity in food and in the body.
3. Although tests have associated this vitamin with the fertility of rats, it has not been proved to have an effect on human fertility.

Effects of Deficiency
None known.

Vitamin E Sources
wheat germ
soya beans
vegetable oils
liver
eggs
pulse vegetables
'Bemax'

It is frequently used as an antioxidant in commercial products such as cooking fats.

VITAMIN F (Essential Fatty Acids)

This is not a true vitamin but is mentioned here because

the essential fatty acids – linoleic, linolenic and arachidonic acids – are sometimes collectively known as vitamin F. They are thought to be essential for the skin and are found in margarine and vegetable oils, particularly sunflower seed oil.

VITAMIN K (Koagulations Vitamin)

This was discovered in Copenhagen and gets its initial from the Danish word for clotting, which is its main function. It is sometimes called the anti-haemorrhagic vitamin.

Functions of Vitamin K

It assists the formation of prothrombin, a protein essential for normal blood clotting.

Effect of Deficiency

Inability of blood to clot, which, after accidents and operations, could be very dangerous.

Vitamin K Sources

1. liver (especially pigs liver)   eggs
   green vegetables   milk
   fish   some fruit
2. It is manufactured by bacteria present in the intestine.

Recommended daily allowance. A good mixed diet supplies adequate amounts, but extra would be required after loss of blood, as in the case of haemorrhaging or surgery.

Water Soluble Vitamins

VITAMIN C (Ascorbic Acid)

Vitamin C is often called the anti-scorbutic (anti-scurvy) vitamin. It was known for hundreds of years that lack of fresh fruit and vegetables caused a disease called scurvy. Sailors, at sea for months, often suffered from this disease as their diet consisted mainly of dried and preserved foods. In the early seventeenth century it was discovered that oranges and lemons cured scurvy, but it was not until this century that vitamin C was isolated. A great deal of research on vitamin C is still taking place.

Composition

Ascorbic acid is a white crystalline substance with the formula $C_6H_8O_6$. It is similar to glucose and has a sweet-sour taste. Although most animals can manufacture this vitamin within their bodies, the primates, that is monkeys and man, must obtain it from food. It is the most unstable of all vitamins and is therefore very easily destroyed. As it cannot be stored for long in the body, it is important to include some in the diet every day.

Functions of Vitamin C

1. It is necessary for healthy tissue, skin, gums, bones and teeth.
2. It strengthens the blood vessels, thus helping to heal wounds and prevent bruising.
3. It prevents scurvy.
4. It helps to prevent infection. Large amounts of vitamin C are said to reduce incidences of colds and influenza.
5. It is necessary for the absorption of iron.
6. It is involved in metabolism (energy release) and in the functioning of the adrenal glands.

Note: It is used in food processing as an antioxidant. It is also added to flour as an improver, to hasten maturity.

Effects of Deficiency

1. Diseases of the skin, gums, teeth.
2. Delayed healing of wounds, excessive bruising.
3. Gradual deterioration of health, leading in severe cases to scurvy – see below.
4. Retarded growth.
5. Susceptibility to infection.
6. Incomplete absorption of iron, leading to anaemia.
7. Tiredness, irritability, a feeling of being 'run-down'.

Scurvy is rare in wealthy nations, although it sometimes occurs among the elderly and those living on limited incomes. Its symptoms are sore and bleeding gums, loose teeth, pains in the limbs and general tiredness.

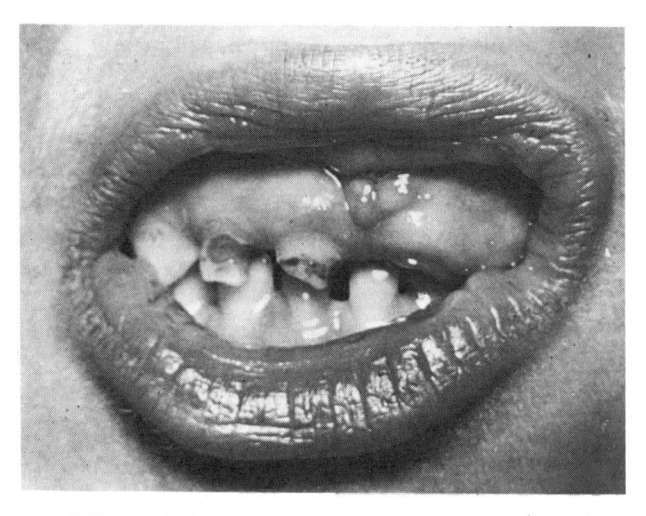

Effects of vitamin C deficiency (C. James Webb)

Vitamin C Sources

Most fruit and vegetables.
Best fruit sources:
rose hip syrup
blackcurrants
citrus fruits
These fruits contain moderate amounts of vitamin C: strawberries, raspberries, gooseberries, tomatoes.
Best vegetable sources:
peppers
parsley
broccoli

brussels sprouts
spinach
watercress
cabbage
bean sprouts
Other vegetable sources: lettuce, onions, new potatoes.

Animal sources: milk in summer (unreliable)

Recommended daily allowance. The minimum is:
    children 30 mg
    teenagers 40–50 mg
    adults 30 mg
    pregnant women 80 mg
    nursing mothers 100 mg

It is important to note that milk, especially cows' milk, is a poor source of vitamin C, so orange juice or rose hip syrup should be included in the diet of a baby four to six weeks after birth.

Allow for 50–90 per cent loss of vitamin C in cooked foods.

Frozen and canned fruit and vegetables are often as rich in vitamin C as the fresh variety but dried foods will lose some of the vitamin in processing.

Properties and Stability

As already mentioned, vitamin C is the most unstable of the vitamins and is therefore liable to be lacking in the

Good sources of vitamin C (John Topham Picture Library)

diet. Great care should be taken in the purchase, preparation and cooking of foods rich in vitamin C to see that

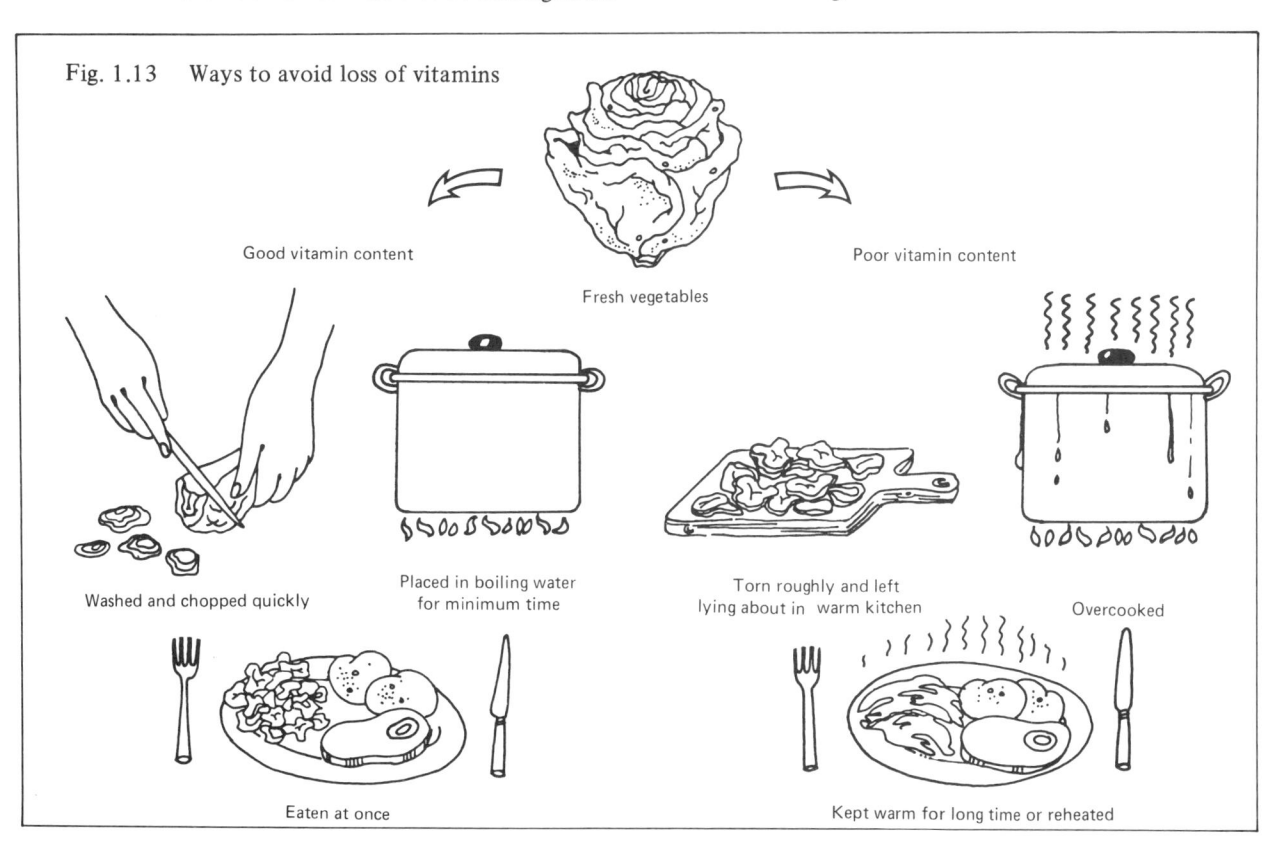

Fig. 1.13 Ways to avoid loss of vitamins

Fresh vegetables

Good vitamin content — Washed and chopped quickly — Placed in boiling water for minimum time — Eaten at once

Poor vitamin content — Torn roughly and left lying about in warm kitchen — Overcooked — Kept warm for long time or reheated

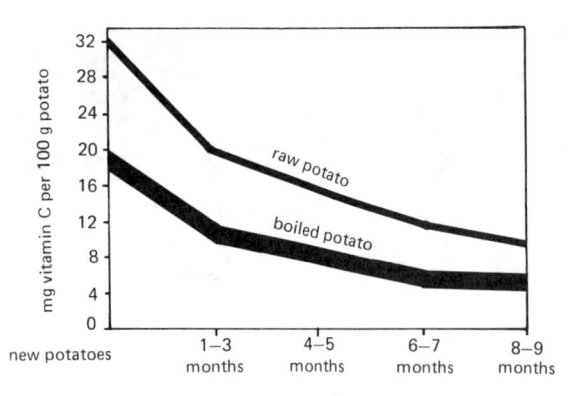

Fig. 1.14 Effect of storage on vitamin C content of potatoes

losses are minimised. Vitamin C is reduced or in some cases destroyed by: heat, oxidation, soaking or cooking in water, chopping, alkalis, storage, processing.

a. *Heat.* Even when carefully prepared and cooked, 50 per cent of vitamin C is lost during cooking of green vegetables. Avoid peeling vegetables such as potatoes before cooking. Cook as quickly as possible, with the lid on to reduce oxidation. Serve quickly (so avoid over-garnishing). Over-cooking, reheating and keeping meals warm are all inadvisable.

b. *Oxidation.* Once vegetables and fruit are harvested the vitamin C content is reduced through oxidation. Tired, wilted produce contains little vitamin C. As warmth speeds up oxidation, fruit and vegetables should be stored in a cool, dark, dry, ventilated place. Bad storage results in a 90 per cent loss of vitamin C.

c. *Water.* Ascorbic acid is highly water-soluble, so the practice of soaking vegetables in water, for any reason, is to be avoided. Cook vegetables in very little water, using up the cooking water for sauces and soups.

d. *Chopping.* An oxidising enzyme (oxidase) which is present in the cell walls of plants destroys vitamin C. If fruit or vegetables are chopped, grated or bruised, the oxidase is liberated. Chopped food such as coleslaw should be eaten soon after preparation. As oxidase is destroyed at a temperature of 85°C, blanching vegetables before freezing them will conserve the ascorbic acid.

e. *Alkalis.* As vitamin C is an acid, the practice of using breadsoda (an alkali) to soften greens and intensify their colour is inadvisable because it destroys *all* vitamin C present.

Processing
Formerly most methods of preserving food resulted in an almost total loss of vitamins B and C. However, constant research has brought about new methods of preserving food and of treating it before it is preserved so that vitamin loss is very slight.

Freezing. As the food is frozen within hours of harvesting and is blanched before freezing, it compares very favourably with fresh food. So-called fresh vegetables are often one day at the market, one day in the shop and one day in the home before they are used.

Canning. Canned vegetables are usually blanched and then sealed and processed quickly, so the maximum amount of vitamins is conserved.

Drying. Although improvements have been made in methods of drying vegetables, including blanching before drying and using anti-oxidants, there is still a considerable loss of vitamin C by this method of processing. Accelerated freeze drying causes the least loss of vitamins. Synthetic vitamins are sometimes added to dried foods to make good the losses caused by processing.

It stands to reason that foods with a high vitamin C content should, where possible, be eaten raw. As this vitamin is present in many salad ingredients, fresh salads should ideally be a daily part of the diet. In winter when lettuces are in short supply, other crisp, green vegetables like cabbage or celery can be used. Potatoes, although their vitamin C content is small (20 mg per 100 g portion), are eaten in such large quantities in this country that even allowing for the loss in cooking, they remain a good source of vitamin C. Remember that the newer the potatoes, the higher the concentration of vitamin C.

Unfortunately fruit and vegetables are becoming very expensive, so that there is an increasing danger of scurvy, especially among those on low incomes. The best source of vitamin C is from fruit and vegetables that have just been picked, so the sooner people realise that it pays both nutritionally and economically to grow their own fruit and vegetables, the better.

Test for Vitamin C
Tablets of the dye 2:6 dichlorophenol indophenol (DC PIP) have been developed to test food for ascorbic acid. This dye is blue in colour when in alkaline or neutral solutions and pink in acid solutions. It loses its colour when it comes in contact with ascorbic acid.

To begin the test dissolve 1 tablet of the dye in 20 cm^3 water; it should now be blue. Crush some food (e.g. an orange segment) and filter. Add some dye solution drop by drop. If ascorbic acid is present, it will begin to turn pink and then eventually go clear.

VITAMIN B GROUP
This group of vitamins was originally thought to be a

single vitamin. After much research, it was found to contain several independent vitamins which share some functions and are present in similar foods. In the study of nutrition they are usually separated into individual vitamins by name. The whole B group plays a part in the release and utilisation of energy and includes the following vitamins:

Thiamine (B_1)
Riboflavin (B_2)
Niacin (Nicotinic acid)
Pyridoxine (B_6)
Folic acid
Cyanocobalamin (B_{12})

Other members of the vitamin B group are: pantothenic acid; choline; biotin (vitamin H); insositol; para-aminobenzoic acid. These are still being investigated so little is known about them.

THIAMINE (B_1)

This is a very complex vitamin which is likely to be deficient in the diet because it is unstable to heat and water-soluble. As is not stored in the body, it should be included daily in the diet. There is considerable loss of thiamine during the milling process, so brown rice and brown flour are better sources than milled rice and flour unless the white flour is fortified.

Functions of Thiamine
1. It is concerned with the release of energy from glucose.
2. It is necessary for growth, good appetite and general health.
3. It is necessary for the nerves, preventing the nervous disease beri-beri.

Effects of Deficiency
1. In mild cases a feeling of being 'run-down', tiredness, loss of appetite, constipation.
2. Retarded growth in children.
3. Neuritis and other nervous diseases. In severe cases beri-beri, a nervous disease common in rice-eating countries, occurs. It attacks the nervous system, causing paralysis and often death. This disease also occurs among chronic alcoholics.

Thiamine Sources
Thiamine is present in many foods and plentiful in:

wheatgerm	pork
cereals (including fortified breakfast cereals)	bacon
	beef
yeast	flour (if fortified)
oatmeal	

Recommended daily allowance. This is related to the amount of carbohydrate eaten. In the average diet, about 1.0 mg is necessary for adults.

Properties and Stability
After vitamin C, thiamine is probably the most unstable vitamin. It is affected by heat and very soluble in water. Many of the rules for vitamin retention in the cooking of vegetables apply also to thiamine: avoid steeping and moist cooking methods, use as little water as possible and do not over-cook, keep warm or reheat. There is some loss in food processing and almost a total loss in the manufacture of breakfast cereals, so synthetic vitamins are often added. In baking, thiamine is destroyed by the use of alkalis such as bread soda. Sulphur dioxide, a preservative, also destroys it.

RIBOFLAVIN (B_2)

This is a yellow water-soluble solid.

Function of Riboflavin
It is an essential link in the process of cellular respiration by which energy is released from food (see p. 176).

Effects of Deficiency
1. Retarded growth in children.
2. The mouth, tongue and eyes become sore and inflamed.

Riboflavin Sources

yeast	eggs
liver	milk
beef	green vegetables
cheese	beer

Recommended daily allowance. This is related to the total calorie intake. The average is 1.5 mg for adults.

Properties and Stability
This vitamin is water-soluble and will leach into the cooking water. Cooking at high temperatures (e.g. pressure cooking, food processing) will destroy it. As ultra-violet light also affects it, avoid leaving milk bottles standing in sunlight.

NIACIN

This vitamin was formerly called nicotinic acid. It is also known as the anti-pellagra vitamin (see below). This is one of the few vitamins which can be manufactured in the body. The amino acid tryptophan can be converted to niacin in the intestine.

Functions of Niacin
1. It helps release energy from food.
2. It is essential for growth and the correct functioning of nerves and skin.

Effects of Deficiency
1. Growth is retarded.

2. Chilblains and skin disorders develop.
3. In severe cases pellagra develops. This is a disease common in maize-eating countries such as Central America, because the niacin present in maize is unavailable to the body. It symptoms are '3 Ds' — dermatitis, diarrhoea and dementia.

Niacin Sources

yeast	white fish
wheatgerm	cheese
bread	eggs
liver	milk
beef	

Recommended daily allowance. This is related to the amount of tryptophan in the diet and total energy expenditure. The average is 15–20 mg daily.

Properties and Stability

Niacin is soluble in cooking water but little affected by heat or acids. There is an 80 per cent loss in milling and a slight loss in food processing. Niacin is the most stable of the B group vitamins.

PYRIDOXINE (B_6)

This is concerned with energy release and prevents convulsions.

Pyridoxine Sources
wheat germ
liver
meat

Daily requirement. 2 mg per person

FOLIC ACID

Essential for the formation of red blood cells and helps to prevent anaemia.

Folic Acid Sources
Offal
beef
cereals
eggs
green vegetables

Folic acid is often prescribed during pregnancy.

CYANOCOBALAMIN (B_{12})

This helps to form red blood cells and prevents pernicious anaemia, a rare disease caused by the malformation of blood cells.

Cyanocobalamin Sources
offal and other animal foods

Vitamin B_{12} is not present in vegetables and is sometimes lacking in the diet of vegetarians.

MINERAL ELEMENTS (OR SALTS)

The human body requires approximately twenty mineral elements if all organs and systems are to work efficiently. These are the most important:

calcium	potassium
sodium	chlorine
iodine	sulphur
phosphorus	magnesium
iron	

Others are only required in minute amounts. These include copper, zinc, cobalt and fluorine and are called *trace elements.*

Inorganic elements do not provide energy. As they are water-soluble care must be taken to avoid steeping foods which contain them. Very little water should be used in cooking and any water remaining after cooking should be used for sauces etc. so that the valuable mineral salts are not lost. Heat and food processing do not usually affect the mineral content of food.

As each mineral salt has a specific function and is found in certain foods, it is difficult to make general statements about the functions and sources of minerals. Some are concerned with the manufacture and upkeep of the body cells, others with metabolism and others with bone formation. As with all food constituents, a good mixed diet should supply all the mineral elements in sufficient amounts.

The two minerals most commonly lacking in the diet are calcium and iron.

CALCIUM

There is more calcium in the body than any other mineral. Most of this is in the bones, but there are small amounts in the blood to control clotting and some is also present in the muscles. Calcium works in conjunction with phosphorus; the proportion should be in the ratio of 1–1.5 calcium to phosphorus for maximum absorption. Vitamin D is vital for the correct utilisation of calcium and phosphorus as it controls absorption and ossification (laying down of calcium in the bones). Vitamin C also increases calcium absorption.

The parathyroid gland controls the amount of calcium in the blood. Certain substances such as phytic acid in the outer layer of cereals and oxalic acid (found in spinach) can interfere with calcium absorption, as can excess fat in the diet. Calcium is excreted in the urine.

Hypercalcaemia is a disease which causes too much calcium to be absorbed as a result of an excess of vitamin D in the diet.

Functions of Calcium

1. It helps the development of strong bones and teeth.
2. It is necessary for normal clotting of blood.
3. It is necessary for the normal functioning of the muscles, including the heart muscle and the nerves.

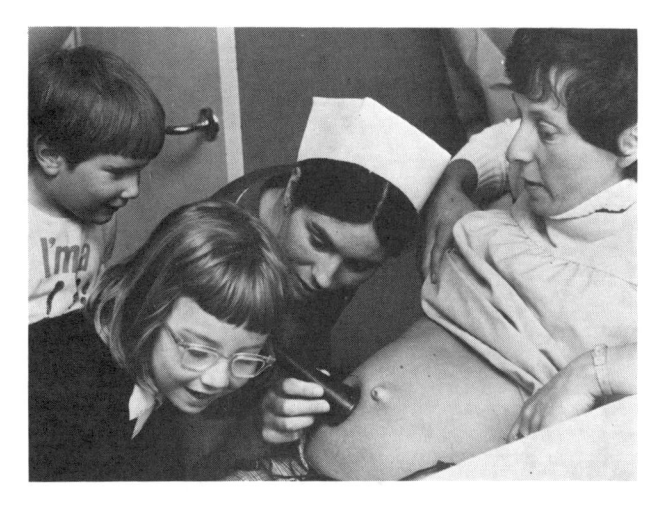

Calcium: vital in childhood and pregnancy (Barnaby's Picture Library)

Effects of Deficiency
1. Rickets and osteomalacia occur in severe cases (see p. 19).
2. Teeth are of poor quality and badly formed.
3. Irritability and muscular spasm.

Calcium Sources
milk
cheese
tinned fish (i.e. fish of which bones are eaten, e.g. sardines, salmon)
green vegetables
hard water
flour when fortified with calcium carbonate

Recommended daily allowance:
>   adults   500–800 mg
>   children   1000 mg
>   pregnant and nursing mothers up to 2000 mg

Note: The importance of sufficient calcium in the diet of children, pregnant and nursing mothers must be stressed. If extra amounts of calcium-rich foods are not taken during pregnancy calcium will be withdrawn from the bones of the mother to supply the foetus, thus endangering the health of the mother, particularly in the case of repeated pregnancies. Lack of calcium in the diet of children can result in stunted growth and rickets.

PHOSPHORUS

This element is present in many proteins and is stored in the bones as calcium phosphate. As with calcium, there are small amounts present in the blood and tissue. Excess phosphorus is removed by the kidneys.

Functions of Phosphorus
1. Bone and tooth formation.
2. Necessary for the reproduction of cells.
3. Necessary for metabolism. It plays an important part in the storage and release of energy in the body as it forms part of the phosphate compounds ADP and ATP (see p. 176).

Effects of Deficiency
As phosphorus is present in most foods, both animal and vegetable, deficiency is unknown.

IRON

Three-quarters of the iron in the body is found in the blood. The remaining iron is stored in the liver, spleen and bone marrow. Small amounts of iron are lost through wear and tear of the body. It is only during haemorrhaging and menstruation that large amounts are lost.

Function of Iron
It is necessary for the formation of haemoglobin in the red blood cells. This transports oxygen from the lungs to the tissues for oxidation.

Absorption of Iron
Only 10 per cent of ingested iron is absorbed by the body. Many different factors can affect its absorption. Iron is eaten in the ferric state but must be reduced to the ferrous state before it can be absorbed. Both vitamin C and vitamin E (reducing agents) help in the reduction of iron, but excess phosphates can inhibit its absorption. The oxalates in spinach also prevent its iron from being absorbed.

Blood cells have a life span of 120 days, after which time they are recycled. The cells are broken down by the spleen and the iron is transported to the liver where it is used in the formation of new red blood cells, made in the bone marrow.

Active teenagers need adequate iron in their diet (Irish Press).

Iron-rich foods should be included in the diets of babies over six months old (W.H.O.)

Effects of Deficiency

The main effects of deficiency are tiredness and lethargy. In severe cases *anaemia* results. There is a high risk of iron deficiency during growth spurts in children and teenagers.

Many women, especially those who have heavy periods, suffer from varying degrees of anaemia. Pregnancy, particularly repeated pregnancies, can increase the risk of anaemia, and severe anaemia can contribute to congenital abnormalities and difficulties at birth. Normal babies are born with enough iron to last six months, but iron-rich foods such as egg yolk should then be introduced. Remember that milk is a poor source of iron.

Iron Sources

    liver, kidney, other meats especially black pudding, corned beef
    whole cereals, brown and white bread, wheat germ
    pulse vegetables, green vegetables, parsley
    treacle, dried fruit, curry powder, wine, soya products

Recommended daily allowance:

adult women	12 mg
pregnant women	15 mg
adult men	10 mg

Remember that more iron is required during puberty, pregnancy and lactation.

Stability

As with all minerals, there will be some loss into cooking or steeping water.

SODIUM

Sodium is present in the fluids of the body and is essential for life. It is readily excreted through the kidneys as urine and as perspiration through the skin. In hot climates

Sodium is lost through perspiration during extreme physical exertion (Barnaby's Picture Library)

and in jobs or sports where physical exertion causes excessive perspiration, extra salt must be taken to make good the loss.

Functions of Sodium

Necessary for maintaining the correct water balance and pH balance (acid/base balance) in the blood. It keeps the blood and some digestive fluids alkaline.

Effects of Deficiency

Muscular cramps.

Sodium Sources

There is usually sufficient sodium in a normal diet. Bacon, smoked fish, cheese, bread and butter are fairly good sources. Extra can be added during cooking or at table.

POTASSIUM

This element is similar to sodium and is present in all body cells.

Function of Potassium

Necessary for the formation of new cells.

Potassium Sources

Usually sufficient is found in a mixed diet. Good sources include prunes, potatoes, brussels sprouts, beef, offal, fish, milk, eggs and cheese.

IODINE

Although iodine is only required in very small amounts, it is absolutely essential to the body.

Function of Iodine

Essential for the proper functioning of the thyroid gland and its hormone thyroxine which controls the metabolic rate.

Effects of Deficiency

1. Enlargement of the thyroid gland, resulting in goitre.
2. Lack of energy, obesity and mental backwardness.

Iodine Sources

Iodine is plentiful in all seafood and sea weed, and most vegetables grown in this country contain iodine. On large land masses such as central Africa, goitre is very common and iodised salt is used in cooking and at table to compensate for the lack of iodine in the soil.

Recommended daily allowance: 150 µg.

Tests for Minerals

Preparation. Use a clean crucible. Add 5 g of the food

Seafood: an excellent source of iodine (Bord Iascaigh Mhara)

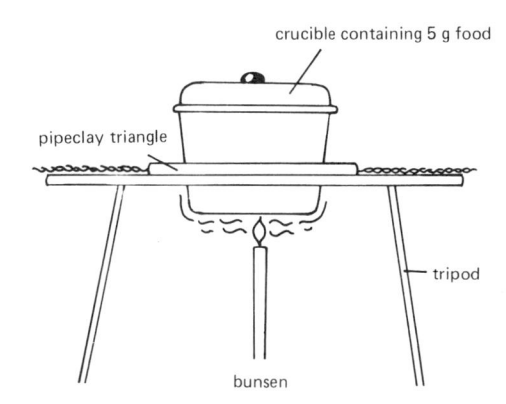

Fig. 1.15   Test for mineral content of food

to be tested. Heat gently over a bunsen burner until only ash remains. Cool.

Iron. Ash is usually a whitish colour. If it has a brown tint, it indicates the presence of iron.

For tests below, mix the ash in a beaker with a little distilled water containing a few drops of nitric acid. Divide into three test tubes.

Sodium. Dip a platinum wire into the solution in one test tube. Heat over a bunsen burner. If it flames yellow, sodium is present.

Potassium. Using the second test tube, repeat the test. A resulting lilac flame indicates the presence of potassium.

Calcium. Gradually add ammonia to the third test tube until it is slightly alkaline. (Test with litmus paper.) Add sufficient acetic acid to acidify the solution. Add an equal amount of 5 per cent ammonium oxalate solution. A

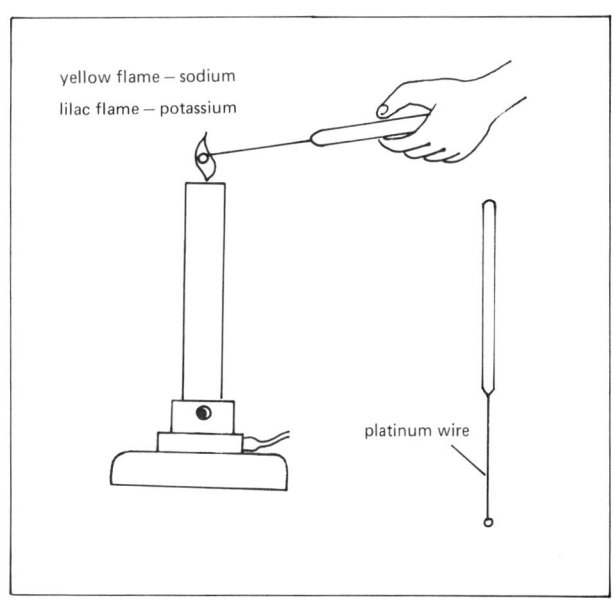

Fig. 1.16   Test for sodium and potassium

cloudy white precipitate indicates the presence of calcium.

Trace Elements
Zinc. A component of many enzymes. *Sources:* seafood.

Copper. Necessary for the utilisation of vitamin C and the absorption of iron. *Sources:* liver, kidney, shellfish.

Manganese. Necessary for enzymic activities. *Sources:* cereals, pulse vegetables.

Chlorine. Necessary for the production of hydrochloric acid in the stomach. *Sources:* as for sodium.

Fluorine. Prevents dental decay. It is added to water supplies in most areas of Ireland in the form of sodium fluoride. *Sources:* offal, fish, tea.

WATER
Water is essential for all life and makes up two-thirds of the body weight. It forms the main part of the cell liquid cytoplasm and the surrounding liquid *extracellular fluid* (ECF). The blood, secretions, digestive juices and lymph are mainly composed of water. Even bone is 25 per cent water.

Composition
Water is composed of hydrogen and oxygen in the proportion H_2O.

Functions of Water
1. Water is the medium for transporting substances from one part of the body to another, e.g. oxygen and carbon dioxide, blood cells, nutrients, waste matter, hormones and enzymes.
2. It distributes the heat generated by metabolism, keeping the body at the temperature necessary for all body functions – 37°C.
3. It keeps lining membranes of organs moist, e.g. pleura, bronchii, joints and eyes.
4. It dissolves food during digestion, forms secretions for enzymes and assists absorption.
5. Its ability to dissolve substances enables chemicals to mix readily with one another, so that hydrolysis takes place.
6. It is necessary for the removal of waste from the body, e.g. urea is excreted by the kidneys.
7. The body is cooled by perspiration, which takes heat from the skin for its evaporation.

Water Sources
1. The main source in the diet is beverages e.g. milk, tea, coffee. The average person drinks 1.5 litres of fluids daily.
2. Water is also obtained from food (.8 litres). Green vegetables and citrus fruits are 90 per cent water. Dried foods, sugar, oils and fats have a low water content, but most dried foods are rehydrated to soak back the amount of water lost in drying.
3. Water is also produced in the body by oxidation.

Daily requirement. 2–2.5 litres of water are excreted daily by the kidneys, skin, lungs and to a lesser extent by the large intestine. About the same amount of liquid is required by the body each day to make good this loss. The human body cannot live for more than a few days without water. If too much is lost from the body quickly, as in the case of gastro-enteritis or severe diarrhoea, dehydration may occur. This can have serious and even fatal consequences, especially in children.

Calorific value. None – thus foods with a high water content, e.g. green vegetables, are low in calories.

Properties of Water
1. Pure water has no colour, taste or smell.
2. The pH value of water is neutral, i.e. it is neither acid nor alkaline.
3. It is very solvent and thus capable of dissolving many substances.
4. It readily absorbs and retains heat.
5. It freezes at 0°C and boils at 100°C.
6. It is often a source of mineral salts, e.g. iodine, calcium.

THE ENERGY VALUE OF FOOD
Energy can be defined as the capacity for doing work. There are various types of energy – electrical, mechanical and chemical. The body uses *mechanical* energy to enable it to function – e.g. move muscles. *Chemical* energy is required for all metabolic activities. Heat, another form of energy, is generated during oxidation and muscular activity to keep the body temperature normal. An analogy is often made between the human body and a motor car: the car uses petrol as fuel to give it the energy to move and the body uses food as fuel to provide energy for muscular movement. The energy is released from the fuel in both cases by a form of burning. In the body this slow burning up of food is called oxidation – oxygen breathed in through the lungs combines in the tissues with the carbon in food, producing heat and energy.

Fats, carbohydrates and proteins all contain carbon and therefore supply energy.

Measurement of Heat
Heat and energy may be measured in different ways:

a. Kilocalorie. A calorie is the unit of energy used in physics. As it is a very small measure, a unit which is 1,000 times greater is used for measuring the energy value of food. It is written with a capital 'C', Calorie, but to avoid confusion this is now more commonly known as a kilocalorie – kc or kcal. *A kilocalorie is the amount of heat required to raise the temperature of 1,000 g water by 1°C, e.g. from 20°C to 21°C.*

b. Joule. The internationally accepted unit for measuring energy is the joule. Again this is a very small measurement so a unit 1,000 times greater is used – the kilojoule (kj). When dealing with very large amounts of food such as the daily intake of food, a measurement of 1,000 kj is used. This is called a megajoule – mj.

1 kilocalorie (kcal) = 4.184 kilojoules (kj)
1,000 kilojoules = 1 megajoule (mj)

To convert kilocalories to kilojoules multiply by 4.2
Example: 20 kilocalories x 4.2 = 84 kilojoules.

c. British Thermal Unit (BTU). Heat from fuels such as coal, heating oil, gas and electricity are usually measured in British Thermal Units. A BTU is the amount of heat required to raise the temperature of 1 lb water by 1°F. 1 BTU is roughly equivalent to 1 kilojoule. A much larger measurement of heat is the therm. 1 therm = 100,000 BTU or 105 mj. The therm and BTU are now being replaced by the joule measurement.

The energy value of food varies considerably. Foods with a high water content, for example, contain relatively few kilocalories/kilojoules.

To measure the energy value of a food, it is burned (oxidised) under controlled conditions in a bomb calorimeter. The heat which is liberated is measured and calculated in kilocalories or kjs per gram.

When each of the three energy producing nutrients are measured, it is found that on average:
1 g protein produces 4 kilocalories/17 kilojoules.
1 g carbohydrate produces 4 kilocalories/17 kilojoules.
1 g fat produces 9 kilocalories/38 kilojoules.
It follows that foods containing a high percentage of fat are a good source of energy.

To calculate the energy value of food, the numbers of grams of each of the three energy-producing nutrients in the food are multiplied by the kilocalories produced per gram.
Example
100 g milk contains 3.3 g protein, 3.8 g fat, 4.8 g carbohydrate.

3.3 x 4	= 13.2
3.8 x 9	= 34.2
4.8 x 4	= 19.2

Total energy value = 66.6 kilocalories

To convert to joules multiply by 4.2 = 280 kj approx.

Many nutrition and calorie tables use 100 g as a basic quantity by which to compare foods. This is roughly equal to one portion of most foods, e.g. meat, fish, eggs, vegetables, cereals and fruit. The amount of kilocalories/kilojoules in some basic foods is shown at the end of the book.

Functions of Energy
Energy is required by the body:

1. For all activity.
  a. Voluntary muscular contractions which occur during any form exercise, e.g. beating a cake, walking, jogging.
  b. Involuntary muscular contractions of internal organs e.g. the heart, digestive organs and those concerned with breathing.

2. For manufacture of body materials, e.g. new cells. This building up process, called anabolism, occurs during growth, repair and the alteration of nutrients for storage.

3. For cell and nerve activity. Every cell requires energy to carry out its functions. Energy is utilised by nerve cells to transmit impulses between the brain and body tissues.

4. To maintain body temperature. The organs of the body function normally at a temperature of 37°C. As the temperature of the environment is usually much lower than this, heat is generated within the body from the fuel foods consumed in order to make good the loss of heat through perspiration, evaporation and convection.

Energy and Activity
Most energy expenditure is used for physical activity. The more active the body, the more energy is used. Running, energetic sports and hard physical work use up far more energy than sitting writing or typing. The speed at which food is utilised for energy production is known as the metabolic rate. This rises as activity increases.

Basal Metabolic Rate
This is the lowest level of energy expenditure maintained by the body. It can be compared to the engine of a car when 'idling' or 'ticking over'. It is the amount of energy required by an individual lying down, relaxed and fasting, in a warm room. The only physical movement is the involuntary activity of the heart, breathing and digestive processes. The basal metabolic rate (BMR) varies considerably between individuals, even those of identical size, age and sex. This accounts for the fact that some individu-

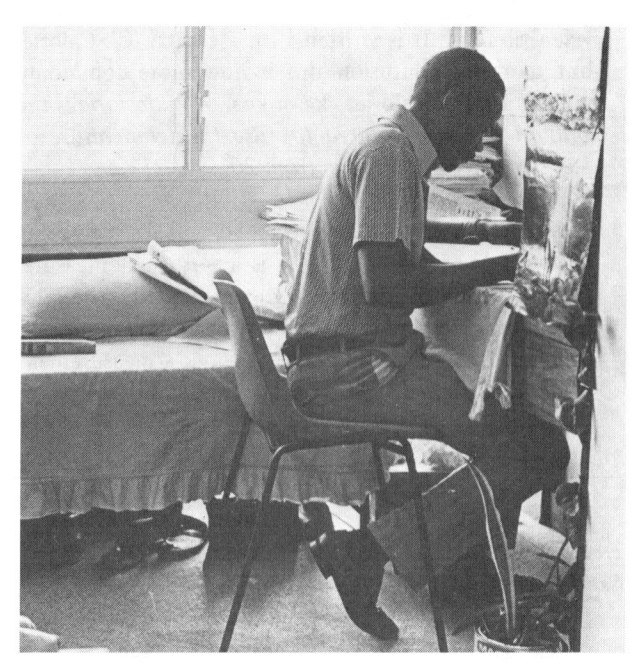

Energetic sports use up far more energy than study (Barnaby's Picture Library and Margaret Murray)

als can eat enormous quantities of food and not get fat, while others have a tendency to obesity in spite of a low kilocalorie intake. Fat-type people tend to store energy, whereas thin-type people metabolise their energy at once as heat rather than store it.

The metabolic rate of an individual shows variations at different ages – for example, during pregnancy the rate increases; with the onset of old age it decreases. From the age of about thirty, most people would benefit from a gradual reduction in kilocalorie intake.

Basal metabolic rate of an average man is 70 kcal (300 kj) per hour

Basal metabolic rate of an average woman is 60 kcal (250 kj) per hour

Any form of activity will increase this rate, as can be seen below.

Activity	kcal per hour	kj per hour
Sitting	85	360
Standing	90	380
Writing	115	480
Walking slowly	185	780
Scrubbing	315	1,325
Cycling	400	1,680
Dancing	450	1,790
Swimming	575	2,415
Walking upstairs	1,000	4,200

Sedentary and mental work uses very little energy by comparison with muscular work.

Daily Energy Requirements

Requirements vary according to:

1. *Body size and composition.* The longer and leaner the body, the greater the amount of energy required: a tall, thin person will require more energy foods than a small fat person.

2. *Age.* During periods of rapid growth – in infancy and childhood – the amount of kilocalories/kilojoules required per kilogram of body weight is greater than at any other time. As the body ages, the kilocalorie requirement gradually lessens.

3. *Climate.* A greater number of kcals/kjoules is required in cold countries than in hot countries. Eskimos have a very high kilocalorie intake which helps to maintain their normal body temperature in the extremely cold conditions in which they live.

4. *Activity.* The greater the physical activity, the more kilocalories are required. Active occupations, sports and hobbies will demand a high kc/kj intake. Sedentary and moderately active occupations will require fewer kc/kj.

5. *Sex.* Males require a higher kilocalorie intake than females of roughly the same size and degree of activity.

6. *Pregnancy, lactation.* The metabolic rate of most pregnant women increases to provide energy for the growth of the foetus. However, care must be taken to avoid gaining too much weight during pregnancy.

Daily Requirement of Kilocalories

	Age	Kilocalories	Megajoules
Children	0–1	800	3.3
	2–3	1400	5.9
	5–7	1800	7.5
Boys	12–15	2800	11.7
	15–18	3000	12.6
Girls	9–18	2300	9.6
Men	18–35 (sedentary)	2700	11.3
	18–35 (moderately active)	3000	12.6
	18–35 (very active)	3600	15.1
	35–65	100 kc less than 18–35 age group	5 mj less
	65–75 (sedentary)	2350	9.8
	75 and over	2100	8.8
Women	18–55 (moderately active)	2200	9.2
	18–55 (very active)	2500	10.5
	55–75 (sedentary)	2050	8.6
	75 and over	1900	8.0
	Pregnancy	2400	10.0
	Lactation	2700	11.3

Note. The average man used for these calculations is a healthy 70 kg (11 stone) man living in a temperate climate. He does not indulge in very strenuous exercise. The average woman weighs 57 kg (9 stone), does some housework and takes moderate exercise. The increase in kilocalorie requirements during puberty is due to a growth spurt, increasing hormonal and chemical activity and because teenagers, like children, are generally more active than adults.

Obesity and Energy Intake

Excess weight is a common problem in developed countries. It is ironic that millions of people are overweight because of overeating while thousands of millions in underdeveloped countries are starving. Obesity can be caused by affluence and ignorance. If more kilocalories are consumed than are necessary for the physical and chemical activity of the body, the excess kilocalories are converted into fat and stored under the skin as adipose tissue. Obesity is usually caused by over-indulgence in fattening foods (i.e. those with a high energy value), combined with lack of exercise. To maintain normal weight, energy intake must balance with energy output. Weight loss will only occur when energy intake is less than energy output.

If a person wishes to lose weight he must either a. reduce his intake of kilocalories; b. increase physical activity; or preferably, do both. This would be the quickest way to lose weight.

Modern eating habits. The average kilocalorie requirements in the modern diet are gradually diminishing as the amount of physical activity in our lives decreases. Increased use of labour-saving machines means less physical work; washing machines, vacuum cleaners and electric mixers do much to reduce the energy output of the housewife, while electric saws, mechanical diggers and pneumatic drills take some of the physical activities out of labouring jobs. Instead of walking or cycling to work many people use cars or public transport. Television viewing has supplanted more active hobbies. These and many other factors have combined to reduce the necessity for as high a kilocalorie consumption as that of our grandparents.

Appetite and energy intake. The part of the brain which controls the appetite is known as the hypothalamus. It is thought that nervous impulses passing between the stomach and this appetite centre control the urge to eat. Low blood sugar, which occurs a few hours after a meal, indicating that reserves of glucose and liver glycogen have

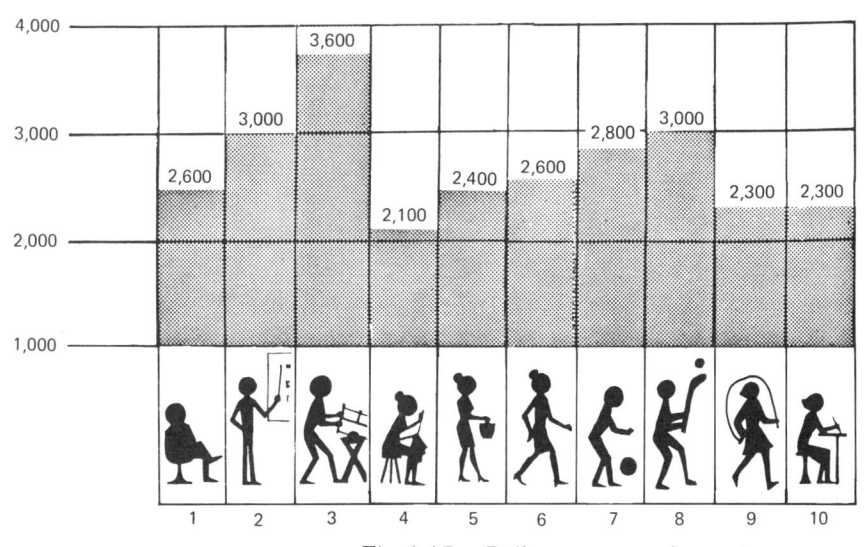

Fig. 1.17 Daily energy requirements

1. sedentary man
2. active man
3. very active man
4. sedentary woman
5. active woman
6. very active woman
7. boy 12–15
8. boy 15–18
9. girl 12–15
10. girl 15–18

Labour-saving machines mean less physical work – and that should mean a drop in kilocalorie consumption (Mansell Collection and Hoover)

been used up, is also thought to stimulate the hypothalamus, producing a feeling of hunger. The appetite controls the balance between energy intake and output and usually indicates when sufficient food has been eaten. If it is ignored and more food is consumed by the body than is required, obesity will occur. Obesity is a definite health hazard and can contribute to heart, liver, kidney and respiratory complaints, high blood pressure, infertility, diabetes and arthritis. Most insurance companies charge a loading fee or refuse to insure adults who are greatly overweight because their life expectancy is shortened.

For more information on obesity see p. 43.

Four international organisations work to promote a better standard of nutrition. These are:
WHO   World Health Organisation
FAO   Food and Agriculture Organisation
UNESCO United Nations Educational, Scientific and Cultural Organisation
UNICEF United Nations Children's Fund.

QUESTIONS

1. Discuss proteins under the following headings:
   a. elemental composition
   b. chemical composition
   c. biological functions

   Explain the term biological value in relation to protein foods and give some examples of high biological value proteins and low biological value proteins.

2. Write a detailed account of each of the following processes:
   a. deamination
   b. denaturation
   List the effects of heat on protein foods.

3. Explain the terms
   a. conjugated proteins
   b. peptide links
   c. fibrous proteins
   Describe a laboratory test which indicates the presence of protein in food.

4. Explain clearly the chemical structure of carbohydrates. List and discuss briefly the properties of **a.** starch and **b.** sugar.

5. Write a note on
   a. cellulose
   b. hydrolysis
   c. photosynthesis
   in relation to carbohydrates.
   Describe an experiment which **a.** identifies the presence of glucose in food and **b.** illustrates the solubility of carbohydrates.

6. Define the term lipid.
   Describe the chemical structure of dietary lipids. Explain the terms saturated and unsaturated fatty acids and describe a test which identifies the presence of each in a food.

7. Write a note on lipids under the following headings:
   a. functions
   b. properties
   c. digestion

   Describe the chemical change known as hydrogenation and explain how this process is utilised commercially.

8. Explain the following terms in relation to lipids: emulsion; rancidity; essential fatty acids; smoke point; plasticity.

9. Classify vitamins.

   Discuss fat-soluble vitamins including references to their sources, functions, effects of deficiency and stability to heat.

   Explain the difference between carotene and vitamin A.

10. Describe briefly four of the following: thiamine; niacin; folic acid; cholecalciferol; tocopherols.

    Mention two tests which may be used to establish the presence of vitamins in food.

11. Discuss calcium or iron under the headings: source; function; effects of deficiency; effects of c vitamin/mineral interrelationships.

12. What are the trace elements? Name three.

    Describe the function of sodium, phosphorus ⌐ iodine in the diet.

    Describe a laboratory test which establishes the presence of a mineral element in food.

13. Explain the term basal metabolic rate. How do the energy requirements of various individuals differ?

    Using the food tables at the back of the book plan a menu for one day for an office worker. Indicate the approximate calorific value of the chief foods in each meal.

14. Why is energy required by the body?
    What is the energy value of
    a. protein
    b. carbohydrate
    c. fat

    Describe how the energy value of (i) eggs and (ii) wheat could be calculated using food tables which show their general composition.

2. The Diet

PLANNING A BALANCED DIET

A well-balanced diet should contain all the necessary nutrients in the correct proportion for the weight and needs of the individual. Although it is possible to calculate from food tables the exact quantities of nutrients eaten each day, the human race has managed to survive for thousands of years with little knowledge of nutrients, simply by eating a wide range of fresh foods.

It follows that if we want to lead healthy, active lives we must eat a *varied* selection of *fresh* foods. These will supply all the nutrients required by the body. It is essential to ensure that the vitamin content of foods is not decimated by over-cooking: fruits and vegetables should be eaten raw where possible. It is important to remember, too, that many modern foods are over-processed, often to the point of being entirely unbalanced. Plenty of fibre or roughage must be included in the diet – in homely dishes such as porridge and brown bread as well as in raw fruits and vegetables.

Convenience foods should be kept to a minimum. Many are bulked out with carbohydrates and laced with additives, and even the least educated tastebuds will accept that their flavour is far inferior to that of the home-cooked or fresh product. Instead of relying on frozen vegetables and canned fruit, use fruit and vegetables in season – when they are at their delicious and nutritious best. It is nice to look forward to the first strawberries or asparagus of the season, rather than eat them as a matter of routine all the year round.

Daily Food Requirements

1. Protein 2 portions (100 g each) of animal protein in the form of meat, fish, eggs, cheese.

(Fish and offal such as liver should each be eaten at least once a week to supply a good selection of minerals.)

2 portions (100 g each) of vegetable protein such as pulses, cereals, bread.

Milk: adults   300–500 ml daily
        children   500–800 ml daily.

Animal protein: two portions a day are required (Anne-Marie Ehrlich)

Recommended daily intake of nutrients

Age ranges	Body weight (kg)	Energy mj	Energy kcal	Protein Recommended g	Protein Minimum requirement g	Calcium mg	Iron mg	Vitamin A (retinol equivalent) ug	Thiamin mg	Riboflavin mg	Nicotinic acid equivalent mg	Vitamin C mg	Vitamin D μg
years													
Infants													
Under 1	8	3.3	800	20	15	600	6	450	0.3	0.4	5	15	10
Children													
1	10	5.0	1,200	30	19	500	7	300	0.5	0.6	7	20	10
2	12	5.9	1,400	35	21	500	7	300	0.6	0.7	8	20	10
3–4	13.5	6.7	1,600	40	25	500	8	300	0.6	0.8	9	20	10
5–6	16	7.5	1,800	45	28	500	8	300	0.7	0.9	10	20	2.5
7–8	20	8.8	2,100	53	30	500	10	400	0.8	1.0	11	20	2.5
Males													
9–11	32	10.5	2,500	63	36	700	13	575	1.0	1.2	14	25	2.5
12–14	46	11.7	2,800	70	46	700	14	725	1.1	1.4	16	25	2.5
15–17	60	12.6	3,000	75	50	600	15	750	1.2	1.7	19	30	2.5
18–34 sedentary	65	11.3	2,700	68	45	500	10	750	1.1	1.7	18	30	2.5
moderately active	65	12.6	3,000	75	45	500	10	750	1.2	1.7	18	30	2.5
very active	65	15.1	3,600	90	45	500	10	750	1.4	1.7	18	30	2.5
35–64 sedentary	65	10.9	2,600	65	43	500	10	750	1.0	1.7	18	30	2.5
moderately active	65	12.1	2,900	73	43	500	10	750	1.2	1.7	18	30	2.5
very active	65	15.1	3,600	90	43	500	10	750	1.4	1.7	18	30	2.5
65–74	65	9.8	2,350	59	39	500	10	750	0.9	1.7	18	30	2.5
75 and over	63	8.8	2,100	53	38	500	10	750	0.8	1.7	18	30	2.5
Females													
9–11	33	9.6	2,300	58	35	700	13	575	0.9	1.2	13	25	2.5
12–14	48	9.6	2,300	58	44	700	14	725	0.9	1.4	16	25	2.5
15–17	55	9.6	2,300	58	40	600	15	750	0.9	1.4	16	30	2.5
18–54 most occupations	55	9.2	2,200	55	38	500	12	750	0.9	1.3	15	30	2.5
very active		10.5	2,500	63	38	500	12	750	1.0	1.3	15	30	2.5
55–74	55	8.6	2,050	51	36	500	10	750	0.8	1.3	15	30	2.5
75 and over	53	8.0	1,900	48	34	500	10	750	0.7	1.3	15	30	2.5
Pregnant, 2nd and 3rd trimesters		10.0	2,400	60	44	1,200	15	750	1.0	1.6	18	60	10
Lactating		11.3	2,700	68	55	1,200	15	1,200	1.1	1.8	21	60	10

Some fruit and vegetables should be eaten raw each day (Anne-Marie Ehrlich)

2. *Fat* A little butter and/or margarine for energy and vitamins A and D.
   Invisible fats are present in meat, eggs, milk and cheese.

3. *Vitamins* ⎫
   Minerals ⎬ 2–3 portions of vegetables at least one of which should be raw (vitamins A and C)
   Roughage ⎭ 1 portion of fruit, preferably raw (vitamins A and C)

4. *Carbohydrates* Fill out the diet with foods containing starch and sugar, e.g. potatoes, cereals, bread, cakes, pastry, puddings. One portion (100 g) at each meal.
   When planning a diet for any individual you should consider:

a. Recommended protein intake. This is related to body size and age. If it is exceeded the extra protein is deaminated and converted into fat. About half of the protein in our diet should come from vegetable sources and at least half from animal sources.

b. Recommended energy intake. This category includes foods containing fats and carbohydrates, and also alcohol. About 50 per cent of our total fat intake should be in the form of unsaturated fats – those of marine or vegetable origin. These will supply the essential fatty acids and are thought to reduce the cholesterol level. When choosing carbohydrate foods, give preference to those which are rich in other nutrients. Oatmeal and pulse vegetables, for example, are rich in vitamins and minerals whereas sugar is pure carbohydrate.

It is important not to exceed the recommended energy intake as excess fats and carbohydrates are stored as fat and often result in obesity. An extra teaspoon of sugar a day can mount up to a staggering 18,250 extra kilocalories every year!

c. Protective foods. Foods with a good quantity of various minerals and vitamins are often called *protective foods* as they 'protect' the body from general lassitude and ill health as well as specific disorders such as rickets, anaemia and scurvy. When people feel 'run down' they tend to ask their doctor for a tonic, which usually consists of nothing more than the vitamins and minerals they would have eaten in a good mixed diet. Protein foods, vegetables and fruit will supply all the necessary vitamin and mineral elements.

d. Roughage. Many dieticians and doctors are now recommending a higher intake of roughage than ever before. This is because tests are indicating that the faster food passes through the intestines the lower the risk of carcinogens forming in the bowel. It seems ridiculous that the best part of the cereal grain is stripped off, complete with vitamins, minerals and fibre, and fed to farm animals, while the incidence of constipation and more serious bowel disorders in humans continues to rise.

e. Special restrictions. Before planning a diet check that there is no restriction on any nutrient (e.g. salt, fat) due to illness or disability. (See p. 40.)

Menu for an Active Man

Breakfast
   Orange juice

   Porridge
   Milk, sugar

   Grilled bacon and tomato
   Toast, butter

   Tea

Packed lunch
   Cheese sandwiches
   Fruit cake
   Apple

   Flask of tea

Dinner
   Roast lamb
   Boiled potatoes
   Cabbage

   Apple tart, custard

Comments

1. The *energy content* of this menu is correct, but if this man were to have taken between-meal snacks or a few

Sample Day's Menu for an Active Man

	Amount	kcal	Protein	Fat	Carbohy.	Vitamins	Minerals	Roughage
Breakfast								
Orange juice	50 g	20	0	0	5.0	A,C	Calcium	trace
Grilled bacon	50 g	225	12.5	20.0	0	B	Cal. Iron	—
Grilled tomato	50 g	6	0.4	0	1.2	A,C	Cal.	medium
Oatmeal	25 g	100	3.0	2.2	18.0	B	Cal. Iron	high
Sugar	25 g	100	0	0	25.0	0	0	—
Milk (⅓ pt)	50 ml	65	3.3	3.8	4.8	A,B	Cal.	—
Tea	0	0	0	0	0	B	0	—
Toast (2 slices)	75 g	190	6.0	1.0	40.0	B	Cal. Iron	low
Butter	50 g	365	0.2	40.0	0	A,D	Cal.	—
Packed Lunch								
Cheddar cheese	50 g	206	13.0	17.0	0	A,B,D	Cal.	—
Bread (4 slices)	150 g	375	12.0	2.25	82.0	B	Cal. Iron	low
Margarine (Flora)	50 g	367	0	40.0	0	A,D	Cal.	—
Fruit cake	50 g	184	2.3	8.0	28.0	A,B	Cal. Iron	medium
Apple	100 g	46	0	0	12.0	A,C	Cal.	medium
Tea, milk, sugar		25	0	1.0	4.0	—	—	—
Dinner								
Roast lamb	100 g	290	23.0	22.0	0	B	Cal. Iron	—
Boiled potatoes	200 g	160	2.5	0	39.0	B,C	Cal.	medium
Cabbage	100 g	15	1.5	0	2.3	A,C	Cal.	high
Apple tart	100 g	281	3.0	14.4	40.4	A,C	Cal.	low
Custard	50 g	46	1.5	1.75	6.5	A,B	Cal.	—
		3,066	84.2	173.4	308.2			

pints of beer on a regular basis he would exceed his daily quota of kilocalories and probably put on weight.

2. The *protein content* is higher than necessary. Salad and egg sandwiches could be substituted for cheese, or a mixed salad eaten instead of sandwiches. These would have a lower protein content.
3. Reducing the protein content would also lower the *animal fats* in this menu. Vegetable fats should be substituted for animal fats, e.g. polyunsaturated margarine instead of butter at breakfast.
4. *Roughage* could be increased without altering the proportions too much, by eating wholemeal bread instead of white bread. This would add valuable iron and vitamin B to the diet.

Menu for a Moderately Active Woman

The day's diet described above can easily be altered to suit a moderately active woman. The intake of sugar, bread and potatoes could be reduced and brown bread could be substituted for white. For breakfast a boiled egg or porridge would be eaten – not both. A nourishing salad at lunchtime would be preferable to sandwiches and would at the same time increase vitamin and mineral intake.

A low-calorie diet would require further reductions in carbohydrate. Sugar would be eliminated, and bread, fat and potatoes drastically reduced. Cake or biscuits would be forbidden, and an orange or natural yoghurt could be substituted for a high-calorie pudding such as apple tart and custard. Throughout the menu portions could be reduced slightly.

For more information on low-calorie diets see pp. 43–5.

BABIES AND CHILDREN

Babies

Breastfeeding. Ideally small babies should be breastfed. Breastfed babies are less likely to be overweight. Breast milk provides the correct nutrients at a perfect temperature. It is sterile and hygienic, and therefore gastroenteritis is less common than among bottle-fed

Breast milk provides the right nutrients at the right temperature – and is sterile and hygenic (W.H.O.)

babies. Because the mother's immunity to many diseases is passed on to the child in breast milk, the risk of other diseases being contracted is also reduced.

Bottlefeeding
1. If babies are bottle-fed care should be taken to sterilise bottle and teat either by using chemical sterilisers or by thorough boiling.
2. Cow's milk, used for babies under one year, should be diluted with water which has been sweetened and boiled.
3. Never make more than one day's supply of feeds together, cool them rapidly and store in a refrigerator. Never keep feeds warm, e.g. in a thermos flask, before going to bed or on a journey as this can cause multiplication of germs.
4. If using a dried milk formula for babies, follow the directions implicitly. The scoop or measure should be lightly filled, neither pressed down nor heaped, and should be levelled with a clean knife. *Never* add an extra scoop or the formula will be too concentrated and may cause dehydration, which can have fatal consequences.
5. Avoid putting solids such as baby rice or rusks into babies' bottles as this may lead to obesity.
6. Babies fed on cow's milk should have their diet supplemented with:
   a. vitamin C – orange juice.
   b. iron – strained green vegetables and egg yolk.
   c. vitamins A and D – vitamin drops or cod liver oil.

Mixed feeding
1. Mixed feeding can be introduced at about four months or when the baby is taking five full 8 oz (225 ml) feeds. As it progresses, bottle feeds can be reduced.
2. The baby should be given small amounts of strained food from a spoon. The food should contain concentrated amounts of nutrients such as protein, iron and vitamins (for example: strained liver and carrot or pulse vegetables) rather than be starchy.
3. Encourage savoury rather than sweet foods.
4. Homemade adult dishes, strained or puréed in a liquidiser, are more nourishing and far cheaper than ready-prepared baby foods.
5. Egg white cannot be digested by babies under 7–8 months and should not be included in the diet before that.

Small Children
1. All foods eaten should be concentrated sources of nutrients. Children should be encouraged to eat a varied selection of foods in order that they grow up to appreciate good food and develop a sensible eating pattern.
2. The atmosphere at meal times should be relaxed and enjoyable. It is usually better to ignore a child's refusal to eat at times rather than make a fuss.
3. Plenty of fresh air and exercise before meals help to build up a healthy appetite.
4. Children should be trained to observe good habits such as washing their hands before eating and using good table manners.
5. Children should not be overtired or too hungry at mealtimes or they may become cross and lose interest in the food. Meals should be served at regular times each day.
6. Avoid too many sweet foods in the diet. Sweets and other snacks between meals are especially bad as they take the edge off a child's appetite and unbalance the diet.
7. Serve food attractively in small portions. Remove bones and gristle from meat and fish, which should be chopped or minced.
8. Adults should set a good example by eating all foods. They should not show their dislike of a food in front of children. They should introduce new foods gradually to a child.

Suitable foods
1. Plenty of animal protein such as meat, fish and eggs, as children are growing quickly.
2. Milk should be drunk rather than tea, coffee, lemonade or squash.
3. A good mixture of vegetables and at least one portion of raw fruit daily to supply minerals, vitamins A and C and roughage.
4. Some bread, butter, cereals and potatoes are necessary to supply energy as children are very active.
5. Milky foods and cheese supply calcium for bone and teeth formation.
6. If oily fish is not eaten, cod liver oil should be taken to supply vitamins A and D. Fresh fruit, orange juice or rose hip syrup will ensure adequate vitamin C.
7. Liver and other forms of offal should be eaten to supply iron. Cocoa, green vegetables, eggs, pulse vegetables and brown bread are also good sources of iron.
8. Hard foods such as carrots, apples, rusks, help to exercise teeth and gums.

Avoid: Fried foods
Highly-seasoned foods
Sugar, sweets, sweet foods, soft drinks

Obesity in Children

This is a frequent nutritional disorder. It usually occurs when a child consumes more kilocalories than are required. It may also be caused by genetic factors; for example, an overweight parent may well have an overweight child. However, it is more likely to be due to the same bad eating patterns as the parent has followed. A lack of interest in sport and physical activity may also be passed on by parents who are overweight and out of condition.

The pattern of obesity starts in early childhood. Most overweight children are bottle-fed and many of them will have had cereals put in their bottles before the age of four months. It is now known that overfeeding in the first year of life not only causes the adipose cells to increase in size, but also to increase in number. Although later weight reduction may cause the fat cells to shrink, the number remains constant, ensuring that the child will have a weight problem for the rest of its life. Eighty per cent of obese children become obese adults.

Obesity may also be triggered off by emotional causes. A mother's natural instinct is to give her baby plenty of food as proof of her love and affection. She may spoil the baby by overfeeding it and bribing it with sweets and biscuits. The overweight child becomes inactive and unhappy and finds comfort in eating, setting up a pattern which may last a lifetime. Overweight children have a tendency to chest infections, are inclined to be small and to suffer from knock knees, flat feet and fatigue.

Treatment. A low carbohydrate diet of 800–1,000 kilocalories daily will help to reduce weight but the child also needs to be re-educated in correct eating habits. Between-meal snacks should be cut out. Carbohydrate consumption should be reduced, sugar eliminated and fresh fruit, raw vegetables and cheese eaten instead of cakes and puddings.

ADOLESCENTS

Teenage boys and girls require a good supply of protein for growth, repair and normal hormone and enzyme synthesis. As energy requirements are usually high, foods with a fairly high kilocalorie content may be eaten if the diet also supplies sufficient vitamins and minerals. Milk and cheese are important as sources of protein, calcium and vitamins. Liver, meat and green vegetables are good sources of iron, a mineral which is often deficient in teenage girls because of menstruation. Too many fried foods, pastry and rich foods should be avoided as they lead to obesity and may aggravate teenage acne.

ADULTS

A normal, well-balanced diet is essential for adults.

Sedentary workers should keep a careful check on their kilocalorie intake if they are to avoid obesity. Foods with a high percentage of cellulose are necessary to prevent constipation, a condition common in those who lead an inactive life. Plenty of fresh air and exercise are necessary to keep the office worker healthy and fit. Rushed lunches with a high kilocalorie content are to be avoided. Salads, cheese and yoghurt should be chosen rather than fried foods such as chips. Fresh fruit should be eaten at coffee breaks rather than cakes and biscuits.

Manual workers can afford a high kilocalorie intake without putting on weight. 3,500 kcal a day is normal in the diet of a male manual worker. Packed lunches should contain protein, fat and carbohydrate: meat or cheese sandwiches are ideal. High energy foods such as fat meat, oily fish, fried food and starchy foods such as bread and potatoes are acceptable in the diet of a manual worker. All other constituents must also be supplied.

Pregnant and Nursing Mothers

Care must be taken that the diet of a woman during pregnancy and lactation is well balanced with particular emphasis on protective food and foods for growth. The old idea of eating for two is incorrect, but an expectant mother needs an extra supply of some food constituents to ensure the healthy growth of the foetus. Extra supplies of the following nutrients are essential:

Protein, both animal and vegetable, for growth.

Vitamins A and D for growth of bones and teeth.

Vitamin B group for energy.

Vitamin C for general health and absorption of iron.

Calcium/phosphorus for bone and teeth formation. Milk and cheese are a good source.

Iron for the manufacture of haemoglobin. Supplies of iron should be sufficient to ensure that enough iron is stored in the child's liver to last six months. Many pregnant women are found to be anaemic. They are usually prescribed iron tablets or folic acid.

Energy needs. The metabolic rate rises slightly during pregnancy, and increasingly (up to 20 per cent) as full term approaches. The appetite usually increases to meet this need, but it is essential that the expectant mother does not put on too much weight. A weight gain of approximately 22 lb (10 kilos) is acceptable and normal. An overweight mother and baby may experience difficulties at birth and the incidences of varicose veins, haemorrhoids and toxaemia in pregnant women also tend to increase with weight. It is often difficult to lose this extra weight when the baby is born.

Nursing mothers. Mothers who are breastfeeding should maintain the same diet as when pregnant with a slight increase in protein, vitamin and mineral intake. The body uses up the reserves of adipose tissue in the mother during milk production, so a mother who breastfeeds is likely to regain her prepregnancy figure more quickly than one who bottle-feeds.

OLD PEOPLE

Protein is essential for the elderly as it repairs worn-out tissues. Meat, fish, eggs, milk and cheese are important protein foods, but they are also expensive. Because of this many old people tend to avoid them and survive on totally inadequate diets of bread, tea, soup and other nutritionally inferior foods. Many elderly people do not make the effort to cook a main meal; they may be arthritic, or they may have poor cooking facilities. Those responsible for the care of an old person should make sure they eat:

a. *Fish* – white fish if they have digestive problems.

b. *Eggs* – 3–4 a week (too many may raise the cholesterol level).

c. *Meat or cheese* at least once a day.

d. *Vegetables* – two portions daily for vitamin C and iron.

e. *Raw fruit* daily to supply vitamin C.

f. *Brown bread or porridge* daily as a source of energy, vitamin B and roughage.

g. *Milk* – 300–500 ml daily to supply calcium and protein.

Some old people may have digestive problems, difficulty in chewing due to loss of teeth and other minor upsets which may restrict their choice of food. For these, steamed, stewed and poached foods are most suitable, and white fish and chicken are particularly digestible protein foods. Most elderly people can continue to follow a normal, well-balanced diet but kilocalorie intake should be reduced because basal metabolism is lower in old age and physical activity also declines. Meals-on-wheels are useful for supplementing the diet of old people.

INVALIDS

Food for the invalid and convalescent should provide the maximum amount of nourishment with the minimum amount of bulk. Foods essential in the invalid diet are:

Protein to repair diseased and wasted tissue.

Vitamin C which helps to heal tissues and wounds and prevent bed sores.

Vitamin A helps to prevent infection and create healthy mucous membranes.

Vitamin B for nervous tissue, vitality and energy release.

Iron to prevent anaemia, which is common in illness.

Roughage to prevent constipation, a common complaint in bed-ridden patients.

Old people often live on an inadequate diet (Source Photographic Archives).

Energy foods should be restricted and used to round off the appetite after more essential foods have been eaten. During illness digestion is impaired. Indigestible foods such as pastry, oily fish, fat meat, fried foods, cheese and highly seasoned foods should be avoided.

Rules for Feeding Invalids
1. Follow doctor's orders.
2. Use best quality fresh food. Never use leftovers and avoid convenience foods.
3. Observe strict hygiene in preparation, cooking and serving of meals.
4. Choose light, easily-digested foods. Steaming, stewing and poaching are suitable methods of cooking.
5. Trim food where possible removing bone, fat, gristle etc. Foods should be cut up or minced if necessary to make them easier to eat.
6. Serve meals at a regular time. Several small meals are better than a few large meals.
7. Meals are usually the highlight of the day for an invalid. The tray should be clean and neatly laid and the food served attractively in small portions in clean, individual dishes. Cloth, glasses, cutlery and napkin should be spotless.
8. The tray should be large enough and comfortable to hold. A tray with legs, or an invalid table such as those used in hospitals, is ideal.
9. Dishes should be served hot and covered to keep them so. Cold food should be well chilled.
10. Remove all food from the sickroom when the meal is finished.

Invalid Diets
1. During fever, e.g. influenza or measles, the temperature is high and much water is lost from the body through perspiration. The diet should consist of plenty of liquid foods to replace the body fluid but should not exclude solid food. If kilocalorie consumption is too low the patient gets very weak and insufficient protein will delay recovery. Drinks and light meals should be given alternately every two hours. *Protein* should be supplied in meat broths, milky drinks and eggs e.g. egg flip. *Fat* may be given in the form of milk and eggs, but other fats should be eliminated. *Carbohydrates* may be given in milk puddings, barley drinks, breakfast cereals, and sugar or glucose in food and drinks. *Vitamins* should be taken in strained citrus drinks, e.g. orange juice, or in blackcurrant juice or rose hip syrup. *Minerals* will be present in most of the foods mentioned.

2. After fever, a light diet can be followed. All foods should be lightly cooked and easy to digest. Plenty of liquids should still be taken.
Suitable foods:
Eggs – poached or scrambled.
Chicken breast, sweetbreads, white fish – poached or steamed.
Fruit, vegetables – stewed and puréed.
Jellies, ice cream, milk puddings, milky drinks.
Raw fruit, vitamin C drinks.

3. In convalescence, the patient should be able to eat small portions of most foods apart from particularly indigestible ones such as fries, pastry and suet. Remember that it is important to supply as much nourishment as possible, but kilocalorie intake will still be reduced through lack of activity.

Suitable foods
Eggs, lightly boiled, poached or scrambled; custards.
Meat – lean beef, minced and stewed; chicken breast; liver.
Fish – white fish grilled or poached with white sauce.
Vegetables – carrots; boiled or creamed potatoes; cauliflower; sprouts.
Milk and milky foods.
Fruit, fresh or stewed; trifle; ice cream.
Bread, lightly buttered; sponge cake.
Drinks – plenty of vitamin-rich, high-energy fluids.

OTHER DIETS
People suffering from certain illnesses are told to restrict their intake of a particular food. In these cases the patient should follow a diet laid down by a doctor or dietician. Such diets include *low salt; low residue* (little roughage); *bland* (for digestive upsets and ulcers); *sugar free* (diabetes); *gluten-free* (coeliac disease); *low fat; high or low protein.*

The following diets are discussed in more detail:
a. Low cholesterol
b. High-fibre
c. Gluten-free
d. Vegetarian
e. Weight-reducing
f. Diets for diabetics and those suffering from certain other illnesses.

a. Low-Cholesterol Diet
Cholesterol is a steroid (solid fat) found in all animal tissue, particularly nervous tissue. As well as being present in food it is synthesised by the liver and occurs in the bile. Excess bile salts cause gall stones. Cholesterol helps to convey fat through the blood. When too much accumulates in the blood it is deposited on the walls of the arteries, particularly those of the heart, causing them to become narrow. This is called arteriosclerosis. A small clot can easily block the narrow vessels, increasing the possibility of coronary thrombosis (heart attack). Research into the levels of cholesterol in the blood points to a connection between animal fats (which are its main source) and heart disease. As polyunsaturated fatty acids are

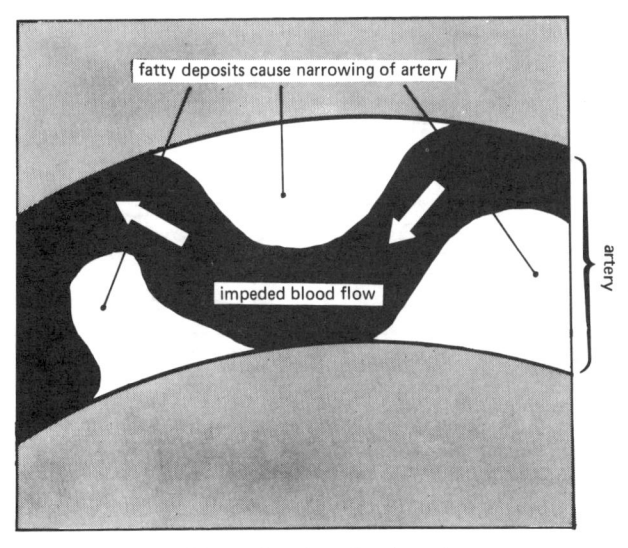

Fig. 2.1 Arteriosclerosis

thought to reduce cholesterol level in the blood, a diet high in these is recommended to people with a high cholesterol level, that is over 260 mg cholesterol per ml of blood. Research into this point is still in progress; it has not been conclusively proved.

Foods containing cholesterol. Egg yolk; milk; butter; cheese; cream; offal; shellfish; suet; lard; dripping; fat in meat.

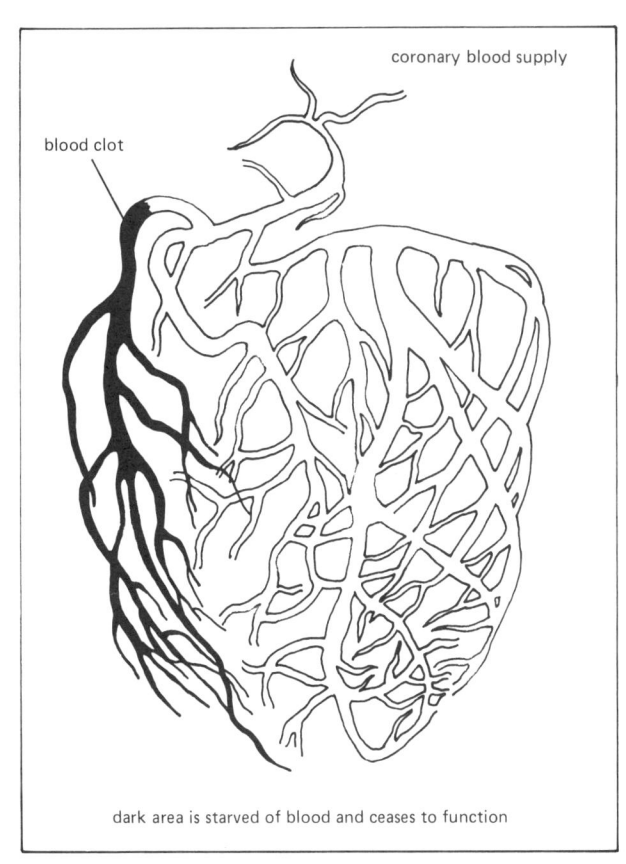

Fig. 2.2 How a heart attack occurs

Foods high in polyunsaturated acids.
1. Polyunsaturated margarines such as 'Flora'. Remember all hard margarines have been hydrogenated, changing the polyunsaturates into saturated fatty acids.
2. Polyunsaturated oils. Sunflower seed oil is the highest in polyunsaturated fatty acids (PFAs). Corn, soya bean and cotton seed oil are also good sources.
3. Vegetables and fruit do not contain cholesterol.
4. Poultry and fish are low in saturated fats.

Other factors which contribute to heart disease are:
1. Emotional stress
2. Cigarette smoking
3. Lack of exercise
4. Obesity
5. Heredity

Rules for reducing cholesterol level
1. Reduce intake of high cholesterol foods and saturated fats (see below).
2. Increase intake of polyunsaturates. Use polyunsaturated oils for cooking.
3. Avoid over-eating. Lose weight if necessary.
4. Take regular exercise.
5. Give up smoking.
6. Avoid worry and stress.

Diet

   Eat plenty of the following foods: fruit; vegetables; skim milk rather than whole milk; cottage cheese rather than full milk cheeses; vegetable proteins as a substitute for animal proteins (many of which are rich in animal fats); fish, chicken and soya protein as substitutes for meat.

   Avoid the following foods: egg yolk (no more than 3 per week); shellfish; meat (5 portions per week or less, with fat removed); offal (once a week or less); fried foods, butter, cream, sugar.

Note: The incidence of death due to heart attack has always been particularly high among white South Africans, whose diet is rich in animal fats, meat and alcohol. Their life style demands little exercise since cheap labour is easily available. The black people on the other hand had a very low rate of heart disease; their diet for cultural and economic reasons was largely vegetarian. Gradually, as the Africans' standard of living improved, they adopted a more European diet and there has since been a steep rise in the incidence of death due to heart disease. This increase is in direct ratio to the rise in the standard of living.

b. High-Fibre Diet

   This is a diet containing large amounts of cellulose or roughage. Many modern foods are highly refined – for

Some low-cholesterol foods (Bill Doyle)

example, white flour and polished rice are more popular in western countries than unrefined brown flour, rice and other cereals. Refining or milling removes the outer husk which contains large amounts of cellulose and vitamin B. Many modern foods, particularly convenience foods, are over-refined, containing large amounts of starch and little or no roughage. As a result the incidence of diseases of the colon such as diverticulosis, constipation, haemorrhoids and cancer of the colon are very common in so-called developed countries. In countries such as Africa, where the diet contains large amounts of roughage, these diseases are very rare.

Most people would benefit from a high-fibre diet, particularly those who suffer from constipation or other complications of the bowel.

Diet

Eat plenty of the following foods: wholemeal bread, porridge, foods containing oatmeal, nuts, fruit, vegetables (preferably raw and unpeeled), brown rice, wheat germ, bran.

Cut down on the following foods: white bread, white rice, white sugar, convenience foods.

Note: It is important that the bowel should be emptied regularly. Failure to do so causes the faeces to harden so that they become difficult to pass, resulting in constipation and haemorrhoids.

c. Gluten-Free Diet

Coeliac disease occurs when gluten cannot be tolerated by the body. It is thought that the lack of the enzyme which breaks down the gluten causes the molecules to be absorbed without being fully broken down. This damages the sensitive lining of the small intestine and much of the food eaten is thus passed through the body unabsorbed, resulting in loss of weight and a failure to benefit from many of the nutrients required by the body.

All wheat-based foods must be eliminated from the diet. These include flour, bread, cakes, biscuits and many convenience foods. Foods containing rye and, in severe cases, oats and barley are also forbidden.

Avoid the following foods:
1. Bread, biscuits, crispbreads, cakes, made from wheat and rye.
2. Meat products containing cereal, e.g. hamburgers, sausages, pies.
3. Fish coated in batter or breadcrumbs, e.g. fish fingers.
4. Many soups, spreads (e.g. mayonnaise), sauces and gravies including those in canned foods.
5. Many sweets and snack foods (e.g. crisps, ice cream).
6. Some breakfast cereals.
7. Baking powder.
8. Almost all convenience foods – they contain wheat-based thickeners.

Suitable foods:
1. Special gluten-free flour and baking powder are available.
2. Many cereals can be eaten, e.g. maize (cornflakes), rice (rice krispies), soya, and in some cases oats and barley.
3. Cornflour should be used as a thickener.

Note: It is thought that some diseases such as multiple sclerosis may be delayed, arrested or even cured by feeding the patient with a gluten-free diet, but no certain proof of this has yet been obtained.

d. Vegetarian Diets

Vegetarianism is the practice of living on vegetable foods. Animal flesh, fowl and fish are not eaten. A large proportion of the world's population is on a vegetarian diet usually through poverty and the unavailability of animal foods rather than from choice. Many vegetarians have religious reasons for not eating animal flesh; others consider it inhumane to kill animals for food. With the current influence of eastern culture and the rising cost of meat, more and more people are following vegetarian diets because they consider it more nutritious and healthy to eat only vegetables and cereals. Health food shops and well-stocked delicatessens now provide many items which used to be difficult to find, enabling vegetarians to follow their diet more easily.

There are two basic types of vegetarian: (a) the *lacto-vegetarian* who refuses to eat animal flesh and fish but will eat animal products such as milk, cheese and eggs; (b) the *strict* vegetarian or *vegan*, who eats no animal produce whatsoever. It is this diet which may produce some nutritional deficiencies unless it is carefully planned. Another type of vegetarian diet, the macrobiotic diet, is dealt with below.

Nutrition. As the diet of lacto-vegetarians is usually well-balanced, the rest of this section deals mainly with the vegan diet. It is accepted that vegetable protein foods contain fewer essential amino acids than animal proteins. In order to obtain sufficient protein vegetarians must eat large and varied amounts of vegetables and cereals to compensate for the lower quality protein in them (see supplementary value p. 5). Textured vegetable protein, a synthetic food made from soya beans, is a useful ingredient in a vegan diet as it can be used in so many ways and is particularly rich in protein.

If a well-balanced supply of vegetables and cereals is eaten it is unlikely that any deficiencies will occur, with the exception of vitamin B_{12} deficiency which used to be common. Most vegans now take supplements of this vitamin. Vitamin A deficiency is sometimes found among those eating mainly rice, and niacin is sometimes lacking amongst maize-eating vegetarians. Surveys have found that obesity is rare amongst vegetarians; blood cholesterol level is lower and owing to the large amount of fibre eaten, 'western' diseases such as diverticulosis and cancer of the colon are also less common.

Handling a vegetarian diet
1. This type of diet must be carefully planned. A basic knowledge of nutrition is necessary to ensure that it is well-balanced, and the use of ingenuity and skill will avoid a dull and monotonous diet.
2. While several small courses are usual, one large one may be served, with an assortment of dishes on the table at one time in Chinese or Indian style. This may be followed by a sweet and tea or coffee.
3. Any vegetable soup may be used but animal stock and meat-based stock cubes should be avoided.
4. Pasta dishes and cereal dishes based on wheat, maize, rice or millet are also substantial and nourishing. *Examples:* risotto, muesli, fried rice.
5. Pulse vegetables should be included for their high protein content. Many dried pulses are available, e.g. lentils, split peas, haricot beans.
6. Nuts are a good source of protein and fat. They may be sprinkled over dishes such as salads, used as the main ingredient of a dish (as in nut rissoles) or put into stuffings and cakes.
7. Whole cereals or ground whole cereals are preferable to processed or milled cereals because of the vitamin B they provide. Use wholemeal flour and bread rather than white flour and bread; brown rice rather than polished white rice.
8. Cheese, milk and eggs, if taken, should be included often as they are well-balanced and nutritious foods.
9. Herbs, spices, and flavourings such as yeast extract should be used to improve the bland flavour of vegetarian dishes. Sauces add moisture and flavour to dishes and improve their colour and appearance.
10. Use soya flour for thickening as it has a high nutritive value.
11. Use vegetable fats and oils only.

Suggested dishes
a. *Lacto-vegetarian:* Omelettes; quiches and savoury flans; curried eggs; pizza; macaroni cheese; milk puddings; cakes and all the dishes listed below.
b. *Vegan:* Vegetables casseroles and pies; salads; vegetable soups; fried rice; nut cutlets; TVP dishes, e.g. spaghetti bolognese; stuffed vegetables; some Indian and Chinese dishes.

Macrobiotic diet. This is the name for a diet made up exclusively of the currently popular 'health foods' — natural foods mainly of vegetable origin, which have been organically grown, i.e. no artificial fertilisers or pesticides have been used in their production. Zen macrobiotic food is the traditional food of ancient Japan. Japanese Buddists who follow such a diet are shown in surveys to be among the longest lived and healthiest peoples in the world. Refined foods, preservatives and chemicals are avoided. Fruits and vegetables are used in season or dried naturally. Grain is the main food eaten, e.g. whole wheat flour, oatmeal, barley, millet and unpolished rice.

The diet is generally a vegetarian one although meat is not forbidden. Sugar is rarely used; honey can be taken instead. Like all vegetarian diets, the macrobiotic one has a high fibre content.

e. **Weight Reducing or Low Kilocalorie Diet**

The so-called slimming diet is what immediately springs to mind when the word 'diet' is used in conversation. A diet is what we eat and drink each day. A slimming diet is one in which the amount of energy-producing foods is reduced to less than that which the body requires for daily heat and energy production. The body then begins to use up its own store of energy – adipose tissue – so that weight is lost. Losing weight is an emotive subject; fads and fallacies abound. A sound knowledge of nutrition is required in order to plan a well-balanced slimming diet.

Remember the following points:
1. It is advisable to consult a doctor before beginning a strict low energy diet. Children, adolescents and expectant mothers should only diet on doctor's orders.
2. The only way to reduce weight is to reduce the intake of kilocalories, or increase activity or, preferably, do both together.
3. There is no quick and easy way to lose weight that is of lasting benefit. A small, steady weight loss is preferable to a sudden reduction in weight which can, in fact, be dangerous. An average loss of 0.5–1 kg (just over 2 lb) each week is ideal and is less likely to be regained than is a sudden loss in weight.
4. Avoid crash diets; most are unbalanced and some (such as bananas and milk, grapefruit and toast), although they reduce weight, may cause illness, e.g. anaemia.
5. Remember that no food is slimming. All foods and liquids with the exception of water, black tea and coffee contain kilocalories or potential energy and, therefore, potential fat.
6. No food helps to burn up calories more quickly.
7. Although exercise alone will not result in marked weight losses, it tones up flabby muscles, improves the figure, is excellent for good health and is better taken regularly than in spurts.
8. It is easier to reduce food consumption slightly before too much weight is gained than to wait until a major diet is necessary.
9. Sauna baths, muscle toners, slimming tablets etc. have little lasting effect.
10. Sufficient motivation is required to keep to a slimming diet. If health and good appearance do not suffice, it may be a good idea to join a group scheme where sympathetic slimmers share and compare problems and compete with one another.

Diet
1. The diet must be well-balanced. The correct nutrients must be supplied in a concentrated form. This is most important in the case of children and adolescents.
2. The purpose of a diet, apart from losing weight, is to *retrain* the appetite into a good eating pattern. This means that the nearer the diet is to a normal, well-balanced set of meals the more likely the dieter is to
   a. keep to the diet
   b. keep the weight down afterwards.
3. It is dangerous to plan a diet on the kilocalorie content of food alone. One could obtain sufficient calories by eating chips for breakfast, dinner and tea! The overall nutritive value of each food must also be considered. This is why cheese, although high in kilocalories, is rarely excluded from well-balanced slimming diets.
4. Restrict carbohydrate intake, especially sugar, sweets and biscuits. Most of these foods contain little nourishment.
5. If carbohydrates are reduced, it follows that there is a reduction in fat intake, e.g. less bread means less butter. Fats may be reduced but should not be eliminated as they are a source of vitamins A and D and essential fatty acids.
6. Protein intake should remain unchanged or increase slightly.
7. Plenty of fruit and vegetables should be eaten. They contain few carbohydrates, are filling and rich in vitamins, minerals and roughage.
8. Eating between meals should be avoided. If it cannot be avoided some raw carrot or celery may be nibbled.
9. Slimming foods such as starch-reduced bread and crispbread are expensive and rarely worth the extra money. Slimming diets based on the use of slimming biscuits or the like are monotonous and cannot be followed indefinitely. Nor do they encourage a new eating pattern.

Causes of obesity
1. Over-eating or eating too much of the wrong foods.
2. Reduced activity resulting in lowered energy needs, e.g. school-leavers may give up sport; a manual worker may take an office job.
3. Boredom, habit or addiction to food – sometimes as a compensation for unhappiness.
4. Entertaining – an active social life may involve too many rich meals and alcohol consumption.
5. Pregnancy – many women gain weight at this time and do not lose it afterwards. Successive pregnancies may mean progressive weight gain.
6. Lack of nutritional education.
7. Poverty, which prompts people to eat cheap, high-carbohydrate foods.
8. A few medical conditions, e.g. hormone imbalance, kidney disease, malfunctioning of the hypothalamus.

Low energy menu
Breakfast. Include a protein food; as this is more sustaining it helps reduce the temptation of mid-morning snacks.
Main courses. Two portions daily from the following: lean meat, poultry, fish, eggs, cheese. Eat offal, e.g. liver, kidney, once a week for iron. Trim visible fat from meat and grill, roast or stew. Avoid frying.
Vegetables and fruit. Raw vegetables or fruit should be eaten daily as well as one or two portions of cooked vegetable.
Milk and milk products. 250 ml (½ pt) to be distributed between food and drinks. Skim milk is preferable in strict diets as it is less fattening. Do not exceed 15 g butter/25 g cheese, preferably cottage cheese, daily.
Starchy foods. 1 slice of bread or 1 small roll daily except on very strict diets. 1 small potato or small portion of rice or pasta.
Drinks. 1 cup of unsweetened tea/coffee after each meal and mid-morning and afternoon. Artificial sweeteners may be used, but it is better to get used to unsweetened drinks.

Suggested weekly plan

	Breakfast	Lunch or Tea	Dinner
Sun.	Whole orange Muesli	Grilled mackerel Roll	Roast lamb Casserole of root vegetables or ratatouille Apple snow
Mon.	Boiled egg Brown bread and butter	Cheese salad Roll and butter	Shepherd's pie Carrots Pear
Tue.	Grapefruit half Porridge, half slice toast	Cauliflower cheese Fresh fruit	Grilled white fish Broccoli, small potato
Wed.	Poached egg on half slice toast Raw tomato	Rollmop herring Green salad Half slice brown bread	Baked chicken Peas Yoghurt
Thur.	Orange juice Grilled rasher and mushroom Half slice toast	Coleslaw Hard-boiled egg	Beef curry Small portion rice Fruit salad
Fri.	Grapefruit juice Scrambled egg on toast	Savoury rice with mushrooms, peas	Liver casserole Fresh fruit
Sat.	Grilled kipper Raw tomato	Cheese omelette small roll	Consommée Mixed salad

Alternative main courses: Grilled steak or hamburger • Kebabs or baked fish • Corned beef

Avoid these	*Restrict these*	*Eat plenty of these*
fried foods	starchy vegetables	lettuce and salad vegetables
tinned fruit	pulse vegetables	mushrooms – not fried
cream	fat meats	green vegetables
sugar, jam	fruit, especially grapes and bananas	liver
biscuits, sweets		white fish
cakes, pastry		fresh fruit juice
mayonnaise	bread, toast	low fat natural yoghurt
alcohol, soft drinks, squashes	breakfast cereals	cottage cheese
convenience foods	rice, pasta	lean meat
nuts	cocoa	
	butter/margarine	

Anorexia Nervosa

This is a psychological condition affecting girls and young women in which they drastically reduce weight by excessive dieting. Although not common it is on the increase. The onset is usually associated with emotional conflict; some doctors feel that the disease is evidence of a rejection on the part of the patient to accept puberty and growing up. It is basically a neurosis, an inability to see oneself as one really is. Patients develop an obsessive aversion to food and will even take laxatives and induce vomiting in order to avoid putting on weight. They lose weight rapidly, become thin and emaciated. Side effects include loss of menstruation and sensitivity to cold. Severe cases result in death. Anorexia always requires specialist treatment in hospital, usually involving a high-carbohydrate diet initially to promote weight gain, and long-term psychotherapy.

In Dublin, a self-help group, Anorexic Aid, has been formed to help people over this illness.

f. Diets for Diabetes and other illnesses
Diabetes

This is caused by a breakdown of insulin production in the pancreas. After the digestion of carbohydrates, insulin enables the glucose produced to be used by the body for energy production. If insulin is not present, or is reduced, glucose accumulates in the blood and is excreted by the kidneys.

Diabetes tends to be hereditary and is more common in people who are overweight and advanced in years.

Diet. A special diet must be worked out to suit each patient individually. It is essential to keep to the diet; too little carbohydrate could cause faintness and coma, while too much would cause excretion of glucose in the urine and the body would become dehydrated.

The diet is basically a low carbohydrate one. Sugar is cut out completely and saccharin may be substituted for it. Many foods are now manufactured for use by diabetics including jam, biscuits and sweets. These usually contain artificial sweeteners.

Consumption of the following foods is reduced: bread, cakes, biscuits, cereals.

Those with a mild form of diabetes can keep quite healthy on a low carbohydrate diet without taking insulin. More severe cases need daily insulin injections.

Note: A patient with diabetes needs to eat food regularly and often.

Anaemia

Iron deficiency anaemia occurs when the red blood cells in the body have a reduced haemoglobin content, caused by insufficient iron in the body. This reduction of iron may be brought about by:
a. Low consumption of iron in the diet
b. Frequent or heavy blood loss
c. Defective absorption of iron from the intestine

Anaemia is one of the most common nutritional deficiencies in the western world. It is very common in women, particularly in those who have heavy periods, and in young children. It is essential for pregnant and nursing mothers to have a high iron intake. Iron tablets are usually prescribed during pregnancy.

Symptoms. Tiredness, breathlessness, headache and lack of vitality.

Prevention. A diet rich in foods containing iron, e.g. lean meat and offal such as liver and kidney; eggs; bread, especially wholemeal; cocoa, dark green vegetables.

Note: Absorption of iron is assisted by vitamin C. The diet should include a daily supply of this.

Severe anaemia requires treatment with iron tablets or injection.

Ulcers

An ulcer is a raw area in the lining of the digestive tract, usually the stomach (gastric ulcer) or the duodenum (duodenal ulcer). They are caused by the action of pepsin on an inflamed section of the stomach, and result in discomfort and pain.

Causes. Stress, overwork, erratic diet, and especially skipping meals.

Treatment. The sufferer must learn to take life easy. He or she must be removed from the stresses which contributed to the condition and have plenty of rest.

Diet
1. Avoid highly seasoned foods and fatty foods.
2. Avoid foods containing cellulose, e.g. cereals, fruit and vegetables.
3. Avoid alcohol and smoking.
4. Eat small meals every 2–3 hours rather than a few large meals.
5. Drink plenty of milk, especially at night.
6. *The following foods are recommended:* milk puddings; white fish; chicken; lean meat; eggs; white bread; potatoes, puréed or sieved vegetables; custard; yoghurt; jelly, sponge cakes.

Note: Dyspepsia or indigestion is often caused by rushed eating, but some people are particularly prone to it. The diet is similar to that for ulcers.

Gastroenteritis

This is a general name for various forms of food poisoning and infections of the gastro-intestinal tract. Symptoms include nausea, severe abdominal cramps, vomiting and diarrhoea.

Diet. Take liquids only for 12–24 hours. These may include glucose drinks, water, carbonated drinks, skim milk, barley water. Gradually include easily digestible foods such as poached eggs, milk puddings, steamed white fish.

Diarrhoea is treated in a similar way.

Acne

This skin condition, common during adolescence, can be improved by a diet rich in fruit and vegetables, preferably raw. *Avoid:* fried foods, cream, pastry, cakes, chocolate.

Gall Stones

These are caused by the accumulation of bile salts in the gall bladder. The main symptom is abdominal pain.

Diet. *Avoid* all fatty foods, e.g. fried foods. Pain can also be induced by foods which cause flatulence, e.g. baked beans. *Suitable foods:* lean meat, white fish, poultry, sieved vegetables.

Treatment. Unfortunately the stones cannot be dissolved by dieting; this merely helps to reduce the discomfort caused to the patient, until the stones or gall bladder can be removed by surgery.

MEAL PLANNING

A good meal should be nourishing and well cooked, with a careful blend of foods and flavours. Our enjoyment of food is increased by many factors such as appearance, taste and smell: a tasty meal well presented is more enjoyable than an elaborate but unattractive one.

One of the most important aspects of any good meal is a relaxed, happy atmosphere; this helps us to enjoy our food and makes a meal a pleasurable occasion. Mealtime is often the only time when the whole family is gathered together and it should be an opportunity for relaxation and casual conversation.

The digestion of food is easily affected by many psychological factors. If we are tense or nervous, if we bolt food in a hurry or if meals are badly cooked without much flavour, our digestive juices fail to function normally. The vagus nerves to the stomach and other digestive organs become so tensed up that the stomach secretes excess acid which leads to indigestion. In the long term tension and rushed meals cause gastric and duodenal ulcers.

Points to Consider
1. The occasion
2. The number of people being catered for
3. The amount of money available
4. The ability of the cook
5. Equipment available, e.g. type of cooker; whether a freezer is available
6. Time available for preparation
7. Time of year: (a) foods in season
   (b) foods suitable to weather – light meals and salads in hot weather and hot puddings in cold weather
8. Individual tastes (within reason)
9. Foods available and their nutritive value
10. Dining facilities

Vary colour, flavour and appearance of courses, e.g. potato soup, Irish stew and rice pudding would be a bad combination.

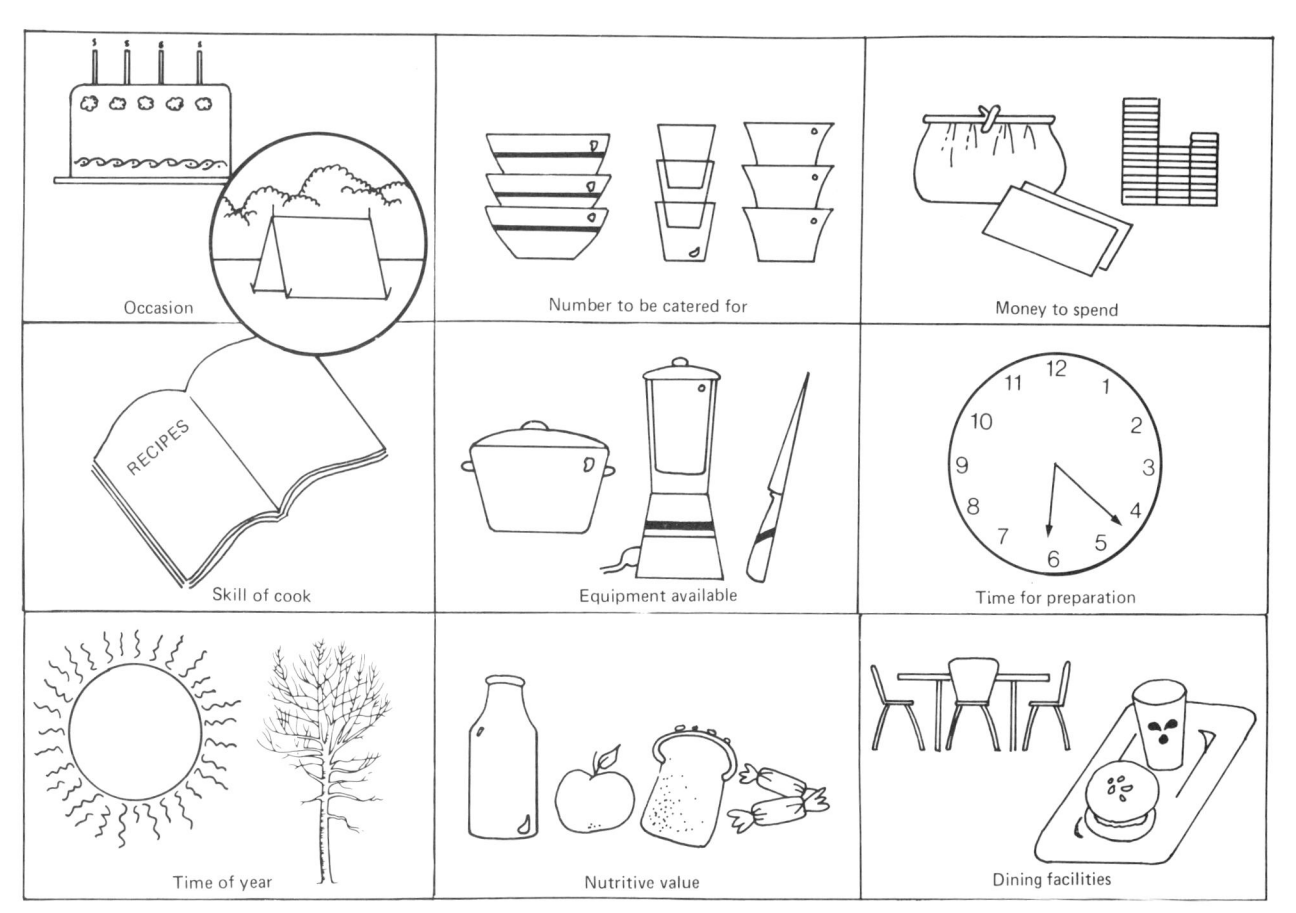

Fig. 2.3 Points to consider in meal planning

Avoid using the same method of cooking in several courses, e.g. fried fillet of plaice; fried chicken and apple fritters would be a bad combination.

Vary the texture – some dishes should be firm, some soft and some crisp. Salads and raw fruit are a good way of introducing texture.

Avoid serving all hot or all cold courses within the same meal; alternate hot with cold.

The Menu

A menu is the plan of dishes available for a particular meal. Family dinners usually consist of two or three courses followed by tea or coffee. More elaborate meals may consist of the following courses:

Hors d'oeuvre
Soup (potage)
Fish
Entrée
Remove or relève
 (main course)
Sorbet
Game
Entremets (hot or cold sweet)
Dessert or fresh fruit
Petit fours or cheese board
Coffee

Weekly Menu Plan

The secret of meal planning is organisation. It is a good idea to plan the main meals for one week in advance. The week's menu will serve as a basis for the weekly shopping. Leftovers are catered for and the preparation and/or cooking of two meals can be done at one time – for example, a double quantity cooked on one day can be used up in a different way the next day (see p. 106). The plan need not be followed too rigidly; a good bargain or a scarcity of a particular food would make it worthwhile to change it. A weekly plan saves time and energy and ensures that the family enjoys a varied and nutritionally balanced diet.

Avoid waste and unnecessary extravagance, and economise on fuel (see p. 106).

Cookery Plan

Meals should be served punctually and regularly at times which are suitable for most members of the family. In order to have each dish cooked correctly it is useful to make a time plan, which takes into account the length of time each dish takes to cook. Those which take longest are put on first and those which cook in a short time are put on last. Following a time plan will reduce the necessity of keeping food hot or keeping a hungry family waiting for a meal. A detailed timetable is particularly useful for a beginner or for a hostess preparing an elaborate dinner party. Large-scale cooking such as batch cooking for a freezer is easier, too, when a timetable is worked out beforehand. With experience, the time plan can become less detailed until finally it can be worked out mentally.

Sample menu and time plan
 Egg mayonnaise

 Chicken casserole
 Baked potatoes

 Apple crumble
 Baked custard

Timetable: Meal to be served at 1 p.m.
11.15 Turn on oven. Prepare chicken and vegetables for casserole.
11.25 Start casserole and put it and potatoes to cook.
11.35 Set table.
11.45 Wash salad vegetables for first course. Chill. Hard boil the eggs.
12.00 Start apple crumble, put to cook.
12.20 Make custard, put in oven.
12.35 Assemble egg mayonnaise, garnish and chill.
12.45 Put water, rolls and butter on table.
12.50 Slit and finish off baked potatoes.
12.55 Tidy up and turn off oven (if electric)
1.00 Serve egg mayonnaise, followed by casserole and potatoes.

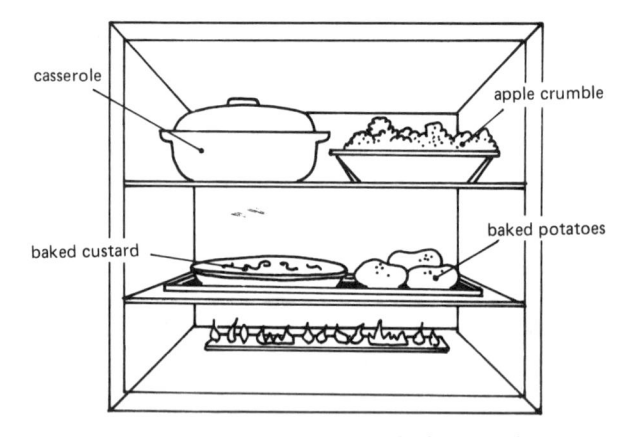

Fig. 2.4 Position of dishes during cooking

All the cooking for this meal is done in the oven at the same temperature. The crumble should be placed on the top shelf to brown the crumbs, with the casserole beside it, and the custard on a tray of water on the lowest shelf with the potatoes beside it. Use the time during which the food is cooking to assemble cold dishes and set the table.

Presentation

Serve meals attractively, on clean dishes with any

splashes wiped off. Hot foods should be served really hot on piping hot plates, and it is best to use a simple garnish such as chopped parsley, a wedge of lemon or a halved tomato rather than an elaborate garnish which will take so much time that the food may get cold. Avoid overhandling.

The table should be attractively arranged, with clean, fresh china, glass, cutlery and linen. The table setting for a dinner party should be carefully thought out, arranging linen and china or pottery to suit the colours of the dining room. The colours of the chosen food should also blend in with the overall plan. A formal setting is suitable for an elaborate meal. A casual homely meal is more suited to informal pottery and tablemats. Floral arrangements should blend in with colour and mood.

ENTERTAINING

The reason why most people entertain is to enjoy a few hours' relaxation in the company of friends or relations. Good food, wine, music and dancing can add considerably to the enjoyment of the occasion.

Many people take entertaining very seriously and feel that a cordon bleu meal and vintage wines are necessary in order to 'win friends and influence people'. The danger in this is that the hostess will wear herself out preparing an elaborate meal and worrying about all the details so much that she will be unable to enjoy herself or put her guests at ease.

The two main ingredients of a successful party are a good mixture of people and a warm and friendly atmosphere. Some of the best parties are impromptu gatherings where everybody takes 'pot-luck'.

On the other hand, it is necessary on certain occasions to entertain more formally. The points listed below should help to make such occasions run smoothly.

Formal Entertaining
1. Try to make as many preparations as possible (e.g. cleaning, extra housework) the day before.
2. Shopping should be done in advance, either the day before or on the morning of the party. This allows for changes in the menu if some foods are unavailable or poor in quality.
3. Sort out tableware, linen and cutlery. China and glasses should be clean and sparkling.
4. Choose food carefully:
   a. Plan menu so that as many items as possible can be prepared in advance. It is a good idea to have a cold first and last course – e.g. paté and fruit salad. This means that only the main course needs last-minute attention.
   b. Avoid dishes which need too much attention or those which must be served the moment they are cooked, e.g. soufflés, fried foods.
   c. It is risky to try out a recipe for the first time on guests. Practise it first on the family!
   d. Remember that a simple meal is often more nutritious and enjoyable than a rich, elaborate one which may well be fattening and indigestible.
   e. Introduce salads, fresh fruit and vegetables into the menu.
   f. Only serve foods which are an acquired taste (e.g. curry) if you are providing a choice (e.g. at buffet parties) as some people may not like them.

Fig. 2.5   A formal table setting

Buffets

A buffet is the simplest and most economical way of entertaining a large number of people. The atmosphere is relaxed and informal and guests can move about, help themselves and sit where they please. It is an ideal party arrangement for teenagers and young adults, but a sit-down meal would probably be more suitable for small children, and more comfortable for older people.

A cold buffet is very convenient as it can be prepared in advance. Dishes should be chosen for colour and appearance as well as taste.

Hot dishes such as curries, casseroles and cheese fondue are more substantial but may be difficult to keep warm if large numbers are to be catered for.

Guidelines for buffets
1. Arrange food, plates and cutlery so that guests can help themselves. (Avoid paper plates and cups as they tend to bend easily.)
2. Make sure there are plenty of napkins to wipe sticky fingers, coasters to protect tables, and ashtrays.
3. Food should be easy to manage. It should be possible to eat it with a fork or with the fingers.
4. Arrange plenty of seats and some small tables around the room.
5. Drinks can be served from a separate table. A wine cup is a pleasant and reasonably economical choice.
6. Have plenty of salads, nuts and dips.
7. Hot dishes should be kept warm over a spirit lamp, hot plate or heated trolley.

Suggested foods: *Cold* – pâtés and terrines, quiches and other savoury flans, smorrebrod (open sandwiches), cold meats, salads, kedgeree, chicken joints, fish cocktails, sandwiches. *Hot* – curry, risotto, vols-au-vent, pizza, cocktail sausages, sausage rolls.

THE FOOD BUDGET

The percentage of housekeeping money spent on food should be generous, since a well-fed family is less likely to spend money on patent medicines and doctor's bills. The amount allocated to food will vary according to income and circumstances. Food will be the largest expense in the budget of a low-income worker whereas it will represent a smaller percentage of the salary of the well-to-do. In both cases the following points should be noted:
1. Buy good quality food.
2. Know the current prices of foodstuffs and shop around for good value.
3. Buy fruit and vegetables in season or grow them at home.
4. Note the prices of meat; cheap cuts are just as nourishing as expensive joints.
5. Avoid overpackaged and convenience foods; they are usually expensive and not very nutritious.
6. Economise on fuel.

Low-Budget Cooking

Many people on low incomes, such as the elderly, the unemployed and the poor, have a very limited amount to spend on food. Many of them have little knowledge of food values and shop unwisely, eking out a miserable existence on foods such as bread, jam and tea.

When money is restricted, remember the following additional points:
1. Use cheaper cuts of meat and fish. Chicken, herrings and kippers are often good buys.
2. Make enough stew for two days, filling it out with dumplings and root vegetables. Make a chicken last for two or three days using leftovers for pies and the bones for soups.
3. Cereals and pulse vegetables are cheap and nourishing. Use porridge instead of the more expensive cereals. Dried peas and beans make stews go further and contain valuable vegetable protein.
4. Salads and green vegetables supply vitamins and iron. Unfortunately fruit, which is another good source of vitamin C, is expensive. Many people on low incomes who are entitled to free medicines could use vitamin supplements such as iron and multivitamin tonics to ensure an adequate supply. Do not neglect roughage: porridge, root vegetables and brown bread are cheap sources.
5. Eggs, milk and cheese are still relatively cheap and provide a good supply of protein, fat, minerals and vitamins.
6. Use margarine instead of butter – it is cheaper.
7. Home made jam, pickles, yoghurt and so on are cheaper than the bought variety.

Breakfast

The importance of a good breakfast cannot be over-emphasised. Research has shown that children who go to school without a breakfast are drowsy, unable to concentrate and aggressive. The efficiency of the body is very low before breakfast as the blood sugar and, therefore, the energy output are low after the long overnight fast.

A good breakfast provides a fresh supply of energy. This means that a person is more efficient, works harder and is also less accident-prone.

Breakfast should supply some vitamin C, protein and mineral elements with carbohydrates and fat for energy. A glass of orange juice or half a grapefruit; porridge or prepared cereal with milk; a boiled egg and a slice of brown bread and butter will supply all these nutrients.

Food Shopping
1. It is more economical and time-saving to shop once a week.
2. Check the larder and make a list of stocks which need replenishing.
3. Add to this a shopping list based on the items in the weekly menu.

4. Avail of genuine reductions, bulk-buying where appropriate.
5. Remember that supermarkets usually sell general groceries cheaply but tend to be expensive for meat, fish, fruit and vegetables.

QUESTIONS

1. List the special points which must be considered when planning meals for **a.** invalids **b.** vegetarians. Plan a menu for one day for either of the above, giving reasons for your choice of foods.
2. What are the usual causes of:
   a. gallstones
   b. diabetes
   c. ulcers
   d. high cholesterol

   Write a three-course dinner menu that would be suitable for a person suffering from one of these complaints, giving reasons for your choice.
3. What are the usual causes of obesity?
   Write a menu for three days (breakfast, lunch, dinner) for a female office worker who wants to lose weight. Refer to the kilocalorie/kilojoule value of each of the foods mentioned and show the total energy intake for each day. List the ill effects of obesity on the body.

3 Protein Foods

MEAT

Meat is considered to be the most important protein food of all. As it is fast becoming a luxury item, some knowledge of the various cuts of meat and methods of cooking and storing them is essential if one is to get good value for money. The term meat covers *carcase meat* – beef, mutton, lamb, pork and bacon; *poultry* – chicken, turkey, duck and goose; and *game* – animals and birds protected by law and killed for sport, such as pheasant.

Nutritive Value

Protein. Meat is rich in high-biological-value protein. The main protein is myosin, but albumin and globulin are also present and the connective tissue contains the proteins collagen and elastin.

Fat. All meat, even the leanest cuts, contains some fat. The amount of fat present in meat depends partly on the animal (chicken, for instance, contains relatively little fat) and partly on the type of cut (streaky bacon will have a higher percentage of fat than gammon).

Carbohydrate. There are no carbohydrates in carcase meat. There is some glycogen in liver but this changes to lactic acid after slaughter.

Vitamins. Meat is a good source of B group vitamins thiamine, riboflavin and niacin; liver, kidney and pork are particularly good sources. Liver is rich in vitamin A and suet contains some vitamin D. Vitamin C is lacking in all meat although there are traces in fresh liver.

Mineral elements. Meat contains iron; liver and kidney are a good source of this. There are small amounts of sulphur in most meats and offal is rich in phosphorus. Calcium is lacking in all meat except tripe, which is a good source owing to the use of lime in its preparation.

Water. Most meat is about 70 per cent water, although the proportion in fatty cuts is less.

Extractives. These are natural flavourings present in the tissue of meat, which dissolve into the cooking liquid or fat and give meat its characteristic succulent flavour. They stimulate the flow of digestive juices and are said to increase the metabolic rate.

Structure

Lean meat or muscle tissue is composed of bundles of very tiny fibres or cells which contain protein, mineral salts and extractives dissolved in water. Each fibre is surrounded by a wall of elastin, a tough connective tissue which is also found in arteries and tendons. The bundles

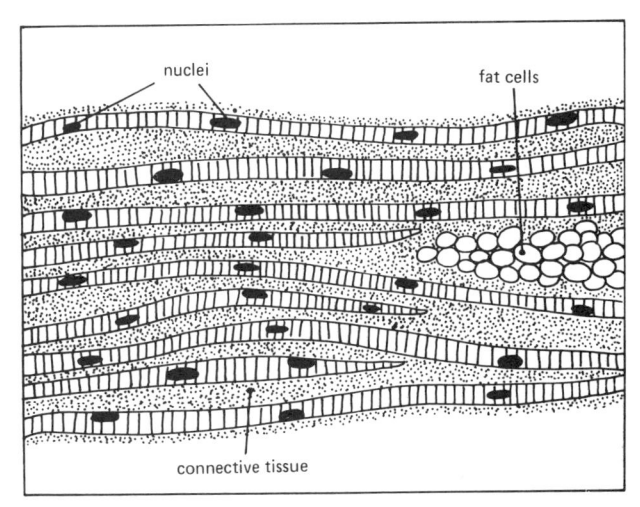

Fig. 3.1 Structure of meat

Composition of Meat (per 100 g)

	Energy kcal.	Value kj.	Water g	Protein g	Fat g	Carbohydrate g	Ca mg	Iron mg	Retinol µg	Carotene µg	Vitamin D µg	Thiamine mg	Riboflavin mg	Nicotinic acid mg	Vitamin C mg
Bacon															
(lean raw)	147	617	67	20.2	7.4	0	9	1.2	0	0	0	0.4	0.16	2.9	0
(boiled joint)	325	1346	49	20.4	27.0	0	13	1.6	0	0	0	0.41	0.21	2.7	0
(grilled rashers)	416	1722	35.2	24.9	35.1	0	12	1.5	0	0	0	0.41	0.16	4.4	0
Beef															
(roast sirloin)	284	1182	54.3	23.6	21.1	0	10	1.9	0	0	0	0.06	0.25	4.8	0
(stewed)	223	932	57.1	30.9	11.0	0	15	3.0	0	0	0	0.03	0.33	3.6	0
Lamb															
(roast breast)	410	1697	43.6	19.1	37.1	0	10	1.5	0	0	0	0.06	0.17	3.4	0
(roast leg)	266	1106	55.3	26.1	17.9	0	8	2.5	0	0	0	0.12	0.31	5.4	0
(stewed neck)	292	1216	52.6	25.6	21.1	0	10	2.2	0	0	0	0.04	0.18	2.7	0
Pork															
(roast leg)	286	1190	51.9	26.9	19.8	0	10	1.3	0	0	0	0.65	0.27	5.0	0
(grilled chops)	332	1380	46.3	28.5	24.2	0	11	1.2	0	0	0	0.66	0.20	5.7	0
Veal (roast leg)	230	963	55.1	31.6	11.5	0	14	1.6	0	0	0	0.06	0.27	7.0	0
Chicken (roast)	148	621	68.4	24.8	5.4	0	9	0.8	0	0	0	0.08	0.19	8.2	0
Liver (lambs)	232	970	58.4	22.9	14.0	3.9	12	10.0	20,600	60	0.5	0.26	4.4	15.2	12
Kidney (fried, lambs)	155	651	66.5	24.6	6.3	0	13	12.0	160	0	0	0.56	2.3	9.6	9
Pork sausages (grilled)	318	1320	45.1	13.3	24.6	11.5	53	1.5	0	0	0	0.02	0.15	4.0	0
Beefburgers (fried)	264	1099	53.0	20.4	17.3	7.0	33	3.1	0	0	0	0.02	0.23	4.2	0
Beef stew	119	498	77.7	9.6	7.5	3.6	19	1.2	0	1600	0	0.04	0.10	1.7	Tr.

of fibres are held together with collagen, another form of connective tissue which also encases the muscle and anchors it to the bone.

Connective tissue. This is a fibrous protein which is insoluble in cold water. Collagen changes to gelatine when subjected to moist heat, and this dissolves in water making the meat tender. Elastin contracts with heat, squeezing out some meat juices and causing the meat to shrink.

Fat. Fat cells are distributed between the fibres; there are more of them in some meats than others, e.g. pork contains many more than chicken. Good quality beef contains visible amounts of fat among the muscle fibres, and this fat is called marbling. Fat is also present on the outer surface of the animal as adipose tissue. As the animal ages there is a build-up of fat. This is why it is uneconomical to buy meat from an old animal.

Tender Versus Tough Meat

1. Age. Meat from an old animal is generally tough because there is a greater amount of connective tissue and there are larger muscle fibres and more gristle. Meat from a young animal has short, fine fibres which have less connective tissue holding them together and little gristle.

2. Activity. When muscle is very active, the fibres become longer and thicker and connective tissue builds up to hold these large fibres together. When a muscle is rarely used, the fibres stay short and little connective tissue is present. This explains why in the same animal neck or leg beef is always tough whereas fillet is very tender.

3. Hanging. Correct hanging can do much to improve the tenderness of meat. After slaughter the protein myosin sets in rigor mortis, making the meat very tough. Gradually muscle glycogen is converted into lactic acid (glycolysis) which softens the meat, assisted by proteolytic (protein-splitting) enzymes present in the meat. Before slaughter it is essential that animals are rested and they should not struggle during slaughter as this will use up the stores of glycogen present, making the meat tough and reducing its keeping qualities.

Tenderising

This can be done

a. Before slaughter by injecting tenderising enzymes into the live animal.

b. Mechanically. Processed meat and meat for caterers is sometimes tenderised with a machine which pierces the meat with thin knives or needles. These break the fibres but release juices, nutrients and flavour. Smaller quantities of meat can be tenderised if they are pounded with a heavy object such as a steak hammer or rolling pin before cooking.

c. Chemically, by sprinkling tenderising chemicals over the meat or steeping the meat in a solution of them. Most of the chemicals contain proteolytic enzymes, e.g. papain (an extract from the paw-paw tree), which help to soften the fibres and are available commercially.

d. In cooking, by using moist, slow methods such as stewing.

Fig. 3.2 Tenderising meat

Dietetic Value

Meat is especially important in the diet of children, adolescents and adults for the animal protein it contains. Its excellent nutritive value makes it suitable for all diets, for it provides every nutrient except carbohydrate and vitamin C. During pregnancy and lactation meat, especially liver, is useful for both its protein and iron content. But contrary to general opinion, meat is not essential in the diet; fish, cheese and eggs make good substitutes and vegetarians can live healthy active lives without eating any animal products at all.

Digestibility

Meat can be digested raw, but it is usually eaten cooked so that pathogenic organisms are destroyed and the appearance and flavour are improved. Cooking develops flavours which increase the secretion of digestive juices, and also makes the tough connective tissue digestible. Stewed meat is particularly easy to digest.

Meat Production

As an agricultural country, Ireland produces most of its

beef, lamb and pig meat, and exports large quantities abroad. Animals are born and bred on farms, many of which specialise in rearing one type of animal. The animals are sold at auctions or livestock marts which take place regularly around the country. There is no longer a centralised 'cattle market'. Those sold for slaughter are transported carefully to the abattoir or meat plant. Care is taken to keep stress to a minimum as, apart from humane considerations, animals lose weight and the quality of the carcase deteriorates if they are frightened.

Slaughter. EEC legislation ensures that animals are slaughtered, bled and processed under ideal conditions. Veterinary checks are carried out at each stage of production to make sure that diseased animals are detected quickly and that scrupulous hygiene is observed throughout the plant. Each beast is stunned before slaughter and bled quickly. The bones and hooves are used to make gelatine or bonemeal. The hides are removed and the animal disembowelled. Edible offal (e.g. liver, heart, kidney, tripe) is removed. Much of the inedible intestine is used for animal feeds and industrial purposes (e.g. oils, glue and fertilisers). Parts of the animal are put to pharmaceutical uses (e.g. the production of insulin and hormones).

The carcase is graded, split in two and quickly chilled to approximately 5°C to retard the growth of bacteria. It is sold in halves or, in case of beef, quarters, to the retailer, who hangs it for the required time.

Hanging. Times vary according to the weather and the temperature of the storage area.

beef	7–10 days
mutton/lamb	2–7 days
pork	2–3 days
poultry	1 day
ducks	1½–2 days
turkeys	4 days
game	until 'high' – usually several days.

Processing

1. Vacuum packing. Much of the meat sold in Ireland goes to meat plants where it is vacuum packed in boneless cuts for the wholesale or export market. These keep about three weeks stored at 0°C.

2. Freezing. Some is boned, trimmed, packed and blast frozen at minus 30°C. Ready-prepared meals such as hamburgers, curries and meat slices in gravy are also frozen and packed in boxes for retail sale. Meat freezes well if quickly frozen and few nutrients are lost, although there is some loss of B vitamins and juices during thawing.

3. Canning. Corned beef, ham, tongue and stewed meat are available canned. Some B vitamins are lost through heat processing and the texture often becomes over-soft and stringy.

4. Drying. Once the only method of preserving meat, it is rarely used now except when meat is accelerated freeze dried (see p. 139). Meat is chopped and used in AFD soups and ready-prepared 'packet' meals. Once reconstituted it must be used up quickly.

5. Curing. Before refrigeration came into use, meat was heavily salted in order to preserve it. It is now possible to use milder cures which are less salty and have a better flavour. Bacon is the cured flesh of a specially bred pig. Sides of the carcass are injected with a solution of preserving salts such as sodium nitrate and potassium nitrate (saltpetre), and subsequently soaked in a solution of brine for about four days. They are then hung in a chilled room to mature for about six days. If smoked bacon is required the meat is subjected to smoke fumes for two to three days. Smoked meats include bacon, ham and sausages such as salami.

Ham. The best hind legs from bacon sides are used to make ham. They are cut off before curing and the rest of the side is cured separately. Ham cures vary. Many are dry salted for one day per 400 g – a 4 kilo ham will remain in salt for ten days. The ham is then hung to mature for a few weeks; it may be smoked before maturing. Mild flavoured hams are cured like bacon. Most ham must be steeped overnight before cooking. Parma and Westphalian hams are hot smoked and eaten uncooked as an hors d'oeuvre.

Corned beef. This is fresh beef which has been soaked in brine in much the same way as bacon. The saltpetre used gives its flesh a bright pink colour when cooked. Usual cuts are brisket, tail end or silverside.

Spiced beef. Brisket or silverside is steeped in a dry marinade of salt, saltpetre, brown sugar, spices and herbs, which are rubbed into the joint each day for ten days.

Note: Smoking and salting preserve the meat by slowing down enzyme action and preventing the multiplication of bacteria. This is why cured meats keeps longer than fresh meat.

Sausages. The food value of sausages varies. They can contain 4–14 per cent protein and 20–30 per cent fat. Sausages may be made from the lean and fat of beef or pork. After these have been minced, cereals and seasonings are added and the sausagemeat is filled into synthetic casings. Continental sausages include frankfurters, Blutwurst, salami and French garlic sausage.

Meat Hygiene

Meat of all types is particularly susceptible to bacterial contamination. Care should be taken at every stage of handling to eliminate the risk of food poisoning.

During production
1. The animal should be tested for disease before slaughter.
2. After slaughter the carcase should be checked for infection from parasites and pathogenic bacteria.
3. Strict hygiene must be observed at all stages of production as one infected animal could contaminate several carcases. All machinery, knives and surfaces should be disinfected regularly and workers should keep their hands clean and wear protective clothing.
4. Raw and cooked meat must never be prepared together because of the risk of salmonella poisoning (see p. 130).
5. Temperatures should be low enough to prevent multiplication of bacteria.

At the butcher
1. The shop should be clean and hygienic.
2. Assistants should not handle both meat and money. This is all the more dangerous if the meat is already cooked, as there will be no further process to destroy bacteria.
3. Raw and cooked meat should not be sold, handled or prepared together (see p. 130).

At home
1. Remove wrapping. Keep fresh meat covered, but not airtight, in a cool place, e.g. directly under the ice box in a refrigerator. Bacon should be stored in an airtight container.
2. The length of time meat should be stored depends on how fresh it was when purchased and the storage facilities available. Most fresh meat should be eaten within two days, but mince and offal should be eaten on the day of purchase.
3. Meat should be removed from the refrigerator at least half an hour before cooking to bring it to room temperature. Cooked meat should also be at room temperature before it is eaten.
4. Cook meat thoroughly, especially pork and mince meats. Bacteria may reach the surfaces of meat during handling, but normal cooking will destroy them. When meat is minced or chopped, however, the bacteria are spread all through it. This means it must be cooked long enough for the heat to penetrate to the very centre of the dish and destroy all bacteria.
5. Avoid keeping meat dishes warm; germs thrive in a warm environment. Avoid cooling boiled meat in its own liquor. Cool leftovers quickly, keep in refrigerator and use as soon as possible.

Frozen Meat
1. Thaw large joints before cooking.
2. Thaw all poultry completely, chickens for 12–24 hours; turkeys for 24–48 hours. If a partly frozen bird is cooked for the normal time, the inside (often a source of salmonella bacteria) will not reach a sufficiently high temperature for the bacteria to be destroyed. Eating a badly contaminated chicken could be *fatal.*
3. *Never* refreeze frozen meat or poultry which has thawed out unless it has been cooked in the meantime.
4. Frozen meat should be used quickly after thawing as any bacteria present before freezing will start to multiply again.

Choice of Meat
The butcher cuts and joints the carcase. Unpopular cuts are sold cheaply and money lost on these joints will be made up by selling popular cuts at a higher price. As preferences vary from one area to another, prices for the same cut may vary too.

Buying Meat
1. Buy meat from a butcher who sells fresh but well-hung meat.
2. The shop should be kept in a hygienic condition with refrigerated storage.
3. Know the cuts of meat and choose one suitable to the proposed method of cooking. Do not forget that the cheaper cuts are just as nutritious as expensive cuts and the flavour is equally good.
4. Choose meat with a small proportion of bone, fat and gristle.
5. The meat should not have an unpleasant smell, should be moist and should have a good characteristic colour. Do not be persuaded to buy unsuitable meat. Many

Hygienic handling of meat (Bill Doyle)

BEEF

Cut	Use
leg	stew, stock, soups, mince
round	stew, braise
tail end	corned beef – boil
silverside	fresh beef – braise
sirloin	steaks, roast
fillet	steaks, roast
flank or lap	stew (corned beef boil)
sirloin and rib roasts	roast
brisket	stew (corned beef boil)
ribsteak chuck	stew, braise
housekeeper's cut	roast slowly, braise, stew
neck	stew, mince, stock
leg	stew, mince, stock

Shaded area indicates expensive cuts

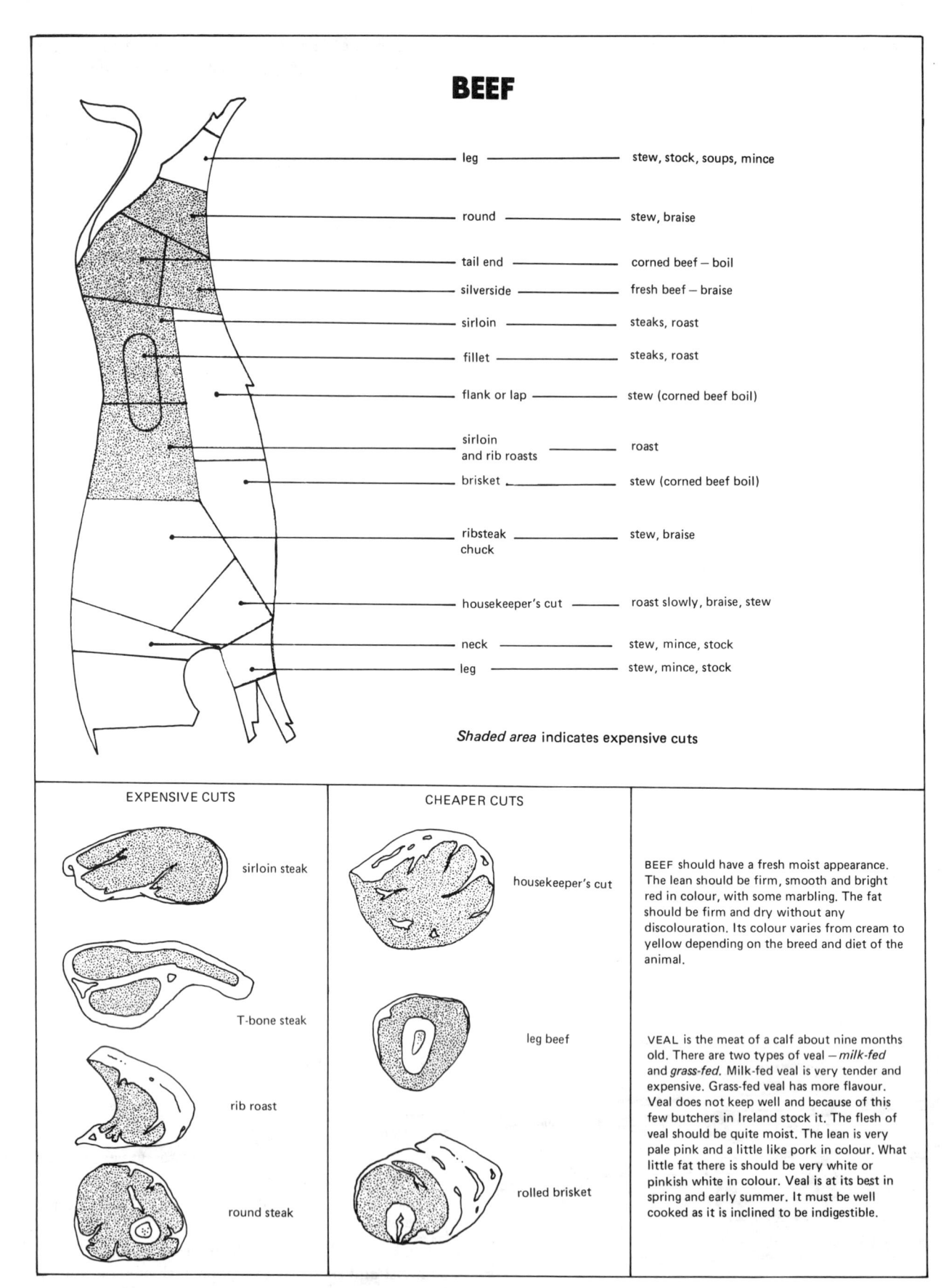

EXPENSIVE CUTS
- sirloin steak
- T-bone steak
- rib roast
- round steak

CHEAPER CUTS
- housekeeper's cut
- leg beef
- rolled brisket

BEEF should have a fresh moist appearance. The lean should be firm, smooth and bright red in colour, with some marbling. The fat should be firm and dry without any discolouration. Its colour varies from cream to yellow depending on the breed and diet of the animal.

VEAL is the meat of a calf about nine months old. There are two types of veal – *milk-fed* and *grass-fed*. Milk-fed veal is very tender and expensive. Grass-fed veal has more flavour. Veal does not keep well and because of this few butchers in Ireland stock it. The flesh of veal should be quite moist. The lean is very pale pink and a little like pork in colour. What little fat there is should be very white or pinkish white in colour. Veal is at its best in spring and early summer. It must be well cooked as it is inclined to be indigestible.

Fig. 3.3

LAMB/MUTTON

- shank ⎱
- fillet ⎰ leg — boil, stew; roast

- side loin (chump) ⎱
- centre loin ⎰ chops: grill, fry, bake; piece: roast

- breast — stew; stuff and roast
- fair (best) end — piece: roast; cutlets: fry, grill
- shoulder — stuff and roast
- gigot — stew
- neck — stew

Shaded area indicates expensive cuts

EXPENSIVE CUTS
- loin
- fillet
- fair end

CHEAPER CUTS
- breast
- shank
- neck

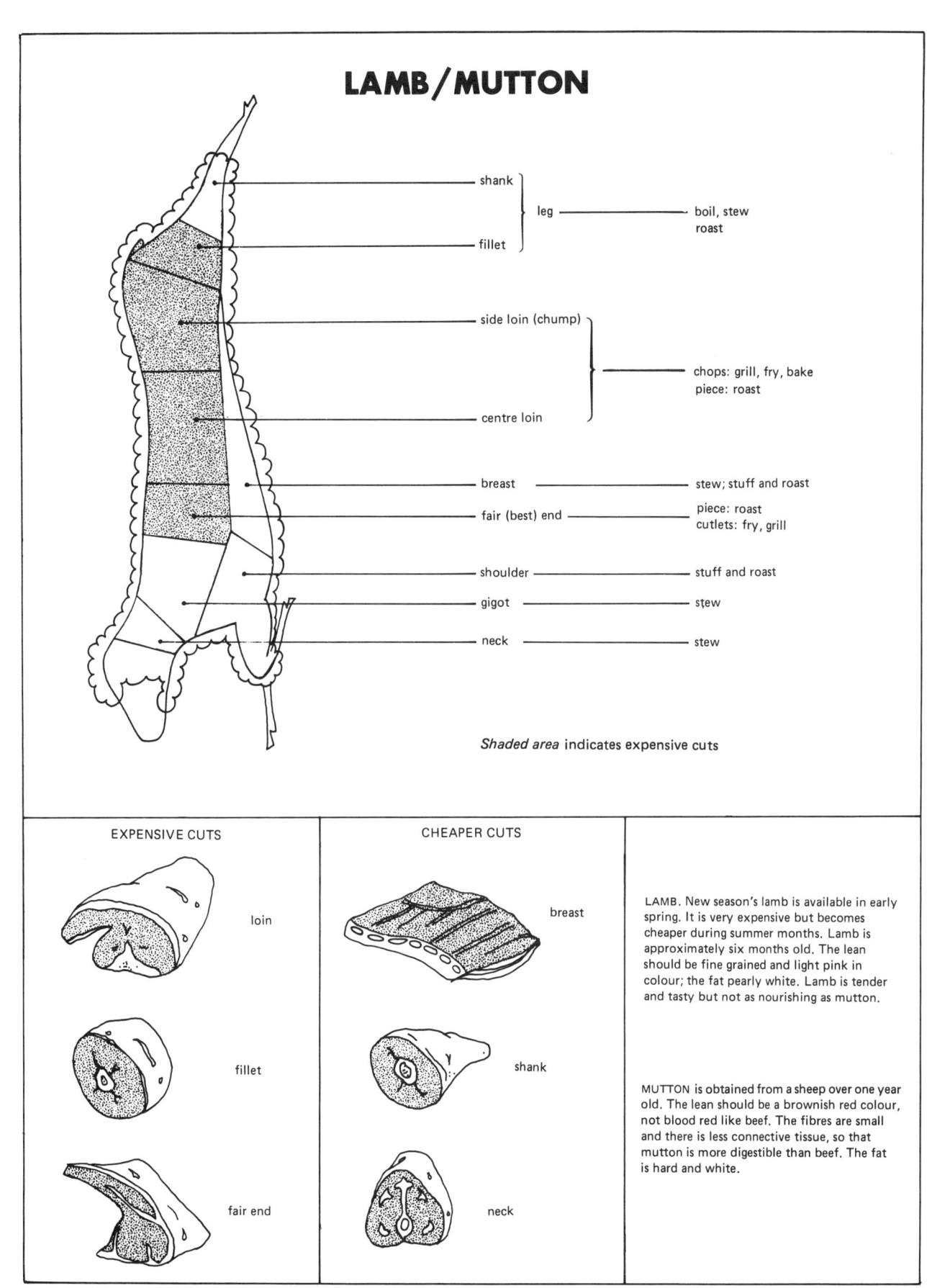

LAMB. New season's lamb is available in early spring. It is very expensive but becomes cheaper during summer months. Lamb is approximately six months old. The lean should be fine grained and light pink in colour; the fat pearly white. Lamb is tender and tasty but not as nourishing as mutton.

MUTTON is obtained from a sheep over one year old. The lean should be a brownish red colour, not blood red like beef. The fibres are small and there is less connective tissue, so that mutton is more digestible than beef. The fat is hard and white.

Fig. 3.4

PORK

Cut	Preparation
feet	boil
leg (shank, fillet)	stuff, roast
loin	chops: grill, fry; piece: roast
porksteak	stuff; roast or braise
belly (streaky)	boil
gigot (spare rib)	roast, bake
shoulder (hand + spring)	stuff and roast; stew
head	boil, use for brawn

Shaded area indicates expensive cuts

PORK is fresh meat from a pig, and the best comes from young pigs. Because of refrigeration pork can now be eaten all the year round. In view of the pig's feeding habits and tendency to harbour parasites, pork must be thoroughly cooked.

The lean of pork should be fairly moist and bright pink in colour, with some marbling. Fibres should be fine and closely packed. The fat should be white and firm without discolouration.

CHEAPER CUTS: belly (streaky), shoulder

EXPENSIVE CUTS: loin, leg, porksteak

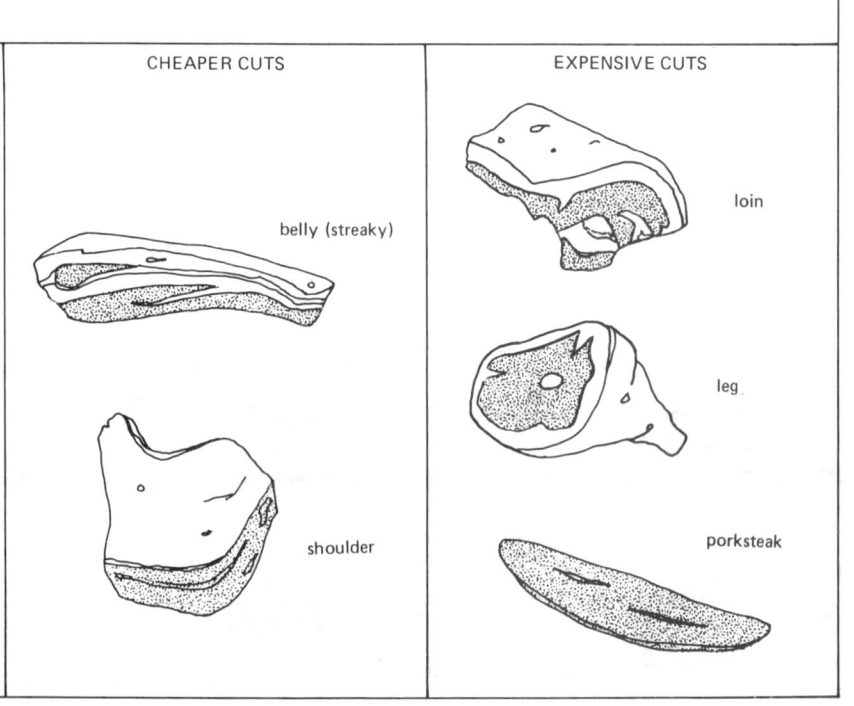

Fig. 3.5

BACON

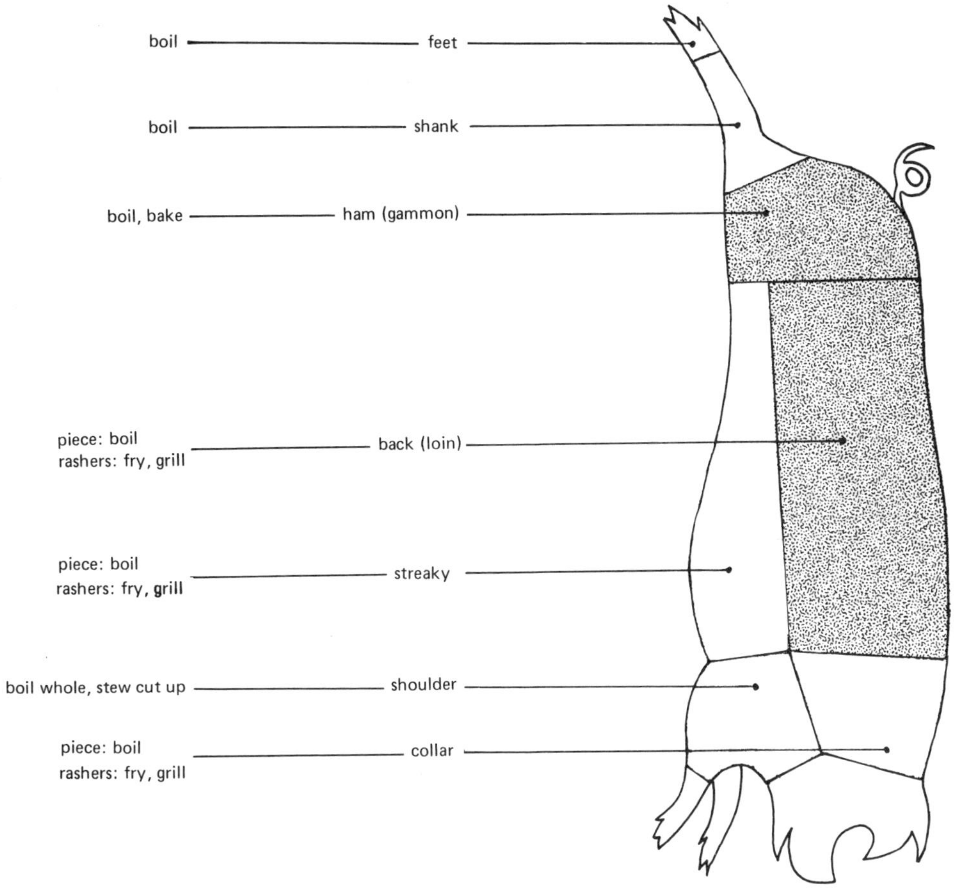

- boil — feet
- boil — shank
- boil, bake — ham (gammon)
- piece: boil / rashers: fry, grill — back (loin)
- piece: boil / rashers: fry, grill — streaky
- boil whole, stew cut up — shoulder
- piece: boil / rashers: fry, grill — collar

Shaded area indicates expensive cuts

CHEAPER CUTS: streaky, collar

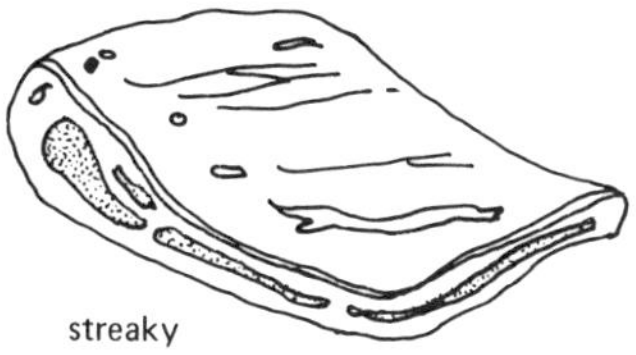

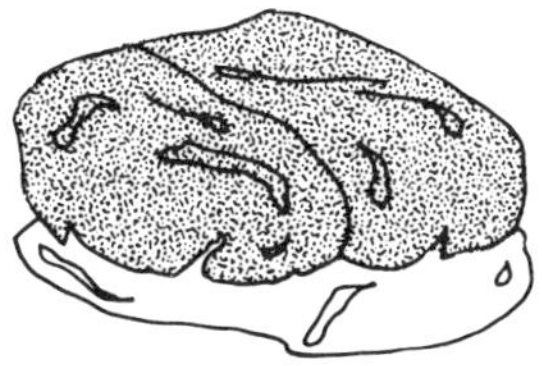

EXPENSIVE CUTS: ham, back rashers

Special lean breeds of pig have been developed for bacon production. Bacon is the flesh of a pig which has been salted and, in some cases, smoked. It should have a thin rind; a thick rind is a sign that it has been produced by an older animal. The lean of bacon should have a bright pink colour and the fat should be a pinkish white and free from discolouration. Mild cured bacon will have a paler pink flesh.

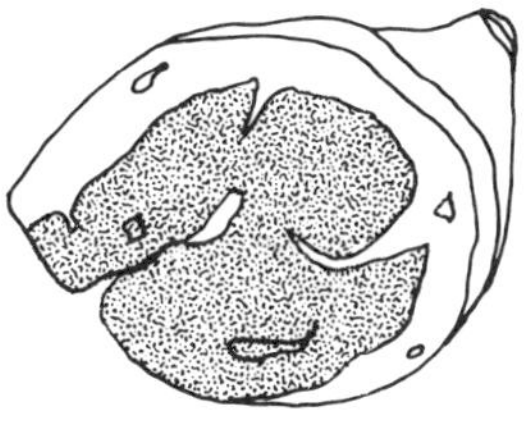

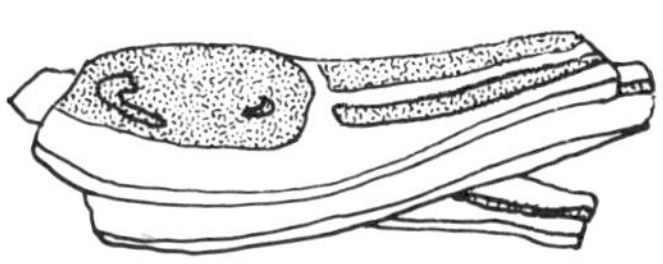

Fig. 3.6

butchers cut up too much meat for display and some of it becomes dark and dried up. Insist on freshly cut meat if necessary and ask to see both sides of a joint.
6. Ask for bones and suet to make stock and dripping. Most butchers make no charge for these if the customer is also buying meat.

Effects of Heat on Meat
1. Protein coagulates; it toughens if cooked too quickly.
2. Elastin contracts and water evaporates, causing meat to shrink and juices to escape.
3. Collagen changes to soluble gelatine upon moist cooking.
4. Haemoglobin turns brown, giving a cooked appearance.
5. Fat decomposes and melts away.
6. Bacteria and parasites are destroyed at high temperatures and decomposition is delayed.
7. Meat becomes tender and digestible.
8. Extractives are released producing appetising odours and flavours and some B vitamins and minerals pass into meat juice.

Offal
This term includes all the edible internal organs of an animal. Structure and nutritive value vary but most organs are a good source of protein, iron and B vitamins. Offal is not hung and should be eaten fresh, preferably on the day of purchase. It is generally cheaper than carcase meat and there is little waste. The flesh should be firm with no unpleasant smell. Wash well in tepid water and cut away vessels and tough parts. Dry in kitchen paper.

Liver. Rich in protein, iron, vitamin A with some vitamin C and a little fat. Lambs' liver is the most tender. Pigs' and calves' liver have a stronger flavour. Ox liver is the most nourishing and cheapest, but tends to be strong-flavoured and coarse-textured.

Kidney. Contains protein, iron, vitamins A and B. It should be very fresh, firm and plump. Lambs' or sheeps' kidney should be surrounded by suet. Remove white core and membrane before washing and cooking. Ox kidney is strong-flavoured and much larger than sheep's kidney.

Heart. A strong muscular organ with little fat. It is inclined to be tough unless cooked very carefully. Rich in protein and B vitamins. Trim well, cut central division, wash, soak for 2–4 hours.

Tongue. Contains protein and fat in equal amounts. Rich in calcium and vitamin B. May be fresh or salted.

Sweetbreads. Usually the pancreas and thymus gland of animals. They are whitish, easily digested and therefore very useful in the diet of invalids and convalescents. They are sold in pairs and must be very fresh. Ox sweetbreads are cheapest but not as tender as those from calves and lambs. Soak in cold water before use, blanch, then remove fat.

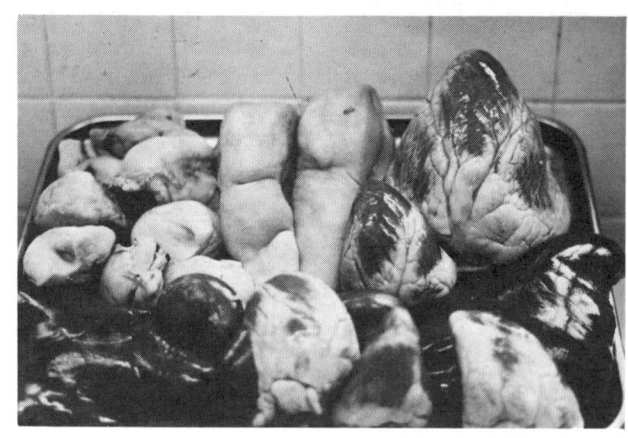

Offal: a good source of protein, iron and B vitamins (Anne-Marie Ehrlich)

Brains. Contain very little protein and considerable amounts of fat. Freshness is essential. Calves' and lambs' are most suitable for cooking and are usually deep-fried.

Tripe. The lining of the stomach of an ox, cleaned and partially cooked by the butcher. It is rich in the protein collagen, which changes to gelatine during cooking, and calcium from the salts used in its preparation. Varieties include honeycomb, blanket and monk's head. It is easy to digest but needs thorough cooking in a well flavoured sauce.

Oxtail. Very bony. Good flavour. Used for soups and stews.

Feet. Calves' feet used for invalid jelly; pigs' feet boiled.

Meat Products
Brawn. Chopped cooked meat from head, feet, etc., set in jelly made from reduced stock and gelatine.
Pâté. A purée of liver, salt-meat, flavourings and fat.
Pâté de foie gras is made from livers of specially bred geese.
Terrine. A cooked loaf of savoury minced meats.
Meat pies. Meat and vegetables cooked in pastry and served hot or cold. Gelatine is added to cold meat pies.

GELATINE
Gelatine is a protein food obtained from the collagen of animals. It is transparent, tasteless and odourless and, as it sets liquid it is used to make jelly, ice cream and other sweets. Powdered gelatine is manufactured from the bone, skin and hooves of animals. The collagen is con-

verted to gelatine by simmering it in water; this is an example of hydrolysis. After purifying, the gelatine is concentrated and dried in granular form.

Gelatine forms a gel at low temperatures and dissolves at higher temperatures, but it must never be boiled as this reduces its setting properties.

Types
Powdered. Available in 15 g envelopes.
Leaf gelatine. Sold in sheets. It is not easily available now, nor is it as convenient as powdered gelatine.
Isinglass. Obtained from the sturgeon and is expensive. It is sometimes used in wine-making.
Agar agar. Obtained from seaweed and not true gelatine. It is useful in vegetarian cooking and is used in laboratories for the culture of micro-organisms.
Aspic jelly. A form of gelatine made from meat stock and used in savoury dishes.

Nutritive Value
Although a protein, gelatine is of little nutritive value because it lacks many essential amino acids, particularly tryptophan. As it is used in very small quantities, it forms an insignificant percentage of total protein intake; nevertheless, if used in conjunction with cereals or high protein foods, it forms a useful supplement to the diet. Gelatine is useful in invalid and convalescent diets as it is very easily digested.

Rules for Using Gelatine
1. Use in correct proportion; 15 g to 500 ml of liquid is the usual combination, but more gelatine may be necessary in hot weather or when a quick set is required. Too much gives an unpleasant flavour and too stiff a consistency, while too little fails to set.
2. Soak gelatine in a little of the cold measured liquid for 10 minutes before use, then dissolve it by placing the bowl in a saucepan of hot water and stirring.
3. Use at once. Pour gradually, in a thin stream, into the prepared ingredients stirring all the time. If it is added too quickly, the gelatine will set in lumps.
4. When it is cold and starting to thicken, pour into a wet mould. To speed up setting, place the mould on ice cubes or in cold water.
5. Allow to set in a cold place overnight or in a refrigerator for 1–2 hours. Prolonged refrigeration will make the food dry and leathery.

TEXTURED VEGETABLE PROTEIN (Synthetic Meat)
As the world food shortage becomes more acute, scientists are spending much time researching alternative or novel protein foods. Animals are a slow, expensive way of producing protein for human consumption; the amount of land used to fatten one ox could, if planted with soya beans, feed more than ten times as many people more economically. Soya beans – containing 40 per cent protein and 20 per cent fat – are now being processed into synthetic meat which is called Textured Vegetable Protein (TVP) or Textured Soya Protein (TSP).

Manufacture
Following harvesting, the oil is extracted from the soya bean and the bean ground into soya flour. The carbohydrate is removed by washing and a protein powder is left. Other ingredients such as vegetable oil, seasonings and flavourings are added. Flavour and colour can be used to give it a beef or chicken-like appearance. It is then pulped and extruded or spun in much the same way as synthetic fibres are produced. The resulting fibres can be chopped to give a mince-like appearance, cut in chunks or woven into large pieces of meat – although this last method has not yet been perfected. The most suitable dishes for TVP are stews, curries, pies and hamburgers – recipes where small pieces of meat are required. TVP can be used as a *meat extender,* that is to increase the volume of fresh meat (usually by one quarter), or as a *meat substitute*. It is sold in dried, canned and frozen form.

TVP has a shelf-life of up to one year, but once reconstituted it must be used up as quickly as fresh meat. Like any new food, synthetic meat will take some time to be accepted, but with the price of meat increasing so rapidly and the sales of meat dropping, it seems likely that it will be part of the staple diet of the next generation.

Advantages
1. Slightly cheaper than meat.
2. Keeps well.
3. Contains almost as much protein as meat.
4. Useful to make meat go further.
5. Little preparation required.
6. No shrinkage or waste.
7. Ideal for vegetarians.
8. Assists in reducing world food shortage.
9. Soya beans are 100 per cent usable, easy to grow in all climates and therefore ideal for feeding people in underdeveloped countries.

Disadvantages
1. Many people dislike the taste, texture and smell. This will probably be overcome as people become accustomed to it.
2. It is not yet available in large pieces, e.g. steaks.
3. Many ingredients are required to flavour it and make it tasty.

Nutritional Value
Although TVP contains only slightly less protein than meat, as a vegetable protein it lacks one essential amino acid, *methionine.* Many manufacturers add this amino acid during manufacture. Vitamin B group and iron are also added. TVP contains more calcium than meat. It contains less fat and therefore has a lower energy value.

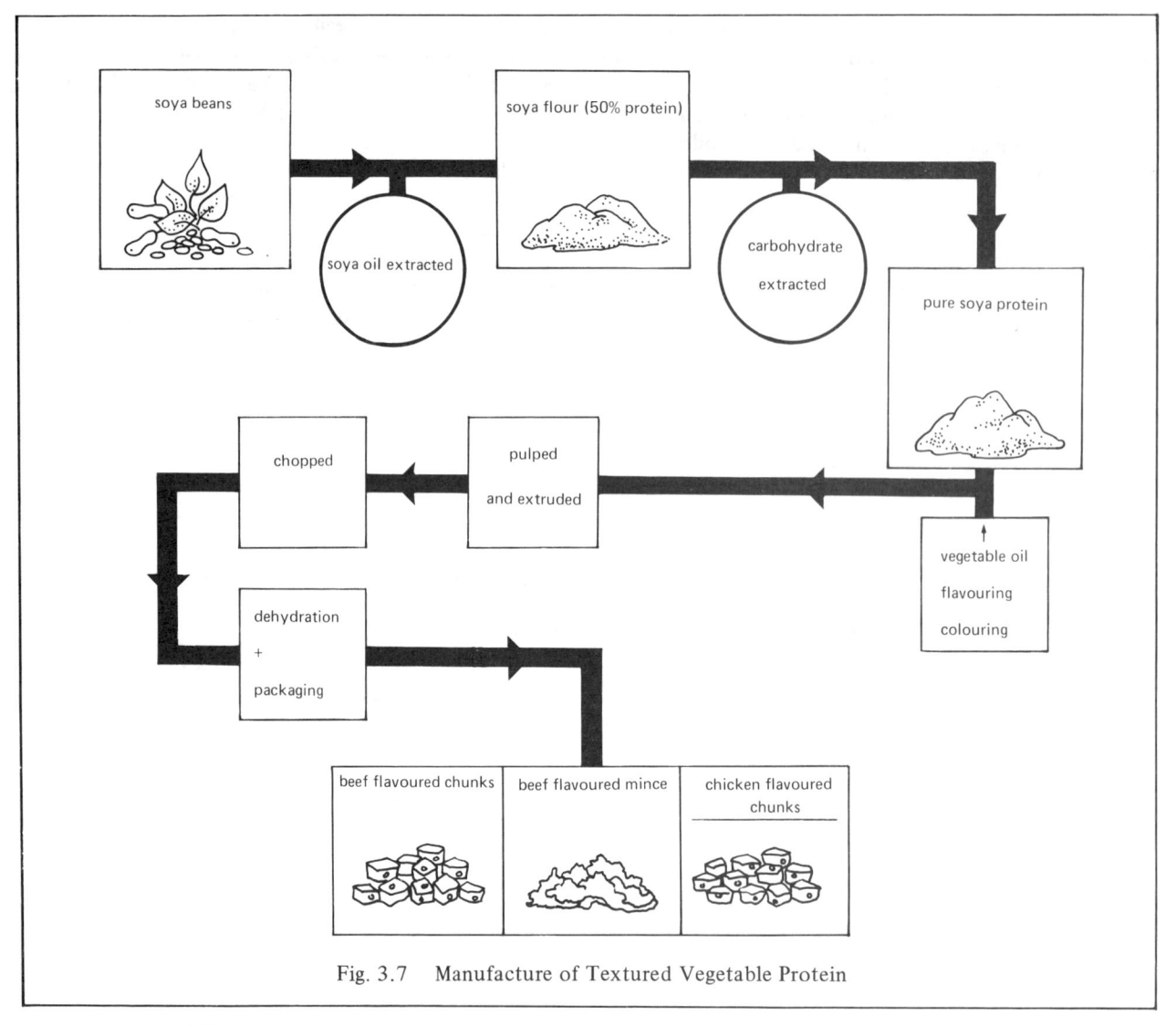

Fig. 3.7 Manufacture of Textured Vegetable Protein

Composition per 100 g

	Protein	Fat	Carbohydrate	Vitamins	Minerals	Water
TVP	16–20%	1–10%	5–10%	B group	Iron/calcium	61–70%
Steak	18–25%	16%	nil	B	Iron 4.3 mg	60–66%

How to Use TVP

Reconstitute by steeping in twice its volume of cold water for 30 minutes *or* hot water for a shorter time. 'Chunks' take longer than 'mince' to reconstitute. Use on its own or with meat in the normal way in stews, hamburgers, meat sauces etc.

Soya protein should be used as a meat extender rather than a meat substitute; if it is used in the ratio of 25 per cent (or less) TVP to 75 per cent meat, it can be almost indistinguishable from meat. If more than 25 per cent soya is used, however, the difference in flavour and texture becomes more obvious. As the strong bean flavour and smell are disliked by many people, seasonings and flavourings should be used to hide them. Frying also helps to improve the flavour.

At present soya is not much cheaper than meat, but as meat prices increase and mass production reduces the cost of TVP, the difference in price is likely to widen.

POULTRY

The term poultry includes chicken, hens, turkey, duck and goose. Chicken is one of the cheapest forms of meat available and can be cooked in many different ways.

Nutritive Value

Chicken is an easily digested form of protein as it has little fat or connective tissue. Ducks and geese have a higher percentage of fat. Fowl provide some B vitamins, calcium and a little iron.

Composition of Chicken

Protein	Fat	Carbohydrate	Vitamins	Minerals	Water	kc/kj (per 100 g)
25%	7%	nil	B_1, B_2	1%	67%	184/770

Dietetic Value

Chicken is particularly suitable for invalid and convalescent diets because it is easy to digest. It is suitable for infants and old people because it is easy to chew.

Production

Modern intensive rearing has made chickens easily available. Most chickens are reared indoors in well-ventilated, heated buildings until they are about eight weeks old. They are slaughtered, plucked and gutted on an assembly-line system, and packed in plastic bags.

Types of Poultry

A poussin is a very small chicken suitable for one portion (600 g).

A broiler is a young chicken specially bred and quickly fattened to produce a good-sized bird in a short time. It is suitable for all methods of cooking including broiling (grilling).

Roasting chickens. Large chickens 1.0–2.5 kg in weight.

Boiling fowl. Usually tough older birds which have completed a laying season. They need to be gently stewed or braised rather than boiled.

Buying Poultry

Poultry can be purchased undrawn or drawn. It may be prepared in one of the following ways:

a. Trussed and packed in a plastic bag, usually with the giblets inside the cavity of the bird. It is essential to keep this sort of bird under refrigeration and to remove the plastic wrapper soon after purchase to allow the free circulation of air.
b. Trussed, packed without giblets in a plastic bag and frozen. It is essential to thaw frozen poultry *completely* before cooking.

Giblets. These are the edible internal organs of the bird. They usually include the neck, gizzard, heart and liver. A quantity of livers can be used to make pâté. The other giblets can be used for stock – and chicken bones should also be saved for stock.

Methods of Cooking Chicken

Whole roasting chickens may be stuffed and roasted in the oven or on a rotisserie, or boiled. If the chicken is to be eaten cold in a salad, for example, it is preferable to boil it as this way the flesh remains moist and plump. As the skin when boiled is pale and rubbery, it is best to remove it. Whole chickens may also be jointed and casseroled, grilled or fried.

Turkeys

Nowadays turkeys have large bodies in proportion to their overall size. The breast should be plump and the flesh firm and white. Frozen turkeys should be thawed for 24–48 hours until no ice at all remains inside.

FISH

In Ireland the consumption of fish is far lower than in other European countries. This is a pity because no part of the country is far from a fishing port. Fresh water fish are also plentiful in our lakes and rivers, although of these only salmon, trout and eels are eaten. Fish is a delicious and nourishing food which can be served in many ways as hors d'oeuvres, soup, entrées, main courses or for snacks, salads and savouries.

Two points must be remembered:
1. Fish must be absolutely fresh.
2. It must be fully cooked but not over-cooked. Most fish cooks within 5–15 minutes.

Classification

Fish may be classified in various ways according to:

a. Physical structure and composition – white fish, oily fish, shellfish.
b. Shape – round fish, flat fish.
c. Habitat – salt water fish, fresh water fish.
d. Sea location: *demersal* – fish which swim at the bottom, and *pelagic* – fish which swim on the surface.

The most usual classification is the first; the nutritional value varies according to structure and composition.

White Fish

1. *Round.* Cod, pollack (black and white), haddock, whiting.
2. *Flat.* Plaice, lemon sole, black sole, turbot, skate, dab, monkfish.

Some white fish contain a little fat. These are hake, gurnet and mullet (round) and brill, bream, halibut and perch (flat).

Oily Fish. Herring, mackerel, trout, salmon, sardines, whitebait, eels (all round).

Shellfish

Crustaceans have legs or claws – shrimps, prawns, lobster, crabs.

Molluscs have no legs or power of locomotion. Some are bivalves (two shells) – e.g. oysters, scallops, mussels. Others, such as periwinkles, have a coiled single shell.

Composition of Fish (per 100 g)

	Energy Value kcal	Energy Value kj	Water g	Protein g	Fat g	Carbo-hydrate g	Ca mg	Iron mg	Retinol μg	Carotene μg	Vitamin D μg	Thiamine mg	Riboflavin mg	Nicotinic acid mg	Vitamin C mg
Cod (steamed)	83	350	79.2	18.6	0.9	0	15.0	0.5	0	0	0	0.1	0.1	2.1	
(fried in batter)	199	834	60.9	19.6	10.3	7.5	80.0	0.5	0	0	0	0.1	0.1	2.0	
Smoked haddock (steamed)	101	429	71.6	23.3	0.9	0	58	1.0	0	0	0	0.1	0.11	1.7	
Whiting (steamed)	92	389	76.9	20.9	0.9	0	42	1.0	0	0	0	0	0	0	
Plaice (steamed)	93	392	78.0	18.9	1.9	0	38	0.6	0	0	0	0.3	0.11	3.2	
Herring (grilled)	199	828	65.5	20.4	13.0	0	33	1.0	49.0	0	25.0	0.18	4.0	3.8	
Salmon (steamed)	197	823	65.4	20.1	13.0	0	29	0.8	0	0	0	0.2	0.11	7.0	
Sardines (canned in oil and drained)	217	906	58.4	23.7	13.6	0	550	2.9	0	0	7.5	0.04	0.36	8.2	
Lobster (boiled) calculated without shell (70% waste in shell)	119	502	72.4	22.1	3.4	0	62	0.8	0	0	0	0.08	0.05	1.5	
Scampi (fried)	316	1321	39.4	12.2	17.6	28.9	99	1.1	0	0	0	0.08	0.05	1.3	
Mussels (boiled)	87	366	79.0	17.2	2.0	0	200	7.7	0	0	0	0	0	0	
Fish Fingers (fried)	233	975	55.6	13.5	12.7	17.2	45	0.7	0	0	0	0.08	0.07	1.4	
Fish Cakes (fried)	188	785	63.3	9.1	10.5	15.1	70	1.0	0	0	0	0.06	0.06	1.1	

Fig. 3.8 Some varieties of fish (not to scale)

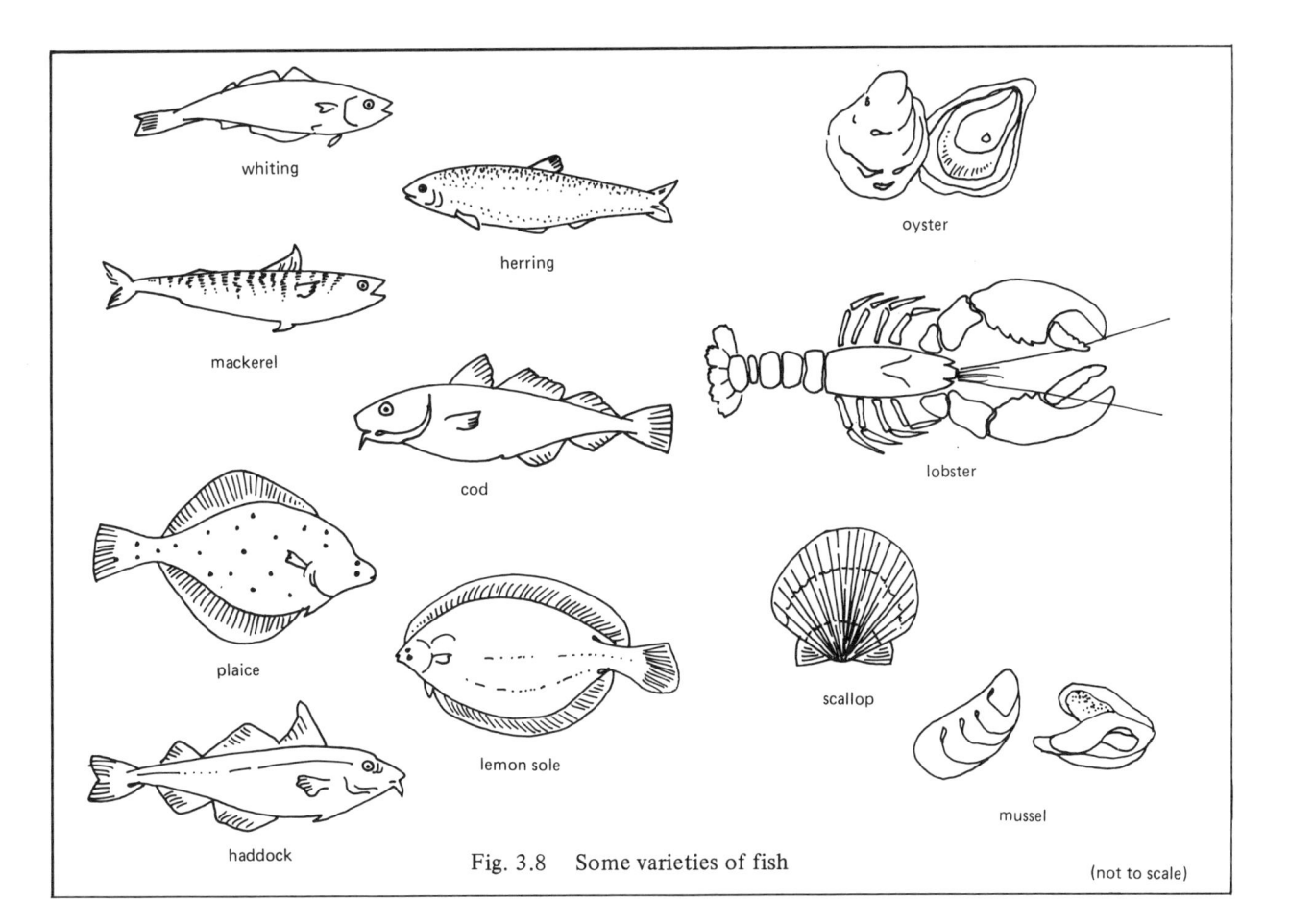

Nutritive Value

Protein. Fish contains protein of high biological value in slightly smaller proportions than does lean meat. White and oily fish contain roughly the same amount of protein. Shellfish has generally more protein (10–15 per cent) but as it tends to be indigestible some of this protein may not be absorbed.

Fat. The flesh of most types of white fish contains no oil; these fish have oil stored in the liver which is removed and used to make vitamin supplements such as cod liver oil. Some white fish, e.g. brill, halibut, hake, contain small amounts of oil (approximately 2–4% per cent) in the flesh.

Oily fish is a good source of oil which contains mainly polyunsaturated fatty acids. It is more nourishing and satisfying than white fish. The amount of oil present varies according to the season, but is higher before spawning. Shellfish contains a small amount of oil.

Carbohydrate. This is lacking in all fish. Carbohydrate foods such as rice or potatoes are generally served with fish to make a balanced meal.

Vitamins. Fish liver oils are the best source of vitamins A and D. Oily fish are also a good source. White fish lacks these vitamins. All fish contain small amounts of B group vitamins; oily fish contain more than white. Vitamin C is lacking in all fish except some shellfish which contain minute amounts.

Minerals. Fish, particularly shellfish, is an excellent source of mineral elements. *Iodine and fluorine* are in sea fish. *Calcium and phosphorus* are in fish of which the bones are eaten, e.g. tinned salmon and sardines. *Potassium and sodium* are present in all fish. *Iron* is found in sardines and tuna, but generally fish is not a good source of iron.

Water. This varies according to the fat content, but is usually up to 80 per cent.

Dietetic Value

Fish should be included in the diet regularly. It contains slightly less protein than meat, but in a very digestible form. White fish, which is deficient in fat, is ideal for invalid and convalescent diets. It is excellent in low calorie diets provided it is not cooked in fat. It is also ideal for infants, young children and the elderly as it is nutritious, has a mild flavour and is easy to eat.

Structure of Fish

The flesh of fish is composed of bundles of short fibres

Fig. 3.9 Structure of fish

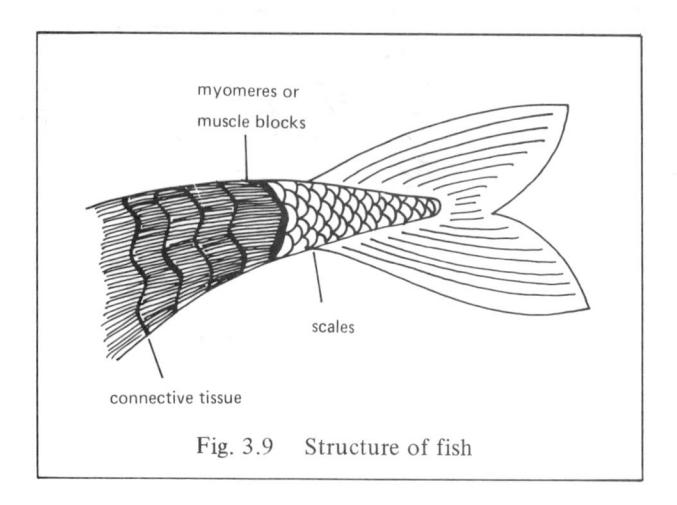

called myomeres, which are held together with thin layers of connective tissue (collagen). On cooking the collagen is converted to gelatine, causing the fish to break apart very easily. Shellfish contains coarse fibres which are difficult to digest. As many shellfish feed in polluted water they can be a source of food poisoning. The bones of most fish contain calcium and following certain cooking or preserving processes, such as canning, they can be eaten.

Filleting on the dockside: fish is ready for retailing or freezing within hours of capture (J. Allan Cash)

Fish Production in Ireland

Ireland has had a long history of fishing and this has always played a part in the economy of the country. In 1931, the Irish Sea Fisheries Association was founded to encourage the fishing industry by providing boats and gear and assisting in the marketing of catches. Bord Iascaigh Mhara was introduced in 1952. This organisation speeded up the boat-building programme, developed modern fish processing plants and expanded the home and export markets. Increased production, training, research and market development followed. By 1970, £10 million worth of fish was being landed by Irish vessels each year.

Modern trawlers have up-to-date navigating and fishing equipment, which includes underwater radar to detect shoals of fish. As the seas around Ireland are well supplied with fish, Irish trawlers need not remain at sea for more than one or two days. This ensures that fish is very fresh when landed.

Many countries have large fishing fleets which remain at sea for weeks at a time. They are accompanied by a 'mother ship' which is a fully equipped fish processing plant, with facilities for gutting and freezing or canning huge quantities of fish at a time. Only periodically does it return to harbour to offload the processed fish. This system ensures that the fish is frozen or canned within hours of capture.

Methods of Fishing

Long lining. Long baited lines are drawn behind the fishing vessel and periodically hauled in.

Drift netting. A straight fence of netting is dropped into the sea with one end attached to the boat. As it drifts with the tide, the fish become trapped in the meshes.

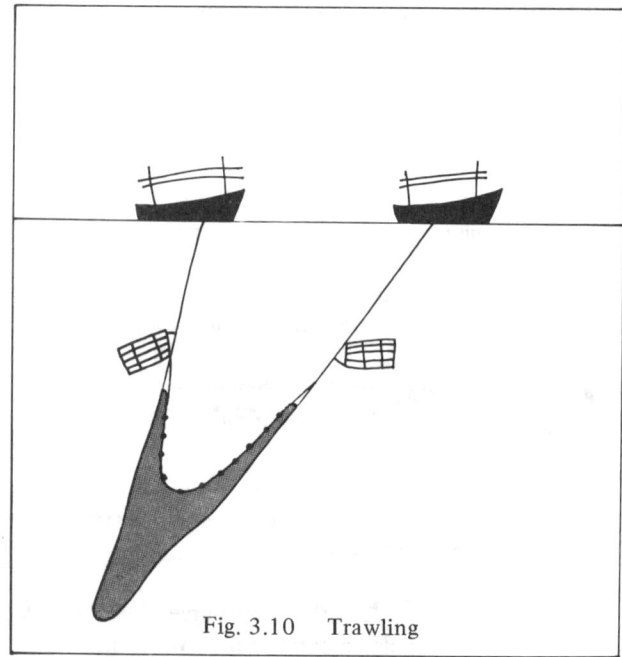

Fig. 3.10 Trawling

66

Trawling. A large conical net is dragged behind the boat.

After landing the fish is boxed in ice and distributed by local fishery co-operatives or is sent to Dublin or Cork fish markets where it is sold by auction to the retailer.

Staling of fish

When fish are caught, they struggle violently using up the stores of glycogen in their bodies. Owing to the deficiency of glycogen, lactic acid, which would have a preservative effect on the fish, is not formed. This causes rapid decay. Bacteria break down the flesh into a nitrogen-based compound called trimethylamine, and it is this which gives stale fish its characteristic unpleasant odour.

Commercial Processing

Because of the perishable nature of fish, all processing must be done as soon as possible after catching.

Freezing. Besides being kept on ice for short-term storage, fish is also frozen on a large scale. Most frozen fish for retail sale is filleted, shaped or trimmed. Sometimes it is coated in breadcrumbs or batter, e.g. fish fingers. Some is packed in 'boil in the bag' form with an accompanying sauce. The fish is blast-frozen at very low temperatures — around $-30°C$ — and then packed. Fish can be frozen and stored for up to one year in a domestic deep freeze, but shellfish should be blast-frozen. There is a slight loss of water, minerals and B vitamins upon thawing.

Canning. This is used extensively for salmon, mackerel, tuna, crab, shrimps, sardines and anchovies. The fish is canned in oil or sauce. In Ireland there are herring canneries in Donegal. Most other canned fish is imported. As the fish most usually canned is the oily variety, canned fish is a good source of vitamins A and D and niacin. The bones, if eaten, are a good source of calcium.

Smoking. Many varieties of fish may be smoked. Before smoking the fish is salted in dry salt or brine. Fish may be cold-smoked at $30°C$ or hot-smoked at $110°C$. Hot-smoked fish needs no further cooking. Smoking is carried out in ovens or kilns. The fish are hung on rods and the smoke from oak or other hardwood chippings is blown over them for approximately four hours. This imparts a distinct smoked flavour to the fish, and the acids, alcohol, creosote and formaldehyde in woodsmoke have a preservative effect and repel micro-organisms. Thanks to improved hygiene and refrigeration in fish marketing, smoking is now more important for its flavour than its preservative effect. At present much 'smoked' fish is flavoured and dyed artificially.

Smoked and preserved fish.

a. *Kippers:* Cold-smoked filleted herrings.
   Bloaters: Whole herrings, lightly smoked. They do not keep as well as kippers.
b. *Smoked mackerel:* Hot-smoked. It is used for paté and hors d'oeuvres.
c. *Smoked haddock or finnan haddie:* Smoked fillets of haddock. True smoked haddock is yellow rather than orange in colour and the skin is not removed, but it is difficult to obtain.
d. *Smoked cod:* These are the large orange sides of fish which are in plentiful supply at most fishmongers. They are rarely cod; black pollock, which is very similar, is usually sold as smoked cod. It is always skinned before smoking.
e. *Smoked trout:* Hot-smoked rainbow trout. It has a very good flavour but does not keep well.

Fish finger production line (Ross Frozen Foods)

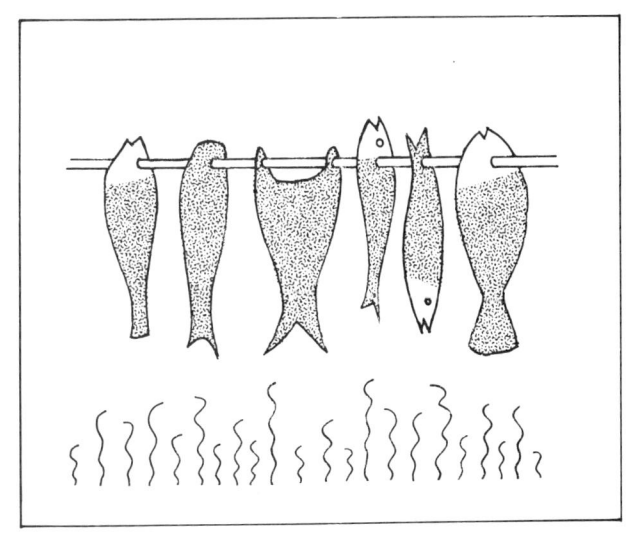

Fig. 3.11 Smoked fish

f. *Smoked eel:* This is uncommon in Ireland.
g. *Smoked salmon:* Cold-smoked. One of the most expensive types of fish available, it should be moist and juicy. Is usually thinly sliced and served uncooked with lemon and brown bread.
h. *Smoked cod's roe:* This is available fresh and in jars. It is used for paté and hors d'oeuvres.
i. *Salted herrings:* These are prepared and preserved in brine or more usually dry salt.
j. *Rollmops:* Rolled herring fillets, marinated in salted vinegar.
k. *Caviar:* The roe of the sturgeon. It is steeped in brine, sieved, drained and packed. It is rich in protein. Black or red in colour, it is served standing in a bowl of ice.
l. *Anchovies:* These are sold in salted fillets or tinned in oil.

Choosing Fish
1. The fish must be as fresh as possible. Ideally it should be eaten within hours of being caught. Remember these signs of freshness:
   a. There should be no unpleasant odour.
   b. Flesh should be firm, the skin scaly and unbroken, the eyes bright and prominent and the gills red.
   c. Cut fish should have a close grain and should not look watery.
2. Buy fish which is in season, as it is the cheapest and has the most flavour.
3. Buy medium-sized fish. Small fish are rather tasteless and have a large proportion of bone to flesh, while large fish may be tough and stringy.
4. Whole fish should be plump.
5. Smoked fish should be glossy and clean, with firm flesh — not sticky.

Buy from a reliable retailer who has a quick turnover of fish, and who buys fresh supplies daily from the market. Make sure that the fish is not frozen overnight and then sold as fresh fish the next day.

If you are freezing fish at home it is essential to order it in advance, stating that it is for home freezing. This will avoid the risk of refreezing frozen fish, which would be highly dangerous.

In rural areas it is often difficult to buy fish. The choice is often restricted to one or two types of fresh fish and a limited amount of frozen fish. It is up to the housewife to ask her retailer to stock fresh fish. Few places in Ireland are more than fifty miles from the sea, so there should be no difficulty in obtaining fresh supplies.

Economic Value
The price of fish varies according to the variety and the time of year. During bad weather, when fishing is hazardous, the price of fish goes up because it is scarce. Its price has little to do with its food value. Oily fish, such as herring and mackerel, which is more nourishing than white fish, is usually cheaper.

There is up to 70 per cent waste in preparing whole fish as the head, skin, bone, gut and perhaps the shell have to be removed; shellfish is particularly wasteful. This is why boned or filleted fish is so much more expensive. But, even allowing for waste, it is still cheaper to buy fish on the bone and to fillet it at home. The fact that fish cooks quickly must be considered as this saves fuel.

Storage
Ideally fish should not be stored at home but cooked as soon as possible after purchase. All fish, especially oily fish, deteriorates rapidly from the moment it is caught. If storage is unavoidable
1. Remove wrappings.
2. Put the fish in a covered container in a refrigerator and if possible cover with broken ice, or wrap in a cloth dipped in vinegar and leave in a cold place. Use within 24 hours.
3. As the temperatures for cooking fish are not usually high enough to destroy all spores of bacteria, any leftover fish must be used up quickly.

Effects of Heat on Fish
1. Protein coagulates and shrinks slightly at 60°–70°C and the fish becomes opaque.
2. The connective tissue changes to gelatine, making the fish tender. Over-cooking causes the fish to disintegrate.
3. Bacteria and parasites such as worms and eggs are destroyed if sufficient heat is used.

Reliable fishmongers buy fresh supplies daily from the market (John Topham Picture Library)

4. Vitamins A and D remain unchanged but there is some loss of the B group vitamins.
5. As some minerals and extractives dissolve into the cooking water, this should be used for sauces etc.

Cuts

Fillets. Thin, boneless sides of fish. The tail end of very large fish such as cod or salmon may be divided into two fillets.

Cutlets. Slices cut across the centre of a large fish such as cod. These contain a hole in the centre from where the viscera were removed.

Steaks. Slices cut across the lower end of a large fish, e.g. cod.

Tail piece. A large, fairly thick cut taken from the end of a large fish. It usually has a bone running through the centre.

Methods of Cooking

Poaching (sometimes called boiling). This method is suitable for whole fish, large centre cuts, thick cutlets and the tail piece. Never boil fish. It should be simmered very gently in just enough boiling liquid to cover it – preferably fish stock. Always add a little salt and lemon juice or vinegar to the water; these improve the flavour and the acid keeps the flesh white. There is a greater loss of nutrients by this method than by any other owing to the large amount of liquid used, but the fish can be served with a sauce made from the poaching liquid.

Steaming. Suitable for fillets, thin cutlets, stuffed rolls. Place the seasoned fish on a buttered plate. Steam over a saucepan of gently boiling water for 10–30 minutes depending on the thickness of the fish. A steamer is used for large pieces of fish. Avoid pressure cooking as the high temperature will make the fish break up.

Stewing. Suitable for fairly solid fish such as chunks of cod, monkfish and smoked cod. The fish is cooked gently in a well flavoured sauce, e.g. white sauce, curry sauce, tomato sauce. All the ingredients in the sauce should be cooked before the fish is added as it will take just 10 minutes at the most to cook. Two well known fish stews are *bouillabaisse*, a Mediterranean dish containing a selection of local fish, and *chowder,* an American shell fish stew/soup.

Baking. Almost all fish can be baked. This method is especially suited to whole fish (which may be stuffed), fillets, cutlets and pies. Avoid using very large fish as the outside will be overdone by the time the centre is cooked. If oily fish is being used, it should be baked with a sharp flavoured sauce or stuffing to counteract the greasiness. To prevent fish from drying out, brush with melted fat or cover with buttered paper, or alternatively pour a little stock, wine or well-seasoned sauce around it. Bake in a moderate oven for an average of 15–40 minutes depending on the thickness of the fish.

Frying. Suitable for fillets, cutlets and steaks and small whole fish such as herrings and whitebait. Fish can be shallow-fried or deep-fried. A coating such as egg and breadcrumbs or batter is necessary when deep frying. Always dry the fish before frying and make sure the fat is at the correct temperature – 180°C. Do not overcrowd the pan. When shallow frying, turn the fish once only as it breaks easily. Drain well and serve with lemon slices. Very few nutrients are lost in frying.

Grilling. This is ideal for whole fish such as herring, mackerel, fillets (e.g. plaice, whiting), cutlets and steaks (e.g. cod, salmon), whole fish on the bone (e.g. trout, sole). Preheat the grill, brush fish with oil or melted butter. Score skin of unskinned fillets (e.g. herrings, whiting). Grill each side for 2–10 minutes turning only once. Serve with maître d'hôtel butter or a suitable sauce.

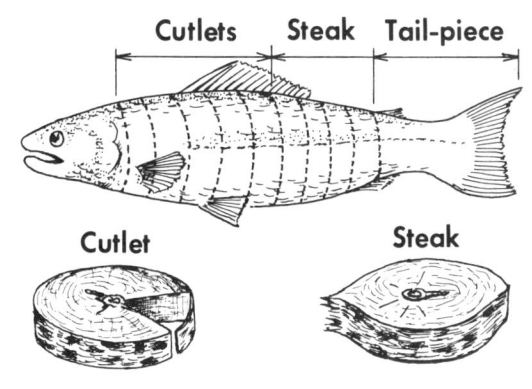

Fig. 3.12   Cuts of fish

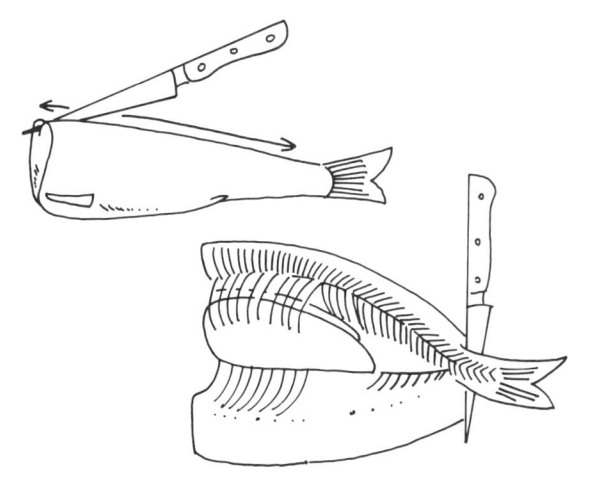

Fig. 3.13   Filleting fish

Smoked fish. Scald by pouring boiling water over and allow to stand for 5 minutes. Stew in a mixture of milk and water. This is then thickened and poured over the fish. Alternatively the fish can be lightly grilled after scalding.

Shellfish can be used for stews, soups or savoury dishes. The fish is usually boiled alive, cooled and the inedible parts removed. It can be served hot in a sauce or cold in a salad. Scallop shells are often used as serving containers for seafood dishes 'au gratin', which are then described as 'scalloped'.

Fish stock. It is well worthwhile making a fish stock from the bones and trimmings of fish. It does not take long and gives a delicious flavour to any fish dish. The stock can be used for cooking the fish or for making an accompanying sauce. To make stock wash the bones and trimmings – the skin, tail, and head of white fish but do not use the head of oily fish. Place in a saucepan and cover with cold water. Add salt, pepper, parsley and a bayleaf together with a small sliced onion and a slice of lemon. Simmer for 20–30 minutes but no longer as the fish flavour will become overpowering. Remove the bayleaf half way through if its flavour is becoming too strong. A *court bouillon* is a stock containing vegetables, herbs and sometimes wine or wine vinegar, which may be used for cooking fish.

EGGS

Eggs are the most versatile of all cookery ingredients. They can be cooked alone, used as a major ingredient in sweet and savoury dishes, cakes and sauces, and also used to bind and coat food. They are a cheap source of many important nutrients. As eggs provide the total food supply of the embryonic chicken, they are a good source of body-building proteins and many essential vitamins and minerals.

Structure

Shell (10 per cent). The outer shell of the egg is made of calcium carbonate. As it is porous, air, bacteria and strong

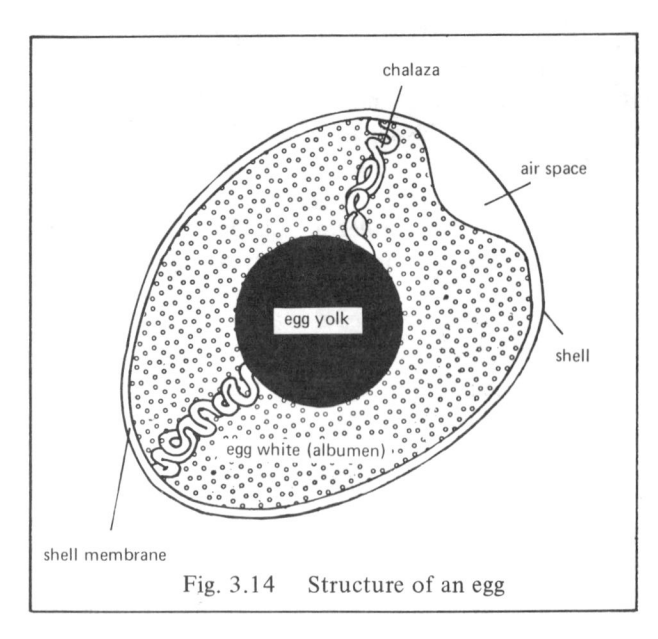

Fig. 3.14 Structure of an egg

flavours can pass into the egg substance. Inside the shell is a strong membrane with an air space between it and the shell at the rounded end. As moisture evaporates through the porous shell, the air space increases and the egg becomes stale; very stale eggs have absorbed so much air that they float in a saucepan of water.

White (60 per cent). The greater part of the white (albumen) consists of water with some dissolved proteins, minerals and vitamins. When fresh, egg white has a thick jelly-like consistency; gradually it becomes thinner as the egg grows stale.

Yolk (30 per cent). Egg yolk deteriorates more rapidly than the white because it contains fat. The egg white helps to protect it and it is kept in the centre of the white by two membranes – the chalazae. If eggs are stored incorrectly with the pointed end uppermost, these membranes break and the yolk rises towards the shell. This makes the eggs go stale more rapidly.

Nutritive Value

Some proteins including ovalbumin are dissolved in the

Composition

	Protein	Carbohyd.	Fat	Vitamins	Minerals	Water	kcal (100 g)	kj
Egg White	12%	nil	0.25%	B	0.75%	87%	45	188
Egg Yolk	16%	nil	32%	A,D,B	2%	50%	350	1470
Whole Egg (shelled)	13%	nil	12%	A,D,B	1%	74%	195	662

white, which also contains small amounts of riboflavin and minerals, mainly sulphur.

In the yolk are two important proteins, vitellin and livetin, which contain all the essential amino acids. Fat is present in a very fine emulsion making the yolk easy to digest. A fatty emulsifier called lecithin and the steroid cholesterol are also present. Vitamins A, D, B_1 (thiamine), B_2 (riboflavin) and niacin are present, together with the minerals calcium, iron, phosphorus. Carotene gives the yolk its yellow colour. The yolk contains less water than the white.

Eggs contain no carbohydrate and lack vitamin C. There is no difference in food value between free range and deep litter eggs. The colour of the yolk or shell has no bearing on the egg's nutritive value.

Dietetic Value

The protein in eggs has 100 per cent biological value and is therefore fully utilised in the body. The fat is in an easily digestible form. As eggs lack carbohydrates, they are usually served with carbohydrate foods, e.g. scrambled eggs on toast, omelette and chips. They are a useful addition to low calorie and invalid diets and can be included as a nutritional extra in many dishes such as soups, puddings, pastry, mashed potatoes and so on. As eggs contain a high percentage of saturated fats, they are usually excluded from low cholesterol diets.

Grading

Although eggs keep for a few weeks in ideal conditions they give better results when fresh. The EEC grading system, now in use in Ireland, has seven grades and three quality groupings.

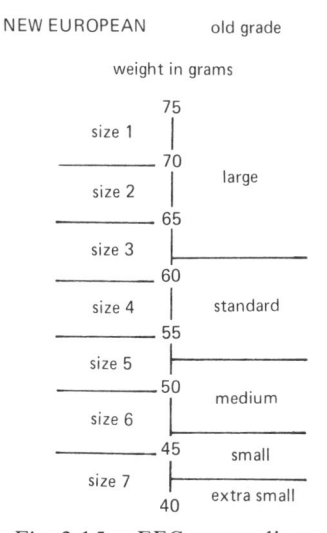

Fig. 3.15   EEC egg grading

Quality

Class A. Top-quality fresh eggs containing a small air space, thick egg white and the yolk in the centre. These are the eggs on general sale in the shops.

Class B. Fair quality. These may have been preserved or refrigerated and may be dirty or washed. Washing removes the natural protective varnish on the egg. As they are less fresh than Class A eggs, they have a flatter yolk and more watery white.

Class C. Used in food manufacture. They may be cracked or opened and beaten.

Extra. Very fresh eggs, packed within the last seven days. The label must be removed when this time elapses.

Packing and Labelling

Candling. Eggs are held up to a strong light and carefully examined for flaws before packing. Originally this was done in front of a candle, hence the term candling.

Labelling. All boxed eggs sold in shops must show:
a. The registered number of the packing station. All Irish registration numbers begin with 8.
b. The name and address of the packing station.
c. The date of the packing. Week 1 is the first full week in January and week 52 the last in December.

To Test for Freshness:
1. Shells should be rough.
2. Fresh eggs feel heavy for their size.
3. Brine test: Put suspect eggs in salt and water. Fresh eggs sink to the bottom; fairly stale eggs stay suspended; a very stale egg floats.
4. Break the egg into a saucer. A fresh egg will have a rounded yolk and thick white, whereas a stale egg will have a flat yolk and watery white.

If the egg is very stale there will be an unpleasant smell due to the formation of hydrogen sulphide as the egg decomposes. Bacteria will also be present.

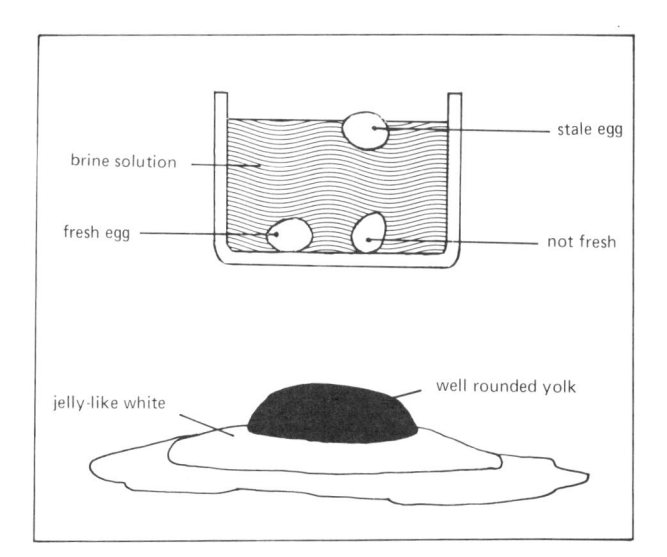

Fig. 3.16   Testing eggs for freshness

Note: Eggs from ducks or fowl other than hens are often contaminated with food poisoning bacteria and must be thoroughly cooked. They must not be used in recipes where the eggs are cooked at a low temperature.

Storage

Eggs should not be washed (see above). They should be wiped with a cloth and stored with the rounded end uppermost, in a cold larder or refrigerator. They should not be placed near strong-smelling foods as the odours will pass through the shell into the egg.

Separated yolks. Place in a cup, cover with cold water and put in refrigerator.

Whites. Place in a screw-top jar and store in refrigerator.
Note: Eggs should be removed from the refrigerator 30 minutes to 1 hour before use to allow them to reach room temperature. If used directly from the refrigerator, they are likely to curdle or, if used whole, to crack.

Preserving Eggs

Eggs are preserved by preventing the evaporation of moisture and entry of bacteria through the shell. This is done by sealing the shell with a commercial varnish or packing eggs closely with preserving liquid (waterglass, i.e. sodium silicate) or dry bran. Eggs for preserving should be very fresh – one to two days old – clean, and without flaws.

Raw whole eggs and hardboiled eggs do not freeze successfully, but separated eggs can be frozen in small cartons.

Dried egg is used commercially. Pasteurised beaten egg is sprayed on to heated rollers and scraped off as a powder.

Economic Value

Improved methods of production, better feeding and disease control have helped to make fresh eggs plentiful throughout the year. The price of eggs varies less between seasons than it did a few years ago, with the result that it is rarely worthwhile preserving eggs at home except when hens are kept by the housewife. Eggs remain a cheap and nourishing alternative to meat and fish. They are quick to prepare and cook, requiring little fuel. Large eggs are better value for money than small.

Effects of Heat on Eggs

1. The protein in egg white coagulates at approximately 60°C. In the yolk it coagulates at approximately 68°C, i.e. far below boiling point.
2. If eggs are cooked at too high a temperature or for too long, they curdle, i.e. the protein shrinks and water separates from the egg.
3. If overcooked the white becomes tough and leathery and the yolk becomes crumbly. Eventually a greenish black rim appears around the yolk owing to the combination of iron and sulphur forming ferrous sulphide. Raw or hardboiled eggs are more difficult to digest than lightly cooked eggs.
4. Egg albumen is soluble in cold water, but becomes insoluble when even slightly heated.
5. Small amounts of vitamin B_2 (thiamine) are destroyed on cooking.

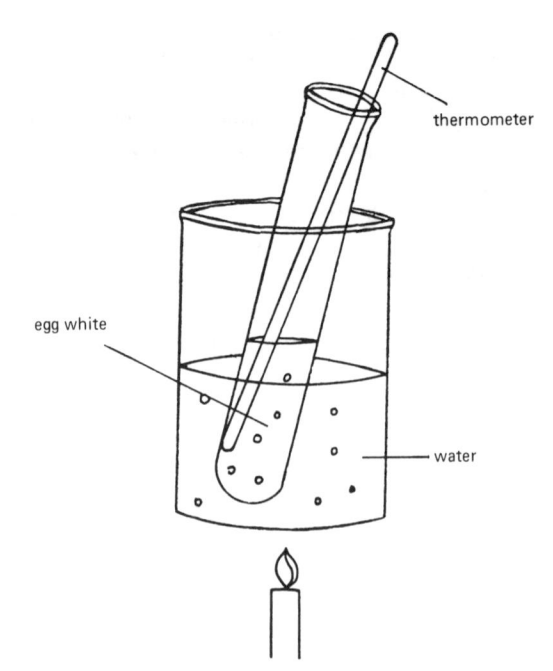

Fig. 3.17 Testing the effect of heat on eggs

Heat test. Place some egg white in a test tube and stand it in a beaker of water. Insert a thermometer in the test tube. Gradually heat the water over a bunsen burner and note the temperature at which the egg begins to coagulate. Bring the water to the boil. Note how the albumin becomes harder as the temperature rises, until eventually it becomes rubbery.

Cooking with Eggs

1. Use alone – boiled, poached, scrambled, in omelettes, baked or as a salad ingredient, instead of meat or fish.
2. To thicken – sauces, soups, custards
3. To bind – hamburgers and reheated dishes such as fish cakes
4. To coat – egg and breadcrumbs, batters
5. To garnish – chopped egg white, sliced hardboiled egg
6. To glaze – pastry, or to frost cakes.
7. To clarify – soups, stock and home-made wines and jellies, using egg whites and shells.
8. To emulsify – lecithin, a natural emulsifier present in egg yolk makes it ideal for stabilising the oil/vinegar mixture in mayonnaise. It also helps to emulsify ice cream and creamed cakes.
9. To enrich – cakes, puddings etc., improving food value and keeping qualities.

10. To aerate – whole eggs or egg white have the property of entrapping air in the albumen. This principle can be used in cake-making, meringues, soufflés and mousses.

Hints for using eggs
1. For boiled and poached eggs the water should simmer very gently for 2–5 minutes depending on the degree of 'set' required. For hardboiled eggs cook in boiling water for 10 minutes. Cool quickly.
2. Always cool hot mixtures before adding to egg. Add warm mixtures to beaten egg; the reverse may cause curdling.
3. When whisking egg whites, make sure that bowl and whisk are clean and free from fat. Even a small amount of fat, e.g. a speck of yolk, will prevent whites from reaching required volume.
4. Use eggs at room temperature. Whisking should be done at approximately 20°C (70°F) to achieve a good volume.
5. When making mayonnaise, oil will emulsify more easily if slightly warmed and added *very* slowly.

Some egg dishes: Scotch eggs; eggs mornay; Oeufs Florentine (with spinach); stuffed eggs; curried eggs; quiche lorraine; eggs *en cocotte* (baked); custards; creme caramel; egg flip; omelettes.

Omelettes. These are quickly made dishes based on eggs with various flavourings and other ingredients added. They must be served immediately they are cooked.

Soufflés. There are two types of soufflé – a baked hot soufflé and a cold soufflé which is usually set with gelatine. A *hot* soufflé is based on a thick panard sauce to which beaten egg yolks and the main soufflé ingredient (e.g. cheese) and seasoning are added. Stiffly beaten egg whites are folded into this mixture and the soufflé is baked in a moderate oven until it is well risen and golden on top. To make a cold soufflé, the eggs are separated and the yolks whisked with sugar and flavouring over a gentle heat until thick. The dissolved gelatine is added and lastly the cream and stiffly beaten egg whites are folded in. The soufflé is set in a refrigerator.

MILK

Milk is often described as nature's most perfect food. It is the ideal food for young children – so much so that babies can live on milk alone for the first six months of life. All female mammals feed their young on milk. Although cows' milk is the type most commonly used in Ireland, goats' and mares' milk is used in many countries.

In 1975 over 2,500 million litres of milk passed through Irish creameries. Only about a quarter of this was used as fresh milk and the remainder was processed or used in the manufacture of other products such as butter, cream and cheese. Bord Báinne is concerned with the export marketing of Irish dairy produce.

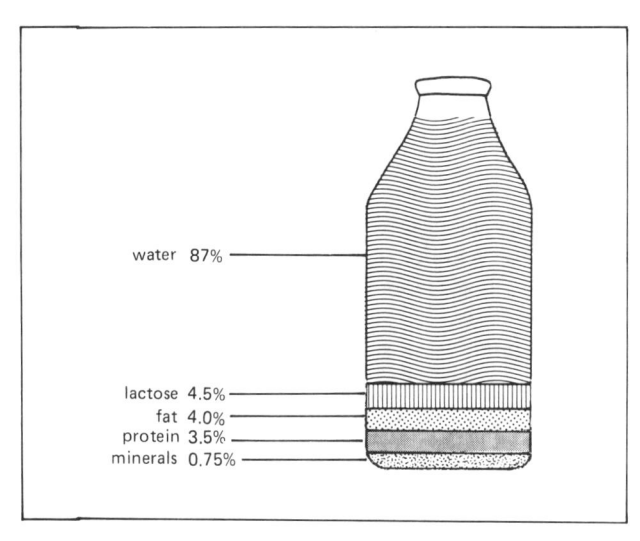

Fig. 3.18 Composition of milk

Composition
The composition of cows' milk varies according to the breed, the length of time since calving and the type of feed the animal eats.

Cow's milk (100 g)

Protein	Fat	Carbohyd.	Vitamins	Minerals	Water	kcal	kj
3.5%	4%	4.5%	A,B,C,D	0.7%	87%	65	272

Human milk

2.25%	3.5%	6.5%	A,B,C,D	0.3%	87.4%	67	280

From these figures it is obvious that human milk contains more sugar and less protein than cows' milk; this is why sugar and water must be added to cows' milk if it is to be used for feeding babies. The protein in human milk is more digestible as it contains less caseinogen than cows' milk. It also contains more vitamin C and iron.

For these nutritional reasons as well as those mentioned on page 36, mothers should breastfeed babies if possible, rather than use commercially prepared formula milk.

Nutritive Value of Cows' Milk
Protein. The proteins in milk are of high biological value, containing all the essential amino acids. Milk contains caseinogen, lactalbumin and lactoglobulin which are easily digested. Caseinogen is acted upon by rennin, a natural enzyme present in the stomach, which aids digestion by breaking down the fats in milk. This forms a clot in the stomach; cheesemaking is based on this principle. Caseinogen is kept in solution by the calcium salts in milk. Lactalbumin and lactoglobulin coagulate on boiling, forming a skin which causes milk to boil over easily.

Fat. Milk fat or cream is easy to digest as it is suspended in a very fine emulsion. Each fat globule is surrounded by a thin layer of protein, which helps to stabilise the emulsion. After standing for some time, the fat globules which are less dense rise to the top of the milk, forming the cream. The main fatty acids present in the milk fat are butyric acid and oleic acid. The emulsifier lecithin and the steroid cholesterol are also present in milk. Milk from Jersey cows contains more fat than that of any other breed.

Carbohydrate. Milk sugar — lactose — is a disaccharide which gives milk a sweetish taste. During souring, the lactose is changed to lactic acid by the lactic acid bacilli present in milk, giving it a sour taste and curdled appearance. These bacilli work more quickly in warm conditions — over 25°C — hence the necessity for keeping milk cold. Pasteurisation (described below) destroys some lactic acid bacteria and thus delays souring.

Vitamins. Milk is a source of all the main vitamins. It contains vitamins A and D in the cream; greater amounts of these vitamins are present in summer when cows are fed on grass. The colour of cream is due to carotene (provitamin A), a natural colouring agent. It also contains vitamin B_2 — riboflavin. Vitamin B_1 — thiamine — and niacin are also present in lesser amounts.

Most milk is deficient in vitamin C, as the amount present decreases from the moment of milking. Much is lost during pasteurisation and more if the milk is left standing in sunlight. This means that the vitamin C content is greatly diminished by the time the milk reaches the consumer.

Mineral elements. Milk is an excellent source of calcium and phosphorus, both of which are essential for bone formation. Because of this, it is important in the diet of pregnant and nursing mothers, babies and children. Milk also contains a little potassium and magnesium, but is deficient in iron. Babies should have their diet supplemented with iron as the amount present in their bodies at birth will only meet their needs for about six months.

Water. Milk is approximately 87 per cent water. This makes it a rather bulky food.

Skimmed Milk

When cream is removed from milk, most of the fat and fat-soluble vitamins present are skimmed off with it. The remaining milk is less creamy but contains protein, lactose, minerals and vitamin B group, with perhaps traces of vitamin C. It can be used in baking and milk puddings and is ideal for low calorie diets. It is also useful for those suffering from digestive upsets.

Note: Skimmed milk must not be used to feed babies and young children as it is lacking in vitamins A and D, which are essential for growth and bone formation.

Approximate composition of skimmed milk

Protein	Fat	Carbohyd.	Vitamins	Minerals	Water
3.5%	0.2%	5.0%	B,C	.6%	90.2%

Dietetic Value of Milk

Milk should be included in every diet. Given its protein, calcium and vitamin content, all of which are related to growth, it is especially important in the diet of babies, children, adolescents and pregnant and nursing mothers. Each of these groups should have 500 ml of milk daily, either in the form of drinks, with cereals or in savoury dishes or puddings. All adults except those on low-calorie or low-cholesterol diets would benefit by taking at least 250 ml of milk daily. Because milk lacks starch, it is suitable for serving with starchy foods such as bread and milk puddings.

Digestion

Milk is an easily digested food. The caseinogen is clotted in the stomach by rennin and then acted upon there by the digestive enzyme pepsin; trypsin from the pancreatic fluid and finally erepsin in the intestine complete the digestion of protein into amino acids. Lactase in the small intestine converts lactose into glucose; lipase and bile convert the fats in milk into glycerol and fatty acids. These are absorbed into the lacteals and amino acids and glucose pass into the capillaries in the villi of the small intestine.

Clotting occurs when the digestive enzyme rennin acts on the milk protein casein, causing it to combine with the calcium salts in milk to form calcium caseinate. This chemical change occurs in the stomach and is utilised in the manufacture of cheese.

Curdling occurs when the lactic acid bacteria in stale milk act on the lactose present, changing it into lactic acid. This makes the caseinogen coagulate into a curd, leaving a liquid which is called whey.

Spoilage

As milk is a liquid food, high in nutrients, it is an ideal medium for the growth of bacteria. Bacteria in milk may be *non-pathogenic,* causing milk to sour, or *pathogenic* — causing many diseases such as tuberculosis, diphtheria, brucellosis (undulent fever), gastro-enteritis and typhoid. It is essential at every stage from cow to kitchen that the utmost care be taken to prevent contamination of the milk. It should always be kept cool, clean and covered.

Quality Control

Scrupulous hygiene must be observed on the farm, in transit, at the dairy and in the kitchen. The quality of milk for human consumption is controlled by legislation under the Milk and Dairy Act 1935 and 1956 and the Food and Drugs Acts 1875–1936.

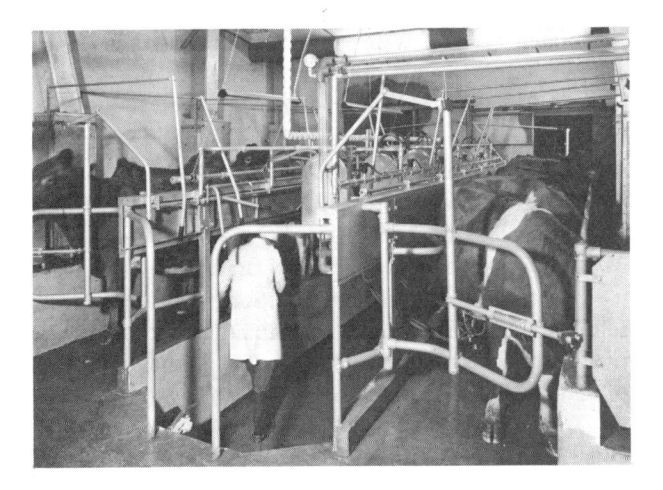

Scrupulous hygiene: milking parlours must be kept clean ... (Milk Marketing Board)

... and sterilisation equipment spotless (Barnaby's Picture Library)

High standards of quality are maintained in the following ways:
1. Inspecting herds: a veterinary surgeon must test all cattle to ensure that they are free from tuberculosis and other diseases. Infected cows are isolated or destroyed if necessary.
2. Dairy farmers are obliged to register with the local authority and submit to inspections of their farms, especially their milking parlours and milking equipment, periodically.
3. Milkers must wash the udder of the cow before milking, wash their hands and wear overalls. The milking parlour must be scrupulously clean. All equipment used for milking must be washed and sterilised (usually with chemicals) after use. Manure must be removed daily.
4. After milking, the milk should be cooled as quickly as possible. It is usually put through a cooling machine or stored in churns.
5. It is transported to the dairy in bulk tankers or churns, where it is tested to ensure it is safe, fresh and free from antibiotics. The usual test employed is the Methylene Blue test; inferior milk will decolourise Methylene Blue within five to six hours.
6. It is then pasteurised, cooled, bottled and sealed. The cap should have the number of the day of the week stamped on it: e.g. 1 – Sunday, 2 – Monday.
7. All dairies must obtain licences from the Department of Agriculture who check and sample milk products regularly.

Note: At all stages of production, milk should be kept at $10°C$ or below to inhibit the growth of micro-organisms.

Processing of Milk
Homogenisation
The milk is pumped through a very tiny valve so that the fat globules are reduced and evenly distributed. This makes the milk more creamy and digestible. Milk is usually homogenised before being treated in other ways.

Heat Treatment
Because of the dangers of infected milk, particularly where large quantities are involved, it has become the accepted practice to heat milk in order to destroy any possible pathogenic organisms. The most popular forms of heat treatment are:

Pasteurisation. This method, discovered by Louis Pasteur, destroys disease-bearing germs without altering the milk to any great extent. The milk is heated to $72°C$ for 15 seconds and then cooled rapidly to $10°C$ or below in a

Milk processing: capping and filling machine with bottle washer in the background (Milk Marketing Board)

heat exchanger. It is essential to cool the milk as quickly as possible in order to minimise the period during which it is warm and therefore capable of breeding bacteria. It is then bottled and sealed. Pasteurisation causes little alteration in the flavour although a little vitamin B_1 (thiamine) and vitamin C are destroyed. All pathogenic bacteria and some souring bacteria are destroyed and as a result the milk keeps longer.

Sterilisation. The milk is homogenised, bottled and sealed. Sterilisation is then carried out at 110°C in an autoclave for about half an hour. As this temperature is higher than boiling point, all bacteria are destroyed, the taste is altered, large amounts of vitamins B and C are destroyed, and the milk becomes less digestible but it keeps for many months. This method of heat processing is not used commercially in the Irish Republic.

Ultra Heat Treatment (UHT). This is similar to pasteurisation, but the milk is heated to a much higher temperature – 135°C – for one second, cooled quickly and packed in sterile polythene-lined containers. As UHT destroys all bacteria, the milk keeps for several weeks. It is ideal for caravans, camping, etc. However, once opened it should be refrigerated and used quickly, like ordinary milk.

Evaporated and Condensed Milk

Following pasteurisation two-thirds of the water is evaporated off at low pressure. The milk is then homogenised and, in the case of unsweetened evaporated milk, canned. The cans are sterilised at 115°C for 20 minutes. In the case of sweetened condensed milk, 15 per cent sugar is added after pasteurising and 60 per cent of the liquid is then evaporated off. High temperature sterilisation is unnecessary as the sugar acts as a preservative.

Note: Neither condensed nor evaporated milk should be used to feed babies as the proportion of nutrients is unsuitable for them. Both have a distinctive taste and can be used in various puddings and savoury dishes. They are often used as a substitute for cream.

Dried Milk or Milk Powder

Milk is homogenised, condensed to 60 per cent of its moisture content and then either spray-dried or roller-dried. In *spray-drying* the milk is sprayed down through a hot air chamber. As the droplets fall, they dry and drop to the base of the chamber as a powder. This form of dried milk reconstitutes easily and is less likely to go lumpy than *roller-dried* milk – that is, milk which is run over heated revolving rollers and scraped off as it dries. This method is used for most baby milk as the high temperature helps to sterilise it.

Both whole milk and skimmed milk can be dried. Skimmed milk lacks fat and fat soluble vitamins, and both it and dried whole milk have reduced amounts of vitamins B and C. All other nutrients remain unchanged when reconstituted.

The standard fat content for skim milk powder is 1.5 per cent maximum. Water content must not exceed 4 per cent.

Storing Milk at Home

1. Do not leave milk bottles standing in sunlight. Ideally milk should be packed in opaque bottles or cartons to prevent loss of vitamins B and C.
2. Leave bottles sealed if possible. When required transfer to clean milk jugs. Never add fresh milk to stale milk.
3. Milk should be kept cool – ideally in a refrigerator. If a refrigerator is not available, bottles or jugs should be left standing in a container of salt water with a clean muslin cloth covering them and trailing in the water.
4. Do not store near strong-smelling foods.

Effects of Heat on Milk

1. There is a loss of some of the B group vitamins and any remaining vitamin C.
2. Flavour is slightly altered.
3. Pathogenic and souring bacteria are destroyed, although the milk will still go bad in a few days as the milk proteins break down and putrify.
4. The proteins coagulate and form a skin on the surface of the milk. When the milk boils it lifts the skin and so the milk boils over.

Spray-drying plant (Milk Marketing Board)

MILK PRODUCTS

Buttermilk

This is a form of skimmed milk which remains after the manufacture of butter. It lacks fat and fat-soluble vitamins, but contains all of the other constituents of milk. It is quite acid due to the concentration of lactic acid bacteria. Buttermilk is used with bread soda as a raising agent.

Cream

When left to stand, the fat present in milk rises to the top as cream. Cream may be skimmed or separated from milk. Skimming is rare nowadays; it necessitates standing the milk for 24 hours and then skimming the cream from the top with a flat paddle. Separating is done on a special centrifugal machine which removes all the cream from the milk. Irish cream is approximately 30 per cent fat, and the legal minimum fat content is 25 per cent. Cream is high in animal fat and vitamins A and D. The use of preservatives in it is illegal, but UHT or long-life cream keeps unopened for many weeks.

Ice Cream

Home-made ice cream is made from milk, cream, sugar and eggs, which act as an emulsifier. Commercial ice cream is a frozen foam made from vegetable fats, dried skimmed milk, sugar and an emulsifier which is usually glycerin monostearate. It is heavily aerated to lighten it and increase its volume. Obviously the food value of home-made ice cream is far greater than that of the commercial product.

Composition of commercial ice cream (100 g)

Protein	Fat	Carbohyd.	Vitamins	Minerals	Water	kc/kj
4%	10%	25%	A,D	Calcium	60%	200/840

Butter

Butter is manufactured from cream. The cream from ten litres of milk is required to make 450 g (1 lb) of butter. The milk is tested, piped to a heat exchanger and then separated in a centrifugal separator. The cream is pasteurised and left to rest in a tank until the cream globules solidify. It is then churned until it turns to butter. After the liquid buttermilk has been drained off and salt added the butter is removed and packed by machine. Butter is checked regularly by Department of Agriculture inspectors for flavour, texture, colour and finish. Water content must not exceed 16 per cent.

Composition of butter (100 g)

Protein	Fat	Carbohyd.	Vitamins	Minerals	Water	kc/kj
1.5%	83%	0.5%	A; traces of D, E.	2 mg cal/ phos. Sodium chloride	12%	795/3339

Unsalted butter is also available.

Butter-making in a Co. Donegal dairy (J. Allan Cash)

Yoghurt

Yoghurt is a fermented milk which originated in south-eastern Europe. It is made by adding a culture of lactic acid bacteria *(Lactobacillus bulgaricus)* to either whole or

Home-made yoghurt: the containers are well insulated in this cake tin (National Magazine Co.)

skimmed pasteurised and homogenised milk. At a temperature of approximately 43°C the bacteria clot or set the milk into a custard-like cream. Flavouring, fruit and sugar may be added. Yoghurt is very easy to digest and can be used as a dessert or unsweetened in savoury dishes or as a salad dressing. Its food value is similar to that of the milk from which it is made, whether whole or skimmed. The energy value is increased when sugar or fruit is added.

Home-made yoghurt

    500 ml milk
    2 tablespoons very fresh unpasteurised natural yoghurt

Sterilise all equipment and containers by boiling if possible. Heat milk to point of boiling. Remove from heat and cool as quickly as possible to 46°C. Add sugar if wished. Place yoghurt in a bowl and blend in the milk until it is well mixed. Divide between 3–6 small containers, cover and put in a warm place (43°C) or pour into a vacuum flask. Allow to stand for about 6 hours. Store in a refrigerator at 4°–5°C for a few hours before use. Flavourings or chopped fruit may be added.

Soured Cream and Cultured Cream

Soured cream is fresh single cream which has been treated with a lactic acid culture, somewhat similar to yoghurt, which thickens and sours it, giving it a piquant flavour. Cream which has been left to sour naturally must never be used instead of soured cream as the souring bacteria produce an unpleasant flavour. Cultured cream is a richer version of soured cream. Both can be used in casseroles, soups and salad dressings to improve the flavour, colour and richness. A home-made version of soured cream can be made by mixing 1 teaspoon of lemon juice with 1 cupful of single cream. It has not the same taste as soured cream but is useful as a substitute.

CHEESE

Cheese is a form of concentrated milk: about five litres of milk are used to make 500 g of cheese. It is usually made from full cream milk.

Cheese is one of the most nutritious foods available as it is a concentrated source of protein. There is twice as much protein in 500 g cheese as there is in 500 g lean beef and it costs less.

The composition, flavour and appearance of cheese vary according to:
1. The *source of supply* – cow, goat, ewe
2. Whether *whole or skimmed* milk
3. The *culture* or starter used
4. The *methods* used in production
5. The *maturing time*

Classification

Cheese is classified as *hard* or *soft* according to the amount of moisture present. Some of the most common cheeses are listed opposite.

Irish versions of many of these cheeses are readily available including Irish Cheddar. Irish Blue, Blarney and Wexford are relatively new Irish cheeses.

Processed Cheese

Most mild cheeses which are foil-wrapped and sold under brand names are forms of processed cheese. Many smoked and flavoured cheeses and cheese spreads are also processed. 'Galtee' and 'Calvita' are processed cheddar cheese. 'Three Counties' is processed gruyère.

Processed cheeses are usually a mixture of ground hard cheese and immature cheese. These are emulsified with water, preserving salts and sometimes milk solids. The mixture is heated and colourings and flavourings may be added before cooling, wrapping and packing. Processed cheeses are more expensive than natural cheese, are less suitable for cooking and are usually milder in flavour with a lower nutritive value. They are, however, more digestible than real cheese.

Average Composition of Hard Cheese (100 g)

Protein	Fat	Carbohyd.	Vitamins	Minerals	Water	kc/kj
27%	33%	nil	A,B_2	6% Cal. Phos.	34%	405/1700

Nutritive Value

Cheese is an excellent food, over 25 per cent pure protein of high biological value mainly in the form of caseinogen. The large amount of fat present makes it an energy-giving food. It is a good source of vitamin A and B_2 (riboflavin). Soft cheeses contain more fat and less water than hard cheeses.

Dietetic Value

Cheese should be a part of every diet. Its high protein content makes it suitable for the diet of pregnant and nursing mothers, children and adolescents. It is an excellent source of energy and, unlike other energy produc-

Familiar cheeses: Irish Brie, Blarney, Irish Cheddar and Irish Blue (Bill Doyle)

Hard Cheese

Country of Origin	Type	Qualities
England	Cheddar	Hard; red or white; excellent for cooking.
	Cheshire	White; hard; milder than Cheddar.
	Wensleydale	Mild; salty flavour; better for table use than cooking.
	Stilton	Cream added; rich with blue veining (white also available).
Holland	Edam (red rind)	Made from partly skimmed milk; not good for cooking; low in calories.
	Gouda (yellow rind)	Pale colour; waxy texture.
Switzerland	Gruyère	Dark rind; small holes; good for cooking and fondues.
	Emmenthal	Large holes (due to special culture); good for fondues.
Denmark	Danish Blue	Strong flavoured, with blue veining.
Italy	Gorgonzola	Strong flavoured and piquant, with greenish veining.
	Parmesan	Very hard and dry; strong flavour; usually finely grated and used for cooking.
France	Roquefort	Made from ewes milk; blue veined; rather expensive.

Soft Cheese

Country of Origin	Type	Qualities
France	Camembert	Small, round; strong flavour; creamy with white crust
	Brie	Large, round; milder flavour; white crust; does not keep well
Italy	Mozarella	White, sold in a pouch shape; cooks well; usually sliced
All countries	Cottage Cheese	Pure, white curd-like cheese; mild flavour.

ing foods such as sugar, it is well balanced providing many other nutrients also. As it lacks carbohydrate it is usually eaten with a carbohydrate food such as bread. The proteins in cheese supplement those lacking in bread, yielding a greater number of amino acids. As many forms of cheese contain cholesterol, care must be taken to use skimmed milk cheese in low cholesterol diets.

Digestibility

Many people find cheese indigestible because it is such a highly concentrated food, containing a very high proportion of fat. Cheese must be chewed well to make digestion easier. Cooked cheese is even more indigestible than raw cheese, but a pinch of mustard or bread soda added to cooked cheese dishes increases their digestibility.

Economic Value

Cheese is still a relatively cheap food. When assessing its cost one must take into account the fact that it is twice as nourishing as almost any other food. There is little or no waste, it keeps well and needs no preparation or cooking. Imported and unusual cheeses are more expensive than plain hard cheese.

Buying and Storing

1. Buy cheese in small amounts as it is inclined to become dry — although vacuum-packed cheese keeps well unopened.
2. Store in a cool place. Wrap loosely in greaseproof paper and then in a polythene bag, but do not exclude air as this makes cheese sweat and become rubbery.
3. Hard ends of cheese may be grated and kept in a screw-topped jar in the refrigerator for use in cooking.
4. Soft cheeses should be used up quickly — within 3–4 days.
5. Try to remove cheese from the refrigerator at least half an hour before use to bring out its flavour if it is to be eaten raw.

Effects of Heat on Cheese
1. The fat melts and after prolonged heating separates.
2. The proteins coagulate and shrink.
3. Prolonged cooking or high temperatures make the cheese stringy and indigestible.

Uses of Cheese
Raw: In salads, sandwiches, savouries, snacks, cheese board.
Cooked: Welsh Rarebit, toasted sandwiches, snacks, cheese soup and sauces (Mornay).
Savouries: cheese straws, aigrettes, macaroni cheese, pizza, omelette.
Au gratin dishes: cauliflower, smoked haddock.
Pastry: cheese flan, quiche lorraine.
Main Meals: lasagne, cheese soufflé, fondue
Puddings: cheese cake, cheese and apple pie.
Garnish: Grated — on soups (e.g. French onion soup), spaghetti.

Manufacture

Hard cheeses
1. Fresh, whole pasteurised milk is poured into vats and a prepared culture or 'starter' of lactic-acid producing bacteria is added to speed up ripening, i.e. the conversion of lactose to lactic acid. This reduces the pH of the milk. The culture develops the colour, flavour and texture in each type of cheese.
2. The milk is heated to 30°C and rennet is added to coagulate the protein casein, forming a thick curd with the fat. The whey is drained off.

Testing cheese (Milk Marketing Board)

3. The curd is then cut and heated slightly. This shrinks the protein and releases more whey.
4. The curd is cut again, salted, drained further, then packed into cloth-lined moulds.
5. The cheese is pressed into the required shape to remove the remaining whey. It is then sprayed with hot water or soaked in brine to produce a skin which preserves the cheese.
6. The cheese is ripened, i.e. the enzymes and bacteria develop flavours within it. The temperature and humidity are carefully controlled. Cheeses may be left to ripen for periods varying from two months to two years.

Quality control extends all the way through the manufacturing process.

Soft cheeses are not pressed after being moulded. The whey drains off naturally, producing a softer cheese.

Blue-veined cheeses. The curd used to make these cheeses is soft and open-textured, still retaining much of its moisture. After it has matured for some weeks the cheese is pricked with needles containing the mould *Penicillium* which develops within the cheese producing the characteristic blue veining.

Cream cheese is made from cream or a mixture of cream and whole milk. It is cultivated with lactic acid bacteria somewhat like normal cheese. It has a high percentage of fat and contains twice the number of calories as cheddar cheese. Cream cheese does not keep well.

Cottage cheese is a white curd-like cheese made from whole or skimmed milk. A 'starter' of rennet is used to

Cheese-making: milk is placed in vats and a starter culture is added (Milk Marketing Board)

make a curd. The whey is drained off, the curd washed and salt added.

Curd cheese is made in the same way as cottage cheese, except that the milk is allowed to sour naturally.

Commercially both of these cheeses are made from skimmed milk and they are useful in low calorie diets.

Home-made cottage cheese

>500 ml milk
>1 teaspoon rennet

1. Sterilise all utensils.
2. Put fresh milk into a scalded jug and leave in a warm place until warmed through.
3. Add the rennet, stir, cover and leave for 12 hours (overnight).
4. Empty curd into double butter muslin and hang over a bowl for a few hours or overnight, until all the whey has drained off.
5. Turn curd into a clean bowl and add 1 tablespoon of milk or cream.
6. Beat until smooth. Add salt and flavouring such as chopped herbs.

Home-made curd cheese

>1 litre milk
>salt

1. Sterilise all utensils.
2. Pour milk into a large bowl, cover and leave in a warm place to sour.
3. Heat gently until the milk separates.
4. Turn into strainer lined with butter muslin.
5. Pour 1 litre of warm, boiled water over the curd. Drain well.
6. Repeat rinsing twice.
7. Tie up muslin and suspend over bowl until all whey is drained off.
8. Add salt and flavouring to taste.

QUESTIONS

1. Discuss meat under the following headings:
   a. structure
   b. composition and nutritive value
   c. economic value

   Explain the chemical changes which occur when meat is hung. Describe three other methods by which tough meat may be made tender.

2. Write a note on textured vegetable protein, stating how it is (a) manufactured and (b) reconstituted and used in cookery and giving (c) its nutritive value and (d) its advantages and disadvantages.

3. Classify the main types of fish and discuss the nutritive value of each group.

   Discuss briefly the following factors in relation to fish: perishability; freezing; smoking.

   List the effects of cooking on fish.

4. Explain the underlying principles involved in cooking with eggs. State how they apply in the making of
   a. mayonnaise
   b. sponge cake
   c. custard sauce

   Write a note on the nutritive and dietetic value of eggs.

5. "Milk is overrated as a source of nourishment." Discuss this statement in relation to
   a. children
   b. the aged
   c. pregnant women and nursing mothers

   Discuss the uses of skimmed milk and state its approximate composition.

6. Describe a project or any research work you have done on milk. Include references to
   a. modern methods of processing
   b. regulations governing the quality of milk
   c. culinary uses

   Differentiate between curdling and clotting and illustrate instances of each.

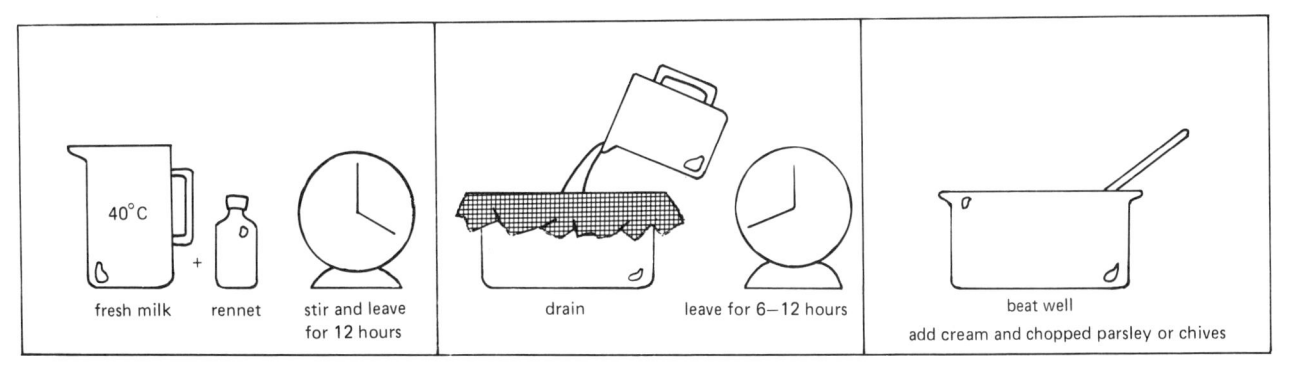

Fig. 3.19 Making cottage cheese

4 Energy Foods

FATS AND OILS

Edible fats and oils are found stored as food reserves in many animals and plants. Fats, which are solid at normal room temperature, are usually obtained from animal sources. Oils generally come from plants and are liquid at room temperature. Fats, of course, can become liquid when heated and most oils will solidify upon freezing. These are physical changes which do not alter the composition of the fat or oil.

Visible fats are foods which are used as fats in the diet, e.g. butter, margarine, oil and the fat on meat. *Invisible fats* include egg yolk, milk, cheese and oily fish.

All fats and oils have a high energy value, are digested very slowly and do not mix with water.

Animal Fats

In animals and indeed humans, fat is found in bone marrow and around nerves and delicate organs. Some fat is present in the liver and varying amounts are laid down under the skin, depending on the age and diet of the animal. In live animals and humans this fat is known as adipose tissue; in carcase meat fat is known as suet. Suet can be used for cooking or it can be rendered down to make other fats such as lard or dripping (see below).

Suet. Cooking suet is beef fat which is trimmed and cleaned. It has a 99 per cent fat content. Suet is a hard fat containing mainly saturated fatty acids such as stearic acid and palmitic acid. It is slow to melt and highly indigestible, so it should be finely chopped to speed up melting, and must be cooked thoroughly. It is possible to buy ready-prepared and chopped suet ('Atora'), which contains a preservative to delay rancidity. It has an 83 per cent fat content. Fats may be obtained from suet by cleaning and melting it down on dry heat (dry rendering) or by melting it with water and removing the fat which floats to the top (wet rendering).

Lard. This is a pure white fat which is rendered down from pig fat. It has no flavour and has good shortening qualities so it is excellent for pastry making. It is suitable for frying as it does not decompose at high temperatures.

Dripping. This is beef or mutton fat which has been rendered down and strained. As it has a slightly meaty flavour it is usually used for frying and roasting, or for making pastry to cover meat pies. It gives a good flavour when used to sauté meat and vegetables for brown soups and stews, and is ideal for gravy-making.

Butter. Butter contains butyric acid, which is a saturated fatty acid. This fat from milk has good creaming and shortening properties and improves the flavour of

Composition of some Lipid Foods (per 100 g)

	Energy Value kcal	kj	Water g	Protein g	Fat g	Carbohydrate g	Ca mg	Iron mg	Retinol μg	Carotene μg	Vitamin D μg	Thiamine mg	Riboflavin mg	Nicotinic acid mg	Vitamin C mg	Cholesterol mg
Margarine (average)	730	3000	16	0.1	81.0	0.1	4	0.3	900	0	7.94	Tr.	Tr.	Tr.	0	*
Vegetable Oils	899	3696	0	0	99.9	0	Tr.	Tr.	0	Tr.	0	Tr.	Tr.	Tr.	0	Tr.
Cooking Fat	894	3674	Tr.	Tr.	99.3	0	0	0	0	0	0	0	0	0	0	*
Butter	740	3041	15.4	0.4	82.0	0	15	0.2	750	470	0.76	Tr.	Tr.	Tr.	Tr.	230
Lard	891	3663	1.0	Tr.	99.0	0	1	0.1	Tr.	0	Tr.	Tr.	Tr.	Tr.	0	70

*Depends on fats used.

cakes and pastries. It can also be used to sauté food but it separates and forms a white scum when heated to a high temperature. As it also burns easily, it is not ideal for frying. (For further information on butter see p. 77.) Butter has a very high cholesterol content.

Cream, cheese, eggs, milk. These foods are rich in animal fats which are high in saturated fatty acids. See previous chapters.

Marine or fish oils. These contain unsaturated fatty acids and are sometimes used in the manufacture of margarine. Whale oil was one of the most common types used until the present world shortage of whales — but oil from any oily fish may be processed in a similar way. It is refined and deodorised, resulting in a colourless, tasteless, odourless oil.

Vegetable Fats

Many vegetables, particularly nuts, have large amounts of oil deposited in their tissues. Olives, ground nuts (peanuts), soya beans, sunflower seeds, coconuts, palm kernels and maize are all sources of vegetable oils. They are used to produce cooking oils and fats, margarine and spreads.

Fatty acids. While all fats contain both saturated and unsaturated fatty acids, vegetable fats usually contain a higher proportion of unsaturated fatty acids (see p. 13). Vegetable oils contain the highest proportion of unsaturated fatty acids, particularly olive, palm, soya, corn and sunflower oils. Coconut oil is an exception, for it contains far more saturated than unsaturated fatty acids. High consumption of polyunsaturated fatty acids is said to reduce the cholesterol level in the blood. These unsaturated fats are usually more expensive than saturated fats, but they are less likely to cling to food, giving crisp, non-greasy results.

Cooking Oils

a. *Olive oil,* obtained from crushed olives, is high in oleic acid (a monounsaturated fatty acid). It is the most expensive oil in general use in the kitchen, has a good flavour and is mainly used for salad dressings.

b. *Corn oil* is made from maize. It is excellent for frying.

c. *Soya bean oil* is cheap and plentiful and is used in the manufacture of margarine and cooking fat.

d. *Sunflower seed oil* (high in unsaturated fatty acids) and *cotton seed oil* are both suitable for salad dressings.

e. *Ground nut oil* (high in polyunsaturates) is used in the manufacture of margarine and cooking oil.

f. *Walnut, coconut, almond oils* are all expensive and not regularly used for cooking in Ireland.

g. *Cooking oils* or *blended oils* are a blend of cheaper oils. They are ideal for frying as most of them contain additives which delay burning and reduce spattering.

Cooking Fats

These are generally made from vegetable oils which have been hydrogenated (hardened), and air is often incorporated to lighten them. They are white, have little flavour and are usually used for frying, although some, such as Cookeen, are also recommended for pastry-making.

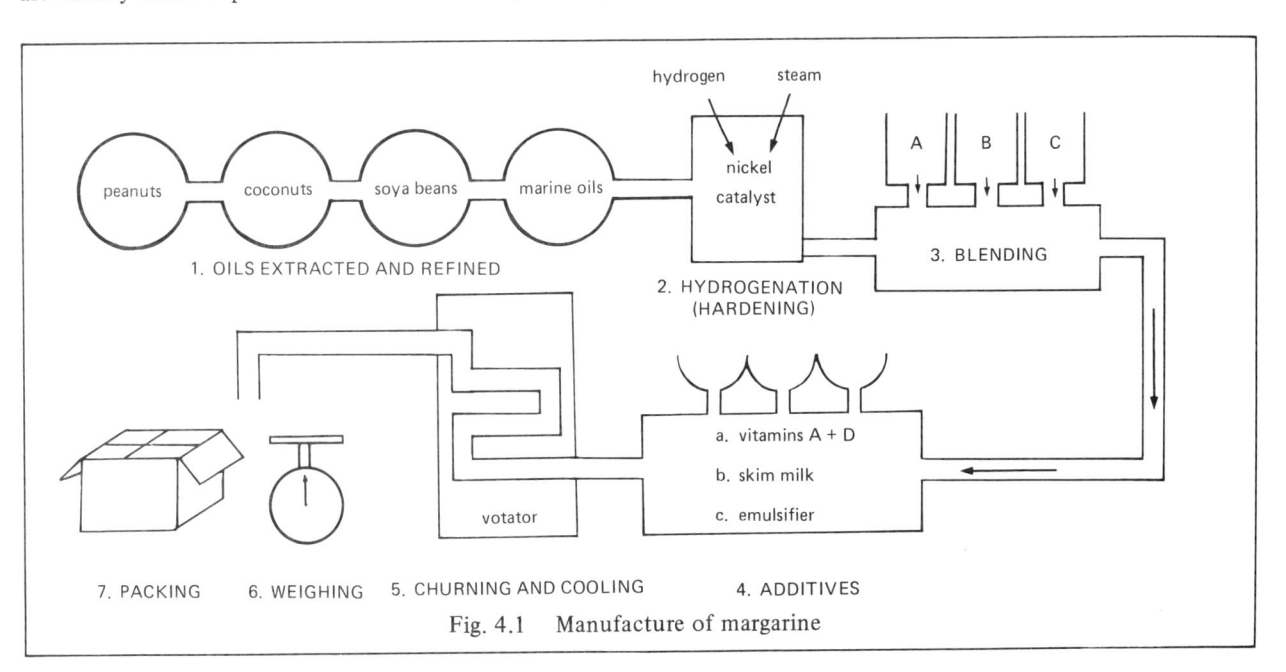

Fig. 4.1 Manufacture of margarine

Widely used fats: butter (animal) and margarine, cooking oil and cooking fat (vegetable) (Bill Doyle)

Margarine

Margarine was originally produced in France in 1861 as a cheap substitute for butter. It is made from a blend of oils — usually ground nut, soya, palm and whale oils. It must not, by law, be more than 16 per cent water.

Manufacture

a. ***Extraction.*** The seeds and nuts are cleaned, crushed and heated slightly to remove the oil. Solvents may be used to assist extraction.

b. ***Refining.*** Caustic soda is used to neutralise the crude oil. It is then bleached and filtered. Steam is passed through it to remove strong flavours.

c. ***Hydrogenation.*** Some of the oils are hydrogenated, i.e. hydrogen gas is forced through the unsaturated oils in the presence of a nickel catalyst. This transforms the unsaturated fatty acids into saturated fatty acids, thus hardening the oils. The hydrogenation process can be used to varying degrees depending on the consistency of fat required, e.g. for easy creaming or hard margarine.

d. ***Blending the oils.*** Cultured skimmed milk, salt, colourings, vitamins A and D and any other additives such as emulsifiers are mixed together and then fed into a votator machine, which has the effect of churning and cooling the margarine until it begins to crystallise and solidify.

e. ***Packing.*** The margarine is weighed, wrapped and packed automatically.

Note. Steps a, b and c above are also used in the manufacture of cooking fats. They are then blended, aerated, chilled and packed.

Composition. Margarine has much the same energy value as butter (see chart). It contains vitamins A and D and if it has a high unsaturated glyceride content, it is considered healthier than butter. However, it must be borne in mind that a large proportion of the oils in margarine are saturated due to the hydrogenation process.

Types

Margarine is a product which can be altered easily to suit the consumer or manufacturer. Various textures, ingredients and additives can be used.

Table margarine is general, all-purpose margarine (e.g. 'Stork'). Butter fats such as butyric acid and often real butter may be added to make it taste similar to butter. It has a medium to hard consistency, creams well and is suitable for cakes, short pastry and sauces, but it is unsuitable for frying as it separates and burns.

Soft or luxury margarines (e.g. 'Blue Band') are more expensive. They are well aerated and spread easily — even straight from the refrigerator. They are used for 'all-in-one' recipes, but are unsuitable for frying as their high water content tends to make them separate.

Slimming spreads (e.g. 'Outline'). As these have a very high water content, the term margarine may not be used to describe them. They are made in the same way as margarine with air, water and an emulsifier whipped into a blend of oils to make them light and give them greater volume. They are useful in low calorie diets but are unsuitable for frying or cooking.

High polyunsaturate margarines (e.g. 'Flora') contain a greater proportion of polyunsaturated fats than other margarines. Many people use these for health reasons as it is thought that they lower the cholesterol level in the blood. Whether or not this is true, they certainly have a much lower cholesterol level than butter or animal fats.

Margarine manufacture: the pre-mix platform where the oils, vitamins and other ingredients are blended (W. & C. McDonnell)

Pastry margarine contains a high percentage of hydrogenated oils or animal fats to give it the very hard, plastic texture which is ideal for pastry making and especially for puff pastry. It contains less water than other margarines and does not blend easily. It is difficult to obtain except through bakeries and specialist food shops.

High-ratio margarines, also called superglycerinated fats, are used commercially. They are specially blended with an emulsifier so that they can be used with abnormally high ratios of sugar and water to produce satisfactory cakes. This makes them more economical than ordinary margarine.

Cholesterol Controversy

The mass media have for some time been giving wide coverage to the question of cholesterol and heart disease — and this in turn has led to a debate on the merits of margarine versus butter.

There is no doubt that cholesterol causes a build-up in the arteries, thereby increasing the chances of heart disease, and there is no doubt that butter has an extremely high level of cholesterol (as do other animal fats). But it must be borne in mind that diet is only one factor which increases the likelihood of heart disease; stress, cigarette smoking and sedentary occupations also contribute largely to the problem. At the same time it is worth noting that Japan — which is highly industrialised and therefore has a similar lifestyle to that of western countries — has a very low level of heart disease, and a diet correspondingly low in animal fats.

Even medical opinion is divided on this question. Detailed scientific studies such as the American McGovern Report have argued convincingly against animal fats, while some highly regarded doctors and nutritionists maintain an opposing stance and claim that overindulgence in polyunsaturated fats may have dangerous side effects.

On the question of butter versus margarine, it is important to remember that while certain soft margarines are high in polyunsaturates, other soft margarines and all hard margarines contain hydrogenated oils (which have been saturated intentionally in order to make them suitable for various cooking processes). Some cheaper margarines also contain lard, an animal fat.

Many people have the mistaken idea that margarine is considerably less fattening than butter, when in fact the two have a very similar energy content. Only certain slimming spreads (e.g. 'Outline') are lower in calories than either margarine or butter.

Both butter and margarine contain vitamins A and D. Whereas in butter the amounts vary somewhat margarine is a reliable and useful source of these vitamins.

Note. More detailed information on fats is found on pp. 11–17. This includes composition, nutritive value, saturation, properties, effects of heat, rancidity and smoke point.

Uses for Fats in Cooking

1. *Frying.* This is one of the quickest methods of cooking due to the fact that the fat can rise to very high temperatures without boiling or decomposing (see p. 115).

2. *Shortening.* When fat is rubbed into flour it should put a waterproof layer on the grains of flour. Fats which do this are said to have good shortening properties making crisp, brittle pastry which melts in the mouth. Hard cooking fats such as 'Cookeen' give the best result but lack flavour. Hard margarines, lard and oil are all satisfactory. It is important that the fat does not decompose at low temperatures, otherwise it will melt in the oven before the starch grains burst.

3. *Creaming.* Fat for creamed mixtures should be soft and easy to cream. During creaming the fat and sugar become emulsified and entrap air, making the cake light. The most suitable fats for cake-making are margarines as they cream readily and have a good flavour.

4. *Flavouring.* The use of fat in plain breads and scone-like mixtures improves the flavour.

5. *Anti-staling.* The keeping qualities of breads and cakes are also improved by fat.

6. *Vitamins.* Adding fat to foods lacking in fat also, in the case of butter or margarine, improves the food value by

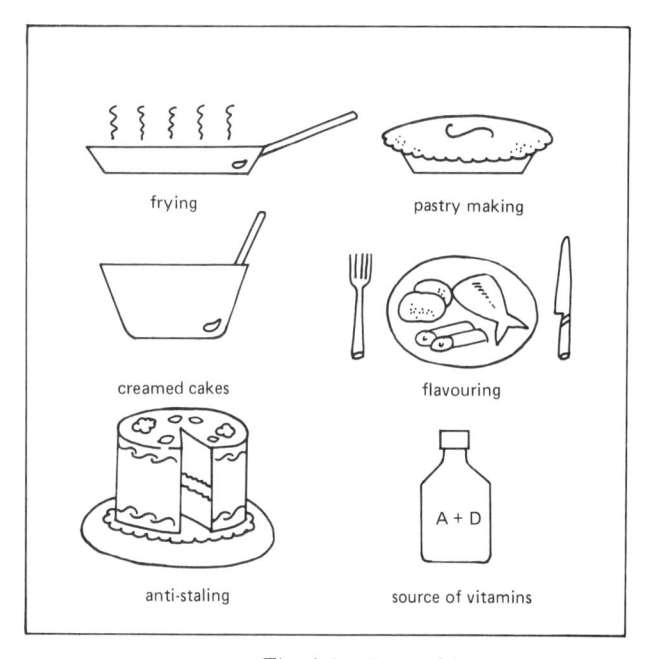

Fig. 4.2 Uses of fat

supplementing the diet with vitamins A or D or both. *Examples:* tossing vegetables in butter after cooking; adding margarine to a rice pudding; buttering bread.

Dietetic Value of Fats and Oils

As these are high energy foods they are useful in the diet of those engaged in strenuous work, but those in sedentary occupations should restrict their intake of fats to avoid obesity. Those on low calorie diets should not reduce their intake of fats too drastically, however, as fats are usually a source of fat-soluble vitamins A and D, and many foods rich in fats (e.g. cheese) are also an excellent source of other nutrients.

Storage

Fats such as butter, margarine and cooking fats should be stored in a refrigerator. Most will keep for six to eight weeks and longer if stored in a freezer, but bulk-buying is not advisable as fats go rancid relatively quickly. Always keep fats covered as they readily absorb odours from other foods. Exposure to air will cause oxidation and hasten rancidity. Oils should keep for six months if stored in a stoppered bottle in a cool, dark place. Too cold a temperature may cause the oil to crystallise or solidify.

CEREALS

Cereals are defined as the edible seeds or grain of cultivated grasses, although the term is often used to include prepared products such as breakfast foods. Cereals are available throughout the world and in one form or another are the staple food of most races: they include wheat, rice, oats, maize, rye and barley. In many third world countries cereals such as rice or maize are the only food available to many people as they are cheap and easy to grow.

Food Value

The main constituent of cereals is starch which often forms 75 per cent of their nutrient content. Small quantities of fat are present in most cereals as well as variable amounts of protein — but this protein is of low biological value as only small amounts of several essential amino acids are present. Lysine, for example, is deficient in many cereals including wheat, and there is only half as much methionine, tryptophan and isoleucine in cereal proteins as there is in animal proteins.

Cereals contain some mineral matter, mainly calcium, iron and phosphorus, and most are a good source of one or more of the B group vitamins. Unfortunately much of the vitamin B is found in the outer husk and germ, which are usually removed during milling.

Because of their low moisture and fat content, cereals store well.

Wheat

Wheat is regarded as the most versatile and nutritious of all cereals. It is probably the most widely-grown cereal as it withstands great variations in climate: wheat is grown in Ireland and Britain, many parts of Europe, the Soviet Union, Africa and both North and South America.

The varieties and composition of wheat differ according to the climate in which the crop is grown. Winter wheat, which is grown in Ireland, is sown in autumn for harvesting the following summer. In countries with very hard winters, such as Canada, the crop is sown in spring and harvested in the autumn of the same year. The quick growth results in a wheat with a higher gluten content than that of winter wheat, producing what is known as *strong* flour. Most flours found in the shops contain a mixture of winter and spring wheat.

Structure of wheat grain. The growing wheat is attached to a stalk which is removed during threshing. The individual wheat grain is small, oval in shape with a tuft of hair projecting from the top. At the point at which it was attached to the stalk is the *germ* which, if the wheat were left in the ground, would germinate and develop into a new plant. The grain is protected with a double layer of *bran.* Inside this is the *aleurone layer* which is largely composed of protein. The central part of the grain consists of the *endosperm* which contains starch cells closely packed and interspersed with the protein known as *gluten.*

The endosperm is the food storage area of the plant. The germ, because it is the embryo plant, contains suffi-

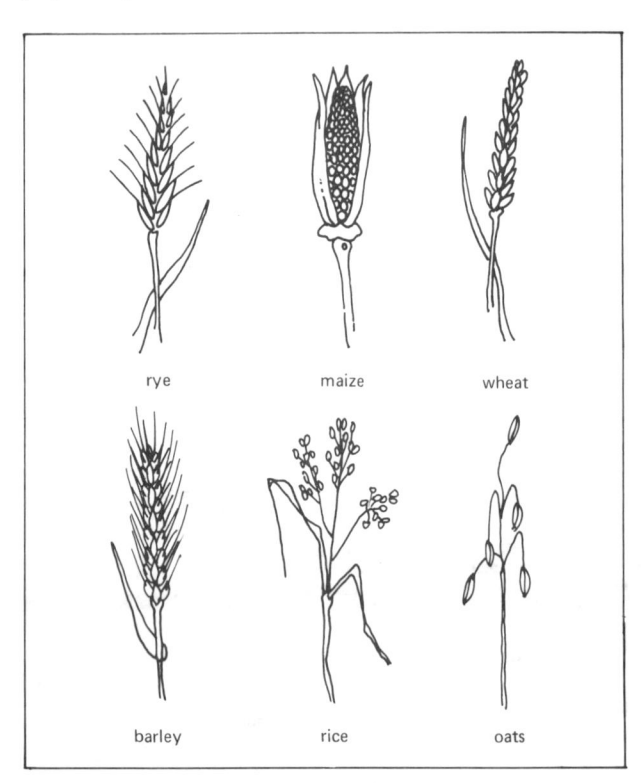

Fig. 4.3 Cereals

Composition of Cereals and Cereal Products

	Energy Value kcal	kj	Water g	Protein g	Fat g	Carbo-hydrate g	Dietary Fibre g	Ca. mg	Iron mg	Retinol µg	Carotene µg	Vitamin D µg	Thiamine mg	Riboflavin mg	Nicotinic acid mg	Vitamin C mg
Flour																
Wholemeal 100%	318	1351	14.0	13.2	2.0	65.8	9.6	35	4	0	0	0	0.46	0.08	5.6	0
Brown (fortified) 85%	327	1392	14.0	12.8	2.0	68.8	7.4	150	3.6	0	0	0	0.42	0.06	4.2	0
White (fortified) 72%	350	1493	14.5	9.8	1.2	80.1	3.0	150	2.4	0	0	0	0.31	0.03	2.3	0
Bread																
Wholemeal	216	918	40.0	8.8	2.7	41.8	8.5	23	2.5	0	0	0	0.26	0.06	3.9	0
Brown	223	948	39.0	8.9	2.2	44.7	5.1	100	2.5	0	0	0	0.24	0.06	2.9	0
White	233	991	39.0	7.8	1.7	49.7	2.7	100	1.7	0	0	0	0.18	0.03	1.4	0
Cornflakes	368	1567	3.0	8.6	1.6	85.1	11.0	3	0.6	0	0	0	1.8	1.6	21.0	0
Porridge	44	188	89.1	1.4	0.9	8.2	0.8	6	0.5	0	0	0	0.05	0.01	0.1	0
Oatmeal	401	1698	8.9	12.4	8.7	72.8	7	55	4.1	0	0	0	0.50	0.01	0.1	0
Rice (boiled)	123	522	69.9	2.2	0.3	29.6	0.8	1	0.2	0	0	0	0.01	0.01	0.3	0
Spaghetti (boiled)	117	499	71.7	4.2	0.3	26.0	0	7	0.4	0	0	0	0.01	0.01	0.3	0
Cream Crackers	440	1857	4.3	9.5	16.3	68.3	3	110	1.7	0	0	0	0.13	0.08	1.5	0

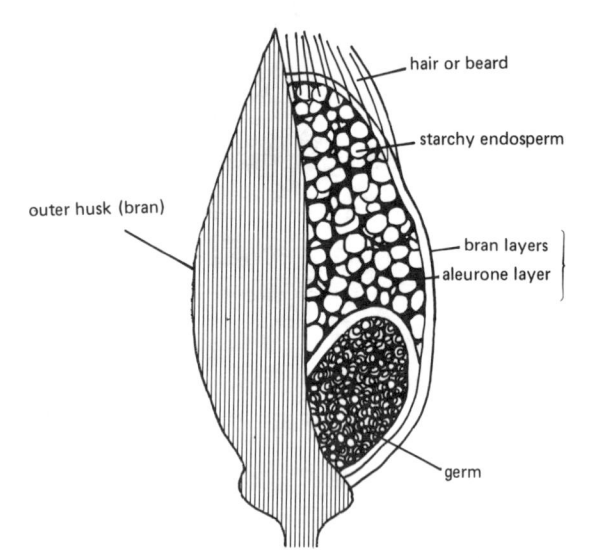

Fig. 4.4 Structure of a wheat grain

Rollermills in a modern flour mill (Henry Simon)

cient nutrients to support the plant during early growth. These include protein, fat and B group vitamins, especially thiamine. The bran, consisting almost entirely of cellulose and B vitamins, is largely indigestible but plays a particularly important part in the diet as a source of roughage. Bran, once regarded as suitable only for animals, is now held in high esteem as a health food and is used in many breakfast cereals, breads and other foods.

As the protein in wheat is of low biological value it should be combined with high value proteins in foods such as milk, cheese, or meat to form a balanced meal.

Milling

Because wheat is an indigestible cereal the outer layers must be crushed before use. This process is known as milling. Since prehistoric times cereals have been ground to flour between two stones. Originally this was done by hand and in many developing areas this is still the case today. In developed countries wind and water mills superseded hand methods and subsequently gave way in the nineteenth century, with the advent of steam power, to the more modern roller milling.

Stages in milling

1. The grain is taken from the storage silos; the foreign matter is removed and the grain is washed, dried and conditioned, i.e. it is given the moisture content most suitable for the rollers.
2. Various types of wheat are mixed to make up the particular grist or mixture to be milled.
3. *Break-rolling.* The wheat is passed through spiral metal rollers which break open the grain.
4. The crushed grain, called semolina, is sieved to remove the germ and bran. These would make the flour dark and reduce its keeping qualities.
5. The endosperm is repeatedly rolled and sieved until a flour of the correct colour and texture is produced. To make flour very light and smooth it is air-classified – in other words, blown about in a special chamber to introduce air.
6. Additives such as bleaching agents, improvers and calcium are usually added at this stage. Improvers help the flour to mature more quickly (maturing time would otherwise be nine months); maturing improves its baking qualities and whitens the flour. Calcium is added for nutritional purposes.
7. The bran and germ which remain are used in the manufacture of breakfast cereals, wheat germ products, health foods and animal foodstuffs.

Varieties of Flour

1. *White flour.* This consists mainly of starch with some protein, mainly gluten, and reduced amounts of B group vitamins. It is milled by the system described above. As the germ which contains the fat is removed, the flour keeps for a long time. White household flour on sale in Irish shops is called plain flour and has a 73 per cent extraction rate; this means that only 73 per cent of the total grain is left in the flour after milling.

2. *Self-raising flour.* This is household flour to which a raising agent such as baking powder has been added for convenience. The disadvantage of it is that while self-raising flour contains the average amount of raising agent for general bread and cakemaking, this may be too little for some recipes. Self-raising flour should not be used for pastry (except suet crust) or for baking where yeast is employed.

Fig. 4.5 Flour production

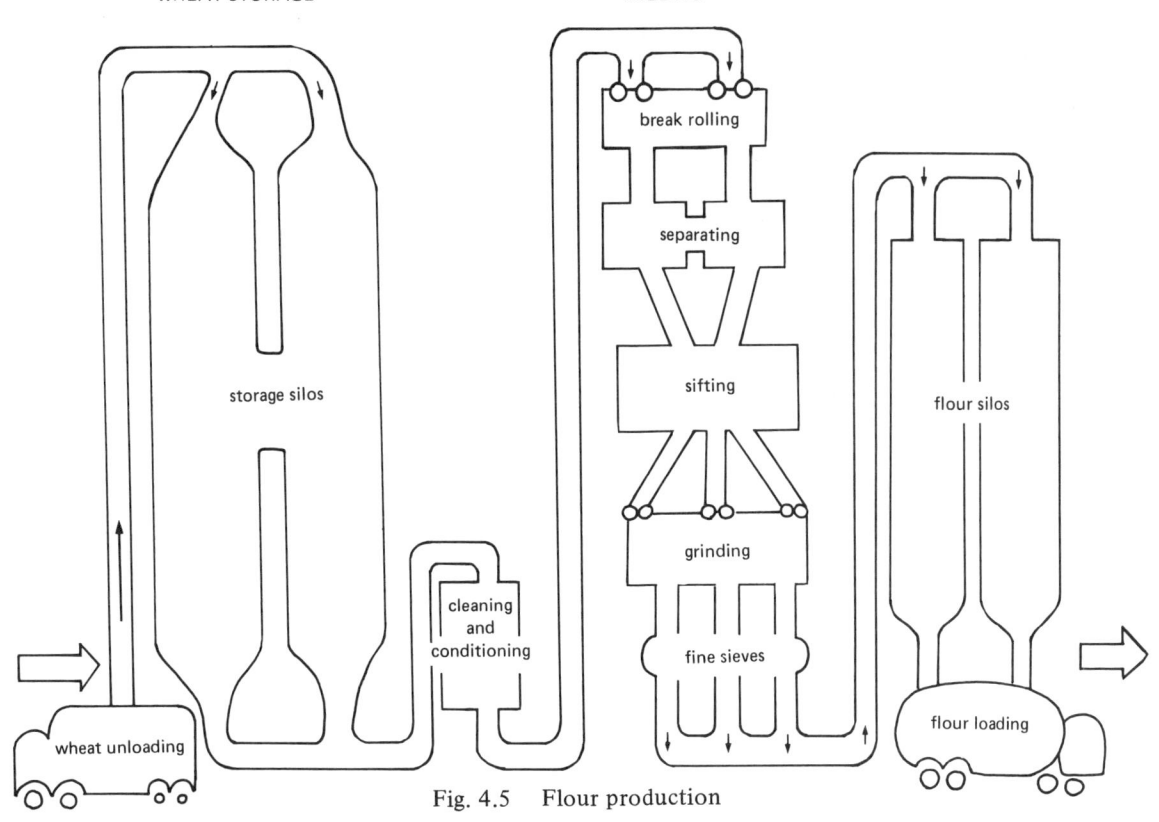

3. **Wholemeal flour.** This contains the whole grain (100 per cent extraction) which is crushed after cleaning. *Wheatenmeal* or *brown flour* usually consists of approximately 85 per cent extraction with added bran.

Wholemeal and wheatenmeal flours are rich in the B group vitamins and contain more iron, calcium, fat and cellulose than white flour. In fact, the more thoroughly the grain is milled the greater the reduction in food value. In Britain legislation ensures that the nutrients lost in high-extraction-rate flours are added after milling. Although there is no such legislation in Ireland, calcium, phosphate and improvers such as ascorbic acid are generally used.

Wholemeal flours do not keep as well as white flours because of the presence of fat and the slightly higher moisture content.

4. **Stone-ground wholemeal.** Wheat ground between stones is said to be superior in taste and food value to flours produced by break-rolling, as smaller amounts of B vitamins are lost. Most stone-ground flour is sold in health food shops and is often ground from 'organically' grown wheat.

The following types of flour are used commercially:

5. **Starch-reduced flour.** Some of the starch is removed by washing, leaving a higher proportion of protein and other nutrients. This is ideal for those who are slimming. 'Energen' products are made from this type of flour.

6. **Wheat germ flour.** A mixture of white flour and cooked germ, which gives a malt flavour to bread. It is used in smooth-textured brown breads such as 'Hovis'.

7. **Gluten-free flour.** This is used by those who suffer from coeliac disease. White flour is 'washed' so that the starch is removed with the water; this starchy liquid is dehydrated to produce a gluten-free flour. The removal of the gluten means that the flour loses some of its nutritive value. To compensate for this soya flour is added. It is rich in protein and gives the gluten-free flour a cream colour.

8. **Strong flour.** This flour has a high percentage of gluten and gives better results in yeast baking and puff pastry. Canadian spring wheat is over 12 per cent gluten. If a high proportion of this wheat is blended with normal wheat (9 per cent gluten), the resulting flour will be strong or hard. Flour made from European wheat is soft and more suitable for cakes. Strong flour, called bakers' flour, is not on general sale in Irish shops but may be bought from bakeries or specialist shops.

Gluten is the main protein present in wheat. When mixed with water it becomes a sticky, elastic substance which makes it ideal for breadmaking. When air or carbon-dioxide bubbles are trapped in the dough, the elasticity of the gluten enables them to expand and rise in the heat of the oven. Once the dough becomes hot enough, the gluten, like all protein, coagulates around the air bubbles and sets the bread. Barley and rye contain small amounts of gluten, but flour from most other cereals, e.g. cornflour, rice flour, does not contain any and cannot, therefore, be used in baking unless when combined with wheat flour.

Other Wheat Products

Semolina. After the initial splitting of the wheat grain during the milling process, the rough endosperm, called semolina, is released. This is used for both sweet and savoury dishes.

Pasta. Pasta is the general term used to describe macaroni, spaghetti, vermicelli, noodles, lasagne etc. It consists of stiff paste made from semolina, water, and sometimes egg. It is extruded under pressure to produce various shapes and is then dried. Wheat used for making pasta must have a particularly high gluten content. A strong flour made from durum wheat (a special variety of high gluten wheat) is used in the best quality pasta.

Barley

Like wheat this cereal is grown in temperate climates. Its principle uses are in the brewing and distilling industries, but it is also widely used for animal feeds. For domestic cooking purposes, barley with the husk removed (pearl barley) is used in soups and stews. Malt and malt products such as vinegar are produced by germinating barley, releasing the disaccharide maltose. Barley is rich in B group vitamins and has a good supply of protein, but is about 80 per cent starch.

Maize

Maize is grown in the United States, Africa and many parts of Europe and Asia. This cereal is the familiar corn-on-the-cob, which is eaten as a vegetable – but maize has many other uses. It produces corn oil for cooking. It is made into breakfast cereals such as cornflakes. It is milled into cornmeal or into the smoother cornflour, which is better known in this country. Cornflour is made by crushing and sieving the grain, washing away the protein and fat and dehydrating the resulting starchy liquid. It is used as a thickener, is the main ingredient in custard powder and is used in puddings and cake making. Popcorn is made from dried maize grains.

Nutritionally maize is inferior to wheat as the presence of phytic acid prevents the minerals calcium and iron from being absorbed by the body. Unlike most cereals, maize is deficient in B vitamins, particularly niacin. In countries where maize is a staple food, there are many cases of pellagra, a chronic disease which can affect the brain.

Rye

This is a hardy cereal which will grow on even the poorest of soils. It is milled like wheat and produces a dark flour, which, because of its gluten deficiency, must be mixed with wheaten flour to make bread. It is used extensively in Central and Eastern Europe and the United States, where it is used to make whisky. It is probably best known in Ireland as the main ingredient in slimming biscuits and crispbreads such as 'Ryvita'. Rye contains some protein including a little gluten, B vitamins and, like most cereals, a large amount of starch.

Oats

Oats are grown in temperate regions and can withstand continuous wet weather. They are extensively grown in Northern European countries including Ireland and Scotland. Oats are produced mainly as an animal feed but can also be used for human consumption. They are first milled into rough pinhead oatmeal; this may then be ground into finer meals. Rolled oats or flakemeal is made by steam-cooking the oats and then passing them between rollers. These flaked oats cook more quickly than pinhead oatmeal. Porridge, muesli and many biscuits are made from flakemeal. Instant porridge is made from more finely milled oatmeal which is treated in the same way. Many baby foods contain finely milled precooked oatmeal.

The food value of oatmeal is high. It is rich in protein, fat, carbohydrates and B group vitamins. Much of the husk is processed with the grain, as it is more difficult to separate than that of wheat, and this makes oatmeal an excellent source of roughage.

Rice

Rice demands a warm, moist sub-tropical climate for its cultivation, hence it is grown and eaten extensively in the Far East and India. It is also grown in the southern United States and around the Mediterranean. Like the wheat grain, the natural rice grain contains a double brown fibrous outer layer, a germ and endosperm, inside which is most of the starch. During processing, the outer husk and germ are removed, leaving the familiar white polished rice. Over 50 per cent of B vitamins, most of the roughage and considerable quantities of protein are lost in processing. Rice is often parboiled commercially in an attempt to prevent further losses of these nutrients. In countries where polished rice is the staple diet of the poor, the disease beri-beri is very common. Brown rice, in which the inner husk and germ are retained, is considerably more nutritious and is becoming increasingly available in Irish shops.

There are two main varieties of rice: *Patna* rice, which has a long grain and is used for savoury dishes, and *Carolina* or *Italian* rice, which is short-grained and used mainly for sweet dishes.

Rice may be parboiled to retain nutrients or pre-cooked

to cut down on the cooking time. It is also available in boil-in-the-bag form.

Rice may be roughly ground to make *ground rice* or further milled to make *rice flour*, which is used together with wheat flour for baking. *Puffed rice* is used in breakfast cereals.

Note. Sago, tapioca, arrowroot are often incorrectly classified as cereals, probably because, like cereals, they have a high starch content (95 per cent). They are processed from the roots and stems of tropical plants and are used in puddings and as thickeners.

Effects of Cooking on Cereals
1. Dry heat causes the starch grains to expand and burst, releasing the starch cells which absorb fat or moisture from the surrounding matter – e.g. fat in pastry.
2. Moist heat swells the starch granules until they burst and absorb the water surrounding them, causing them to gelatinise. This makes the starch digestible.
3. Cereals absorb large amounts of water, often tripling their weight by the end of the cooking time.
4. Protein such as gluten in its expanded form coagulates and sets baked products such as bread.
5. Cellulose absorbs water and softens.
6. Surface starch is converted into dextrin which browns baked products such as bread or toast.
7. Thiamine, which is plentiful in most cereals, is reduced by heat. The degree to which it is affected depends on the temperature and the cooking time.
8. An alkali (e.g. bread soda) combined with cereals destroys the thiamine present.

Breakfast Cereals
These are cereals which have been processed – usually by some form of heat treatment such as puffing or flaking. Other ingredients such as honey and sugar are usually added to improve the flavour. Owing to the processing methods used, much of the thiamine and smaller quantities of other B group vitamins are destroyed. Few breakfast cereals retain the germ of the grain and many also have the bran removed, with the result that most are insipid and nutritionally unsound. Some manufacturers replace the vitamins which are lost and emphasise this in their advertising slogans, but it should be borne in mind that the chief nutrients eaten with breakfast cereals are in the milk which accompanies them. Porridge and muesli are two of the most nutritious breakfast cereals. Their higher protein and fat content makes them more filling and satisfying than more highly processed varieties.

SUGAR

Sugar is a carbohydrate composed of the disaccharide sucrose which, in turn, is made up of two monosaccharides: fructose and glucose. It is a cheap, concentrated and easily digested form of energy.

Sugar may be extracted from two plants which have a very different appearance: the sugar cane, which resembles bamboo, and the sugar beet, which looks rather like a turnip. Sugar cane grows in a sub-tropical climate in such countries as the West Indies, whereas sugar beet is grown in temperate climates like our own. Almost all the sugar eaten in Ireland comes from sugar beet grown in this country. The main sugar processing plants in Ireland are in Tuam, Carlow, Mallow and Thurles.

Cane and beet sugar are identical in appearance and chemical composition – $C_{12}H_{22}O_{11}$ – and give exactly the same results in any cookery process.

Breakfast cereals: most are far less nutritious than the milk that accompanies them. Fruit improves their vitamin C content. (National Magazine Co.)

Sugar cane (J. Allan Cash)

Field of sugar beet: most Irish sugar is home-produced from this (Irish Sugar Company)

Production

Sugar is manufactured in the growing plant by photosynthesis (see p. 7). This raw sugar is extracted, crystallised and purified at the sugar factory.

Beet is harvested in autumn and brought to the processing plant where it is washed and sliced. The slices are soaked in hot water and the sugar diffuses from the cells of the sugar beet by the process of osmosis. The exhausted beet slices are dried and processed into animal feeds. The sugar solution, which now contains about 14 per cent sugar, includes many non-sugars (impurities) and is black in colour. This raw juice is mixed with lime and carbon dioxide which causes the impurities to precipitate out of the solution. These are filtered off and the juice is boiled to evaporate much of the water, bringing the sugar solution to 60 per cent concentration.

The sugar is crystallised by further evaporation in special vacuum pans. The crystals are then separated from the remaining liquid by centrifuging the mixture two or three times in a perforated drum similar to a large spin dryer. The juice which runs out is known as molasses and is used in the manufacture of animal feeds and alcohol — mainly rum. The sugar is bleached with charcoal and finally dried in hot air cylinders called granulators. It is then stored until required for packing and distribution.

Varieties of Sugar

Granulated sugar. Pure white sugar made up of fairly large crystals. It is used at table, for sweetening food, and in preserves and homemade sweets.

Caster sugar. Very fine crystals, used in all types of baking. Caster sugar is ideal for creamed mixtures as it blends well.

Icing sugar. Obtained from granulated sugar ground to a fine powder. It is used for icings and fillings and some homemade sweets.

Cubed sugar. Moistened granulated sugar moulded and cut into squares.

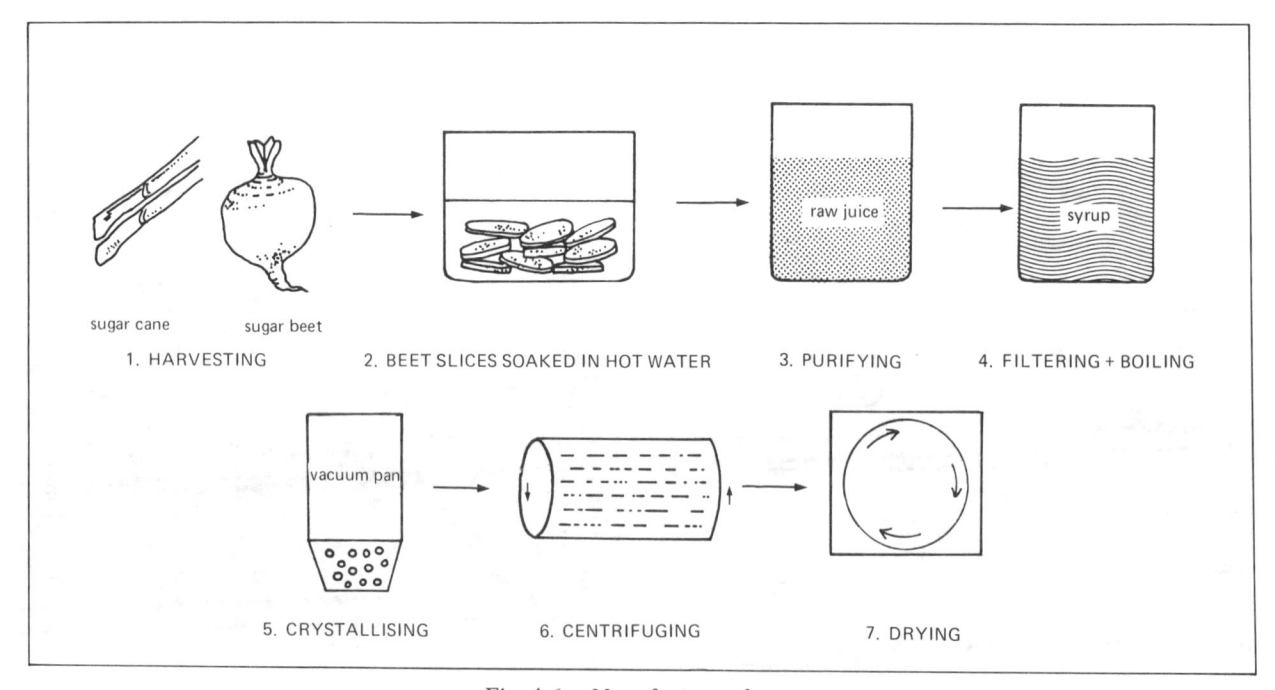

1. HARVESTING — sugar cane, sugar beet
2. BEET SLICES SOAKED IN HOT WATER
3. PURIFYING — raw juice
4. FILTERING + BOILING — syrup
5. CRYSTALLISING — vacuum pan
6. CENTRIFUGING
7. DRYING

Fig. 4.6 Manufacture of sugar

Brown sugar. True brown sugars are made from sugar cane, but most brown sugar on general sale is granulated beet sugar to which molasses has been added to give it flavour and colour. Brown sugar helps to darken and flavour fruit cakes and gingerbread. *Demerara sugar* is a large-crystalled light brown sugar which is unrefined, while *Barbados sugar* is a darker brown sugar with a finer grain.

Golden syrup and treacle. Products of sugar refining, manufactured from the waste liquor of the final crystallisation process. They are sweet viscous liquids containing colouring and minerals such as calcium and iron. Iron gives treacle its dark colour.

Glucose. Usually bought in powder form, although liquid glucose is available. Glucose is used for making home-made sweets as it prevents sugar crystallisation. It provides energy quickly and is sometimes used during illness and convalescence.

Honey. A mixture of glucose and fructose which can be used instead of sugar in some recipes. It is made by bees from the nectar of flowers, and the flavour depends on the type of flowers from which it is collected. It can be sold in the honey comb or pasteurised and strained. Many products sold as honey are diluted. Honey has fewer calories than sugar, and contains traces of vitamin B, iron and many other minerals.

Nutritive Value of Sugar

Sugar consists of 99.9 per cent sucrose and is broken down into its components fructose and glucose during digestion. Sugar is pure carbohydrate. It contains *no* proteins, minerals, vitamins or other food constituents except traces of water. In fact, it needs B vitamins for its utilisation in the body, solely as a source of energy. As we in Western Europe get more than enough energy from our normal diet, it is evident that sugar is superfluous. We need fewer energy-giving foods than our ancestors, who had more strenuous occupations, fewer labour-saving machines and little transport — and yet the average person in this country eats as much sugar in two weeks as his ancestors ate in one year in the seventeenth century. The current tendency to overindulge in sweets, biscuits, soft drinks and sugar in other forms displaces other, more nutritious foods from the diet. Too much sugar results not only in severe dental decay but also contributes to obesity, heart disease and many associated complaints.

Uses of Sugar in Cooking
1. As a *sweetener.*
2. As a *preservative.* In concentrations of more than 60 per cent it inhibits the growth of micro-organisms, e.g. yeasts and moulds.
3. *Syrups and glazes.* When sugar solutions are boiled, the water is driven off and the concentration becomes greater until it eventually forms a thick, sticky syrup which solidifies on cooling.
4. *Caramel.* If a sugar solution is boiled to a temperature of 194°C (380°F) it will change colour and caramelise. *Note.* Sugar must be dissolved slowly and completely before boiling or crystallisation will occur.
5. In *creamed cakes,* the fat and sugar form an emulsion which traps air, making the cake light.
6. Sugar, when moistened, softens the gluten in flour making a *more elastic dough.* This is useful in cake-making.
7. When *cooking with yeast,* sugar must be present to supply food for the fermenting yeast cells. Flour contains a small percentage of sugar. Adding a little sugar to a yeast dough causes fermentation to take place more quickly, but remember that too much sugar can destroy yeast cells.
8. *Meringues.* Sugar when added to egg-white helps the albumen to retain air. The eggs must be stiffly beaten *before* the sugar is added, otherwise the meringue will collapse.

Sweetening Without Sugar
1. *Glucose* or *honey.*
2. *Sorbitol,* a sweetener made from glucose, is about two-thirds as sweet as sugar. As it is still high in energy value, it is not useful in slimming diets but it is used extensively in diabetic preparations. It can also be used for sweetening diabetic foods at home.
3. *Cyclamate,* an artifical sweetner based on sodium cyclamate. Its use is banned in many countries as the result of tests carried out in the United States which associated it with cancer.
4. *Saccharin,* a synthetic substance produced from coal tar and with no carbohydrate content, is over five hundred times sweeter than sugar. It is ideal for those on slimming diets and is used extensively in low-energy foods and drinks. When concentrated it has a bitter aftertaste. In the United States saccharine is being thoroughly researched at present as there are indications that it may have carcinogenic side effects.

VEGETABLES
Vegetables are both nutritious and tasty. They may be served as an hors d'oeuvre, as an accompaniment to meat and fish dishes or as a separate course, and add colour, flavour and variety to a meal. They are valued chiefly for their mineral and vitamin content, mainly vitamin C, so as far as possible they should be eaten raw or lightly cooked. In this country, unfortunately, we tend to overcook vegetables in too much water until they are a soft, tasteless pulp. Vegetables should be eaten as soon as possible after picking. Their appearance is a good indication of their nutritional content — for instance, a crisp green lettuce has a much higher vitamin content than a yellow, wilted plant, and it tastes a lot better too.

Classification Fresh vegetables may be divided into several groups:

Green Vegetables	Roots and Tubers	Pulses	Fruits	Stems and Bulbs
Brussels sprout	carrot	broad bean	courgette	celery
cabbage	beetroot	runner bean	vegetable marrow	asparagus
curly kale	parsnip	French bean	cucumber	leek
endive	swede turnip	pea	tomato	onion } bulbs
spinach	white turnip		sweetcorn	garlic
lettuce	kohlrabi		pepper (green and red)	
broccoli, cauliflower } flowers	potato, Jerusalem artichoke } tubers			

For culinary purposes edible fungi such as mushrooms are also considered vegetables.

Nutritive Value

The composition of vegetables varies. Most contain a high percentage of water. Besides being valued for their mineral and vitamin content, they are a good source of roughage.

Protein. Traces of protein are found in many vegetables such as potatoes. However, pulse vegetables such as peas and beans and especially soya beans, and also nuts, are the best sources of vegetable protein. Root vegetables, cauliflower and Brussels sprouts contain small amounts of protein.

Fat. Many vegetables are deficient in fat although nuts and soya beans are a good source of natural oils. Maize and other cereals also contain fat as do many tropical plants such as olives, palm kernels and cocoa beans. To improve the fat content and flavour of vegetables they can be tossed in butter when cooked or can be served with a sauce.

Carbohydrate may be present in the form of starch or sugar. Potatoes, carrots and pulse vegetables contain starch. Beetroot, carrots, tomatoes, leeks, onions and pulse vegetables contain sugar. Green vegetables contain little digestible carbohydrate, but are a good source of cellulose (see below).

Cellulose is a type of indigestible carbohydrate which forms the cell walls of vegetables. It is most plentiful in the skin or outer husk of vegetables and cereals. It is not digested but stimulates the peristaltic movement of the intestine, preventing constipation and associated diseases. When vegetables are peeled or cereals milled, much of this roughage is lost.

Vitamin A. Many vegetables are a source of provitamin A (carotene). It is plentiful in dark green and most red or orange fruits and vegetables such as carrots, red peppers, spinach, broccoli, cabbage, lettuce and tomatoes.

Vitamins E and K. Present in green vegetables in small amounts. As these vitamins are fat-soluble and are stable to heat, little is lost during preparation or cooking processes.

Vitamin B. Thiamine is plentiful in pulse vegetables and, to a lesser extent, in green vegetables. Small amounts of B_2 (riboflavin) and niacin are also present in many vegetables.

Vitamin C. Vegetables are one of the best sources of vitamin C (ascorbic acid). It is plentiful in all green vegetables, particularly dark greens such as cabbage, spinach, broccoli and sprouts. Onions, some root vegetables (especially swedes), cauliflower and pulse vegetables are also sources of vitamin C. Tomatoes and potatoes contain smaller amounts. Watercress and parsley are very rich in this vitamin, but because they are used in such small amounts, they cannot be counted as a good source. Potatoes, because they form a major part of the Irish diet, supply us

Well-loved vegetables (Barnaby's Picture Library)

with a considerable amount of vitamin C.

Vitamins B and C are water-soluble and both, especially vitamin C, are unstable in heat. This means that care must be taken to avoid steeping vegetables containing them; as little water as possible should be used during cooking and they should be cooked for the minimum time. Young new potatoes, carrots and other vegetables supply more vitamin C than older vegetables.

Calcium is present in green vegetables, some root vegetables and tomatoes. Some types of calcium such as calcium oxalate are present in a form which the human body is unable to assimilate.

Iron. Most vegetables contain a small amount. Greens and pulses are rich in iron. Oxalates prevent the iron in spinach from being absorbed. Some greens also contain sulphur.

Water. Most vegetables contain large amounts of water — many over 90 per cent (see p. 211).

Dietetic Value

Vegetables are important in the normal diet. Their mineral and vitamin content provides substances essential to the proper functioning of the body. Minerals, for instance, help to keep the blood alkaline. Roughage (or dietary fibre), which is plentiful in all vegetables, is of major importance to balance the large quantities of highly-processed, over-refined foods in the modern diet.

Babies should be fed sieved vegetables such as carrots and dark green vegetables at about three months to supplement their vitamin and iron supply. *Children* should be encouraged to eat a variety of vegetables in order to attain a balanced diet. The alkalinity of vegetables is useful in the diet of the *adolescent* as it counteracts greasy skin and helps prevent acne. Vegetables (with the exception of pulse vegetables, potatoes and some root vegetables) are excellent for *weight-reducing diets* due to their low energy value. But those on *low residue diets* may have to reduce their intake of vegetables.

Buying Vegetables

Ideally, vegetables should be eaten straight from the garden, so that they are young and fresh. But as this is not always possible, remember the following points about buying vegetables:
1. Choose vegetables that look fresh, unbruised and insect-free.
2. Buy vegetables in season when they are at their best, plentiful and cheap.
3. Avoid buying them in polythene bags, as these can cause condensation, mould growth and, eventually, decay. Choose instead loose vegetables or those in net bags.
4. Avoid bulk-buying unless time and facilities for deep freezing are available.

Storage

Even in ideal conditions vegetables lose much vitamin C through oxidation during storage. Air, light, heat and damp speed up their deterioration.
1. Store all vegetables, especially greens, for as short a time as possible.
2. Store in cool, dry, dark, ventilated place such as a larder, clean, dry garage or utility room. A kitchen is the worst place to store vegetables as it is usually warm and damp. A ventilated cupboard is a useful storage place.
3. Greens can be washed, shaken dry and stored in a polythene bag in the base of the refrigerator, or kept unwashed in an airtight saucepan or biscuit tin.
4. Large quantities of roots and tubers can be stored in a clamp, i.e. covered with straw and earth or in a sand-pit.
5. Preserving techniques such as freezing, bottling or pickling may be used to preserve vegetables for up to one year.

Economic Value

Although many vegetables have become more expensive since Ireland joined the EEC there is still a good choice of cheap, nourishing produce available. Prices tend to fluctuate depending on weather, demand, scarcity, production costs and quality.

Many people are now growing their own vegetables, and this makes good economic sense. In addition, tastier varieties can be grown than the large-cropping commercial varieties and any surplus can be used for freezing. Because there is no delay between harvesting and cooking, home-grown vegetables are rich in vitamins and minerals.

Quality and Grading

To maintain high quality produce the EEC standards for both fruit and vegetables are now in use in Ireland.

This ensures that they are sold clean, free from soil and contaminants such as pesticides and fertilisers. Each pack should contain produce of uniform quality and size. It should be clearly labelled to show quality, origin and, where appropriate, variety.

Classes

a. Class Extra – best quality produce
b. Class I – good quality
c. Class II – marketable quality but with defects of shape, colouring and blemishes allowed
d. Class III – marketable but inferior.

Effects of Heat on Vegetables

1. The cell walls are softened and broken, making the food more digestible.
2. The starch grains burst, releasing starch.
3. Water-soluble vitamins and minerals dissolve into the cooking water.
4. About 50 per cent of vitamin C and some vitamin B are lost in cooking. More is lost if the vegetables are chopped beforehand.
5. Some water is absorbed.
6. There is a loss of green colour.
7. Over-cooking causes vegetables to disintegrate and lose flavour.

To Retain Vitamins:

1. Only use fresh vegetables.
2. Trim sparingly if necessary and wash under running water. Avoid chopping and peeling if possible, as this liberates oxidase from the cell walls destroying the vitamin C, and avoid steeping.
3. Do not prepare vegetables in advance, as vitamins and minerals will be lost and discolouration may occur.
4. Cook quickly in the minimum amount of boiling water, in a saucepan with a tight-fitting lid, until *just* tender.
5. Do not use bread soda as it destroys the vitamin C. Avoid the use of copper or brass pans; these also reduce the vitamin C content of food.
6. Drain thoroughly, retaining vegetable stock for sauces etc., and serve at once.

Preserved Vegetables

Methods of preserving have improved in recent years and many vitamins formerly lost in processing are now retained, thanks both to increased knowledge of nutrition and to technological advances.

Dried vegetables. Pulse vegetables are among the most popular forms of dried vegetable. Most of these are preserved by air drying in warm ovens. Many varieties of dried beans and peas are available in delicatessens, making it possible to cook foreign dishes using traditional ingredients. *Varieties:* Lentils, marrowfat peas, split peas, chickpeas. Beans – broad, haricot, kidney, red, soya, butter. The nutritive value of reconstituted dried vegetables is similar to that of fresh pulse vegetables. They are a good source of protein, but there is some loss of vitamin B in the drying process and almost complete loss of vitamin C.

Freeze-dried vegetables (see p. 139). Flavour and appearance are reasonably good and reconstitution quick. They are easy to store as they are light and no refrigeration is required. There is little loss of food value. *Varieties:* Instant mashed potato (vitamin C is added to replace that lost in the processing); onions, carrots, peas, French beans, cabbage, vegetable soups.

Canned vegetables. Quick and easy to use: as they are already cooked they only need to be reheated. There is a slight loss of vitamin B group and vitamin C. Some minerals and vitamins dissolve into the canning liquid so this should be used. Many vegetables lose texture and flavour in canning.

Frozen vegetables. These retain natural colour, flavour, texture, shape and vitamins. Frozen vegetables often contain more vitamin C than fresh vegetables because of efficient harvesting, rapid preparation and blanching which seals in the vitamin C before goodness can be lost. They are quick and convenient to use and very little cooking is required. Follow the directions on the packet *exactly*. Frozen vegetables may seem very expensive but there is no waste and cooking time is shorter than for fresh vegetables.

Less Common Vegetables

Many vegetables which were formerly almost unobtainable and very expensive are now in plentiful supply in many green grocers. Special varieties of some, such as sweet corn, pepper and soya bean, have been developed to grow in cooler climates. These give an authentic flavour to many continental recipes and most are delicious and nourishing.

Globe artichoke. A round green vegetable with spiky leaves, which is a member of the thistle family. The base of the leaves and the heart are eaten.

Jerusalem artichoke. This looks like a knobbly potato and has a smoky taste. It is prepared and cooked like a potato, or made into soup.

Asparagus. A thin, pale, green stem with a leafy tip. It is scarce and expensive, and has a delicate flavour. It can be served as a vegetable or used as a garnish.

Aubergine (egg plant). An elongated pear-shaped vegetable with glossy dark purple skin. It can be stuffed and baked or used chopped up in some recipes such as Moussaka.

Less common varieties: avocado, fennel, artichoke, courgettes, chicory and bindi (ladies' fingers) (Anne-Marie Ehrlich)

Avocado pear. Pear-shaped with a rough dark green skin. When ripe it yields to gentle pressure. Used raw, sliced in two, as an hors d'oeuvre with French dressing or stuffing in the cavity. One of the few vegetables with a substantial oil content.

Broccoli – purple sprouting. A member of the cauliflower family, rich in vitamin C, carotene, calcium and iron. It is usually boiled.

Chicory. Long pointed leaves tightly packed and graduating from white to yellowish green. Chicory tastes like slightly bitter lettuce. It is usually served raw in salads or cooked in a minimum of water and served with sauce.

Courgettes (zucchini). These specially-grown baby marrows are long, dark green, cucumber-like vegetables. They can be served halved and stuffed, or cut up in salads or in vegetable casseroles.

Leeks. A member of the onion family, they have long white leaves with green tops. They should be split and washed thoroughly to remove grit. Leeks can be served with béchamel sauce, melted butter or vinaigrette; in casseroles; or in soups. Vichyssoise is a cold soup made with leeks.

Red and green peppers (capsicums). Green peppers have a shiny bright green skin and are almost hollow inside, with a few seeds which are removed. Red peppers are called pimentoes. Green peppers, as they ripen, turn yellow and then red, and their flavour gradually mellows. *Chillies* are small, very hot peppers. Large peppers can be stuffed and used in casseroles, or raw in salad and for garnishes.

Sweet corn. This is readily available frozen in cobs or grains, but it is now possible to buy fresh corn-on-the-cob in summer. The cobs are cooked in boiling salted water and eaten with butter as a first course.

Kohlrabi. A green, turnip-like vegetable with a delicate flavour. It can be eaten both raw (finely shredded) in a salad or cooked.

SALADS

A salad is a mixture of savoury or sweet ingredients which are usually raw and which, after assembly, need no further cooking. Ingredients include vegetables, fruit and in many cases a protein food such as meat, fish or nuts. Salads add colour and variety to meals and have a good nutritive value, as many of the raw ingredients are an excellent source of vitamin C. They provide a contrast in flavour, texture and appearance to most cooked dishes, are quick to assemble and require little or no fuel.

Nutritive Value

This varies according to the ingredients used and their freshness.
1. Raw vegetable salads such as a green salad are rich in vitamin A (carotene) and C and water-soluble minerals iron and calcium. Most vegetables and fruit contain potassium.
2. Pulse vegetables, if used, provide protein and carbohydrate.
3. Root vegetables and fruits supply carbohydrate.
4. If meat, fish, eggs or cheese are included they will provide both protein of high biological value and fat.
5. Salads add bulk to the diet by providing water and are a good source of roughage.

Salads are nutritious, colourful and quick to assemble

6. As salads tend to be deficient in fat a dressing such as mayonnaise or vinaigrette is usually served to make good this loss and to improve the flavour.

Classification
Green salads. Combination of green vegetables, e.g. lettuce, chives, spring onions, cucumber.

Mixed salads. Lettuce, spring onions, egg, tomato, cucumber and possibly cheese.

Meat salads. Cold meat, e.g. ham, chicken, salami with usual salad ingredients.

Fish salads. Salmon mayonnaise, rollmop (herring) salad, Salade Nicoise (made with tuna fish).

Jellied salads. Beetroot or a mixture of vegetables set in aspic or gelatine.

Winter salads. Diced root vegetables, rice and vegetables in mayonnaise.

Fruit salad. A mixture of fresh and perhaps canned fruit in syrup.

American salad. Usually a mixture of savoury ingredients and fruit.

Nuts. Flaked or chopped and added to many salads, they increase the protein value and provide a contrast in flavour and texture.

Seasonings. Use discriminately. Half a garlic clove rubbed into the salad bowl gives a subtle flavour. Use fresh herbs to suit the vegetables — e.g. basil goes well with tomatoes. Parsley, which should be washed, can be finely chopped and mixed into dressings or salads. Pickles, gerkins, olives and capers also provide extra flavour.

Rules for Salad-making
1. All ingredients must be absolutely fresh and clean. They are usually washed and shaken dry.
2. To avoid loss of vitamin C use a sharp stainless steel knife to cut vegetables. This causes less damage to cell walls, releasing less of the oxidising enzyme oxidase.
3. Avoid making salads too far in advance. If this is unavoidable, cover with cling wrap or foil and store in the refrigerator.
4. Avoid shredding vegetables. If this is necessary, do so at the last minute.
5. Dressing improves the flavour and nutritive value of salads. Either serve the dressing separately or toss salad in dressing immediately before serving. Remember a dressed salad becomes stale and limp very quickly.

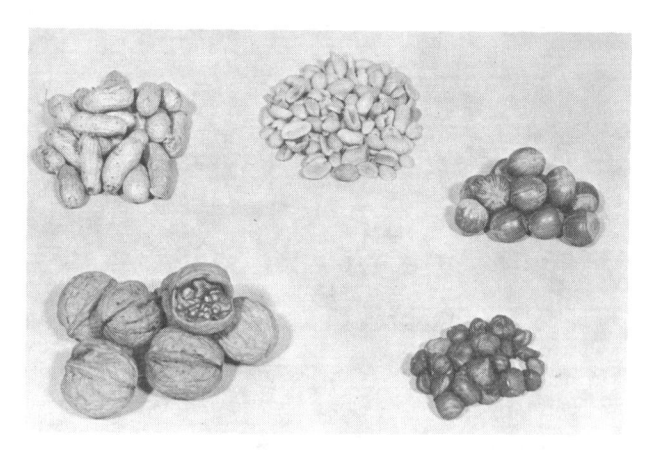

Nuts: rich in protein and vegetable fat (Bill Doyle)

NUTS
Nuts are a useful addition to the diet and can be used whole, ground or chopped, in both sweet or savoury dishes. They are important in the diet of vegetarians because of their relatively high protein content.

Nutritive Value
Nuts are rich in protein and vegetable fat. The carbohydrate present is usually in the form of cellulose. They contain small amounts of vitamins C and B and some iron and calcium, which is usually not assimilated by the body.

Uses of Nuts
Cakes: walnuts, almonds, desiccated coconut.

Sweets: marzipan and almond icing.

Biscuits: almond slices.

Salads: chopped and sprinkled over a salad.

Main course: nut cutlets in vegetarian diets.

FRUITS
Fruit, like vegetables, contains a high percentage of water and is an important source of vitamin C. Raw fruit is refreshing and palatable. It adds colour and texture to the diet.

Classification	Examples	Start of season
citrus fruits	oranges, lemons, grapefruit	January
hard fruits	apples, pears	September
stone fruits	plums, damsons, cherries	summer
berries	strawberries, raspberries, blackberries, gooseberries, black, white and red currants.	summer
other	rhubarb	summer
	grapes	September
	bananas	imported all year round

Nutritive Value

Most fruits contain large proportions of water, usually 80–90 per cent, and few, if any, proteins. All fruit is deficient in fat.

Carbohydrate may be present in many forms — as the disaccharides fructose or sucrose, which give fruit its sweet taste, or as the polysaccharide pectin, which is found in some fruits and is essential in the setting of jams. Another polysaccharide, cellulose, is present in the skin and cell walls of fruit and necessary in the diet to provide roughage.

Many fruits contain traces of mineral elements. Some contain minute amounts of B group vitamins, and apricots, peaches and blackcurrants have a good supply of pro-vitamin A (carotene). However, fruit is mainly valued for its ascorbic acid (vitamin C) content. The following fruits, listed in descending order of importance, contain large amounts of vitamin C:

- rosehips
- blackcurrants
- citrus fruits
- berries
- tomatoes
- apples
- bananas

Vitamin C is essential for the absorption of iron, prevention of scurvy and general health. Fruit has a high acid content which makes it refreshing and helps to retain its ascorbic acid content.

Ripening

Underripe fruit contains considerable quantities of starch, which changes to sugar as the fruit ripens. The pectose changes to pectin, the substance necessary for gel formation and eventually pectic acid. After ripening, enzymes begin to break down the fruit until eventually with the assistance of natural yeasts and moulds in the atmosphere, the fruit decomposes. Various methods have been used to delay decomposition, including impregnating wrapping papers with preservatives such as phenols and keeping the fruits in an atmosphere where the oxygen is reduced and carbon dioxide increased. This method is sometimes used in shipping when the large containers holding the fruit are pumped with carbon dioxide.

Effect of Heat on Fruit

Fruit is affected in a similar way to vegetables by cooking. There is some loss of vitamin C but, because of the high acidity of many fruits, this loss is considerably less than in vegetables which have been cooked by similar methods.

There is a 10 per cent loss of vitamin C in cooked fruit compared with a loss of around 50 per cent in cooked vegetables. Fruit should be eaten raw if possible as it tastes better and has a higher food value.

Storage and Preservation

Fruit should be stored in a cool, dry, well-ventilated place. Decay can be delayed for a considerable time if it is stored under suitable conditions. Fruit can be successfully preserved by freezing, canning or drying.

a. Freezing. The appearance and texture may be slightly altered but the food value remains similar to that of fresh fruit.

b. Canning. There is some change in texture and appearance: most fruits soften considerably, so that they bear little resemblance to the fresh product. The vitamin C content is reduced and the carbohydrate content is greatly increased because of the high concentration of sugar in the syrup.

c. Drying. When reconstituted most dried fruits such as prunes and apricots have a similar energy value to fresh fruit, but substantial amounts of vitamins B and C are lost in the drying process.

Economic Value

As most fruit is expensive, it makes sense to grow some in the garden if possible. For maximum nourishment avoid peeling where possible, as it is wasteful, but be careful to wash fruit thoroughly to remove deposits of pesticides.

Buying Fruit

The rules for buying vegetables (p. 95) also apply to fruit. Choose only very fresh, sound, ripe, unblemished fruit; it should have a good colour and feel firm and heavy for its size. Buy in small quantities and use quickly.

QUESTIONS

1. Compare animal and vegetable fats under the headings
   a. physical properties
   b. sources
   c. culinary uses
   State which type you think is preferable in the diet, giving reasons for your answer.
2. List four different types of margarine, stating the purposes for which each is used.
   Describe in detail how margarine is manufactured. Make a brief comparison between butter and margarine in the diet.
3. State the reasons why you think vegetables are important in the diet. What points must be observed when preparing and cooking vegetables in order to achieve maximum vitamin retention?
   Name three modern methods of preserving vegetables and describe the effect of each method on the nutritive value of vegetables.

4. Classify fruits and give the approximate composition of the fruits in each classification.
   What is the effect of heat on fruit?
   Discuss the value of nuts in the modern diet and list some ways in which they could be included in it.
5. Describe a project which could be carried out on a cereal of your choice. Include the following information in your answer:
   a. basic structure
   b. processing
   c. nutritive value
   d. effects of cooking
6. Describe the culinary uses of sugar. Discuss the use of sugar in the modern diet. Write a note on artificial sweeteners.

5 Food Preparation

BASIC METHODS OF COOKING

Cooking does much to improve the appearance, taste and aroma of food. It makes many foods more digestible while at the same time making them safer by destroying bacteria and parasites.

Reasons for Cooking Food
1. The *appearance and flavour* of foods are improved.
2. The taste and aroma stimulate the *digestive juices*.
3. *Flavours* in the food (e.g. extractives) are accentuated and new flavours are developed by blending different ingredients.
4. Cooking food breaks it down and makes it *more digestible*.
5. *Parasites and bacteria* are destroyed.
6. Cooking has a *preservative effect*.

COOKING BY DRY HEAT
Baking
This involves cooking food in dry heat in an oven. Bread, cakes, pies, meat, fish and vegetables can all be baked. The oven is heated by convection and radiation: the hot air rises and then falls as it cools. For this reason the top shelf is the hottest part of an oven and the bottom shelf the coolest, except in the case of a fan-oven where the temperature is even throughout.

Roasting
Strictly speaking, this means cooking food on a spit which rotates over or under a source of radiant heat. However, the term has come to mean cooking meat, fish or vegetables in hot fat in an oven.

a. Quick roasting is only suitable for tender joints of meat and for poultry. Meat is cooked at a high temperature – 230°C (450°F)/Gas 7 – for 20 minutes to seal the juices. The heat is then reduced to finish cooking at 205°C (400°F)/Gas 5. This method gives a crisp, roasted finish and flavour to the meat.

b. Slow roasting. The meat is cooked at 175°C (350°F)/Gas 3 for up to twice the quick-roasting cooking time. The meat becomes more tender and shrinks less when this method is used, but it lacks a good 'roasted' flavour and appearance.

Autotimer roasting. Contrary to popular belief, meat may be put into a cold oven when an autotimer is being used. The timer is set and the meat will gradually begin to cook as the oven heats up. It should be brushed well with fat beforehand. Results are satisfactory, but not as good as when the oven is preheated.

Spit roasting. Many modern cookers have a rotating spit (rotisserie) on which joints of meat and poultry may be fixed. As the joint rotates the fat melts over the roast, basting it. This prevents the meat from drying out and improves the flavour.

Meat thermometers. These are useful to determine accurately whether a roast is sufficiently cooked. They are inserted into the joint before cooking and will register a certain temperature on the dial when the meat is cooked.

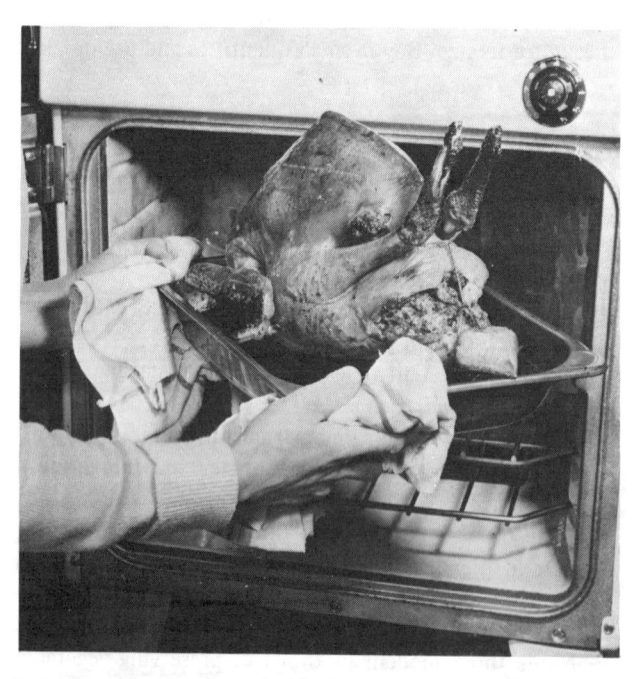

Quick roasting: only suitable for tender meat and poultry (Barnaby's Picture Library)

Rules for roasting
1. Meat should be at room temperature.
2. Preheat oven to correct temperature.
3. Prepare, trim and weigh meat, and wipe clean.
4. Place meat on grid and baste with smoking hot fat.
5. Cook according to table below.
6. After removing from oven allow meat to 'set' in a warm place for 10 minutes before carving. This makes it easier to carve and allows time to make gravy.

Suitable cuts of meat	Slow Method 175°C (350°F)/Gas 3	Quick Method 230°C (450°F)/Gas 7 (reducing after 20 min)
Beef: ribs, sirloin, fillet	35 min. per 500 g	20 min. per 500 g plus 20 min. extra (15 min. per 500 g plus 15 min. extra for rare beef)
Lamb: loin, leg, shoulder, breast	35 min. per 500 g	25 min. per 500 g plus 25 min. extra
Pork: leg, loin.	50 min. per 500 g	30–35 min. per 500 g plus 30 min. extra
Veal: leg, shoulder, loin, breast.	50 min. per 500 g	30–35 min. per 500 g plus 30 min. extra
Fowl: chicken, turkey, duck, goose, game (well buttered and wrapped in foil)	20–30 min. per 500 g	10–15 min. per 500 g plus 10–15 min. extra

Note. If stuffing is used or meat is covered with aluminium foil or a roasting bag add an extra 20 to 30 minutes to cooking time. But remember to remove foil or roasting bag for final 30 minutes to brown meat.

Pot roasting. This is an economical method of cooking a small roast or chicken, in a heavy saucepan or cast-iron stewpan with a close-fitting lid. The prepared meat is browned in hot dripping. The heat is then reduced and meat cooked using the same times as for conventional roasting. It should be turned often. Vegetables may be included.

Grilling

This is a quick method, by which food is cooked under or over radiant heat. The temperature is high enough to seal the juices quickly, thus retaining food value and flavour. Grilling is only suitable for thin pieces of food and tender cuts of meat. Broiling is another name for it.

Rules for grilling
1. Preheat grill so that when the food is placed underneath the heat is sufficiently intense to seal surface protein. Insufficient heat will result in escape of juices and subsequent loss of food value, flavour and moisture.
2. For best flavour and moist results continue grilling at a high temperature. For thick pieces of meat, reduce heat to finish cooking.
3. Brush grid of grill pan and food to be grilled with melted fat. Baste once or twice during cooking.
4. Season meats with pepper only. Salt causes juices to escape.
5. Never prod or turn food with a pointed implement

Grilling: for thin pieces of tender meat (John Topham Picture Library)

such as a fork as this, too, releases juices. Use tongs or two knives.
6. Length of cooking time depends on type and thickness of food, distance from grill and degree of 'doneness' required.
7. Marinating improves flavour and tenderness.

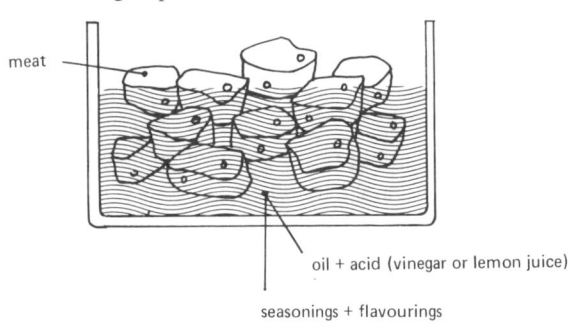

Fig. 5.1   Marinating

Marinating. Braised, grilled and stewed meats all benefit from this process, which involves soaking meat, poultry or fish in a mixture of oil, some form of acid (e.g. wine vinegar, lemon juice), herbs, seasoning and vegetables for flavouring (e.g. chopped onion, carrot, celery). These may be used to form the mirepoix (see p. 104). The meat or fish is left standing in the marinade mixture for a few hours and is turned often. This improves its flavour and moistness, and also tenderises it, as the acid present softens the meat fibres.

COOKING IN FAT OR OIL
Frying

Frying is a method of cooking in hot fat or oil. As frying temperatures are high, it is quick and suitable only for fairly thin pieces of food. Foods which have been properly fried should be crisp and tasty. When the frying temperature is too low the results will be greasy and sodden, whereas when it is too high the outside may burn before the inside is cooked.

The two basic types of frying are:
1. Shallow frying – cooking thin pieces of food in a small amount of hot fat.

2. Deep fat frying – cooking food by complete immersion in hot fat or oil.
Note. Particular caution must be observed during deep frying as it is probably the most dangerous culinary process.

Frying fats. These should be tasteless with a high smoking point (the point at which blue smoke can be seen rising from them). Food must be cooked at below the smoking point or it will be burnt.

Types	Smoking Point
lard	200°C or over
dripping	190°C
cooking fats	225°C or over
cooking oils	230°C

Butter and margarine are unsuitable for frying as they contain salt and water, which make them spatter and separate. As they have a low smoking point they burn easily.

Coating. Many foods are coated (e.g. in flour, egg and breadcrumbs, batter) before frying. This protects the surface from the hot fat and reduces the loss of flavour and nutrients into the fat.

Dry frying is a method of shallow frying used for foods containing a certain amount of fat e.g. sausages, rashers, steaks, chops. They are placed in a heated frying pan which has been brushed with a little melted oil or fat, and some fat from them melts on to the pan. The results resemble grilled foods. Do not use a lid when frying as it traps the steam and prevents the food from becoming crisp.

COOKING BY MOIST HEAT
Boiling

By this process food is cooked in liquid at 100°C. Water, stock, wine or a combination of the three may be used. The food should be covered or almost covered with the liquid, which should be bubbling rapidly when food is immersed in it in order to:
1. Coagulate the surface protein.
2. Destroy the oxidising enzyme in plant foods.
3. Set the colour of green vegetables.
The food is then simmered until cooked.

Foods suitable for boiling
Salt meats: Ham, bacon, corned beef. Cooking is started in cold *unsalted* water.

Shallow frying: thin pieces of food cooked in a small amount of fat (John Topham Picture Library)

Fresh meats: Leg of mutton, chicken. Immerse in boiling *salted* water.

Vegetables: Almost all vegetables may be boiled. Green vegetables are particularly suitable as boiling retains their colour. For *maximum* vitamin retention, the *minimum* amount of liquid should be used.

Rice, pasta: Most forms are cooked for about 12 minutes in boiling salted water.

Advantages. Boiling is an easy, quick, clean method of cooking. The food remains moist and juicy.

Disadvantages. Vitamins, particularly vitamin C, mineral salts and extractives leach into the water. This major loss of nutrients may be partly overcome if the cooking liquid can be used for soups, sauces and gravies.

Poaching

This method differs from boiling in that the temperature is lower, i.e. the water in which the food is cooked is at 85–90°C, just below simmering point. The water should be 'trembling' but not bubbling. The food is partly covered by the liquid.

Simmering

Food is cooked in liquid which is at 90–95°C, just below boiling point.

Stewing

The basic principle of stewing is to cook the food very gently in a little liquid. This usually takes a long time. The temperature should not rise above 90°C, and the cooking pot should be tightly covered to prevent evaporation. Stewing may take place in a covered saucepan on the hob or in a covered casserole in the oven. The second method takes about 20–30 minutes longer than the first.

Foods suitable for stewing
  Meat: Tougher cuts of beef, mutton, pork, rabbit
  Poultry: Chicken, duck, game
  Fish: White fish and shellfish
  Vegetables: Vegetable casseroles
  Fruit: Most fruit – whole or sliced.

A stew is a mixture of meat or fish and vegetables cooked slowly in liquid which may later be thickened by liaison or by reduction. Stewing is an ideal method for cooking the tougher cuts of meat, as the low temperature and moist heat soften the connective tissue, turning it into gelatine. The fibres are thus separated and extractives are released into the cooking liquid. As this is eaten with the stew little flavour or nutrients are lost, except for small amounts of vitamin B group and vitamin C, both of which are affected by heat. A very low temperature is essential to prevent toughening of meat fibres and evaporation of liquid.

Simple stews or cold water stews. This method is ideal for very tough meat (e.g. mutton, in Irish stew). Meat (usually tossed in seasoned flour), vegetables and seasoning are placed in layers in a casserole or saucepan, moistened with stock and very slowly brought to simmering point. The stew is simmered gently for 1–3 hours. This type of stew is particularly suitable for auto-timer cooking.

Brown or fried stews. The prepared, cubed meat is tossed in hot dripping until browned, followed by onions and other vegetables. Flour is added to thicken the stew and cooked for 3 minutes; then stock is gradually stirred in and brought to simmering point. The stew is cooked very slowly for 1½–3½ hours.

Examples
  Ragout – a meat stew, usually brown.
  Salmis – a brown stew made from poultry or game.
  Navarin – a mutton or lamb stew containing root vegetables.
  Exeter stew – a brown stew with dumplings.
  Curry – a very highly spiced Indian stew served with rice.
  Goulash – a Hungarian stew containing paprika and tomatoes.

Rich white stews. Meat, e.g. lamb or veal, is soaked overnight and blanched to whiten it and remove the strong flavour. Vegetables, seasonings and bouquet garni are added, with sufficient cold stock to barely cover the meat. The stew is simmered for 1½–2 hours and thickened with a roux. Egg yolks and cream may be added to enrich it.

Examples
  Fricassée – a white stew made from meat, chicken or fish.
  Blanquette – a white stew enriched with egg yolks and cream.

Pressure cookers and electric slow cookers are particularly useful for cooking stews. The electric slow cooker consists of an earthenware pot which sits in an electrically heated base. Brown stews must be started in an ordinary saucepan and transferred to the slow cooker for simmering. Stews cook very slowly by this method, using between 70–160 watts of electricity (about the same as an electric light bulb). They can be turned on in the morning and left to stew for up to 8 hours, and they are also useful for cooking milk puddings, porridge and steamed puddings.

Braising

This is a combination of stewing, roasting and steaming: the food is cooked slowly by moist heat in a tightly covered saucepan. Braising is suitable for cuts of meat which are too tough to roast but are more tender than stewing cuts. Meat for braising is usually left in the piece. Very lean meat is sometimes larded with strips of bacon fat to make it more juicy and tender. The meat is first

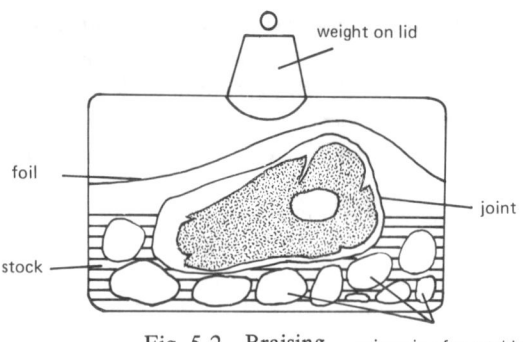

Fig. 5.2 Braising   mirepoix of vegetables

browned in hot fat to seal in the juices and is then placed on a base of diced vegetables *(mirepoix)*. Stock is added to come half-way up the meat and it is cooked until tender – 35 minutes per 500 g and 35 minutes over. The meat should be basted frequently with the stock. When cooked, the sauce is strained, reduced or thickened and poured over the meat. The mirepoix may be served with it or used in soup.

Foods suitable for braising

Meat: Topside, brisket of beef (unsalted); breast or leg of lamb; veal, pork, ham.

Vegetables: Celery, onions, root vegetables.

Note. The term braising is sometimes used for stewing – for example, beef stew is sometimes called braised beef – but this is not strictly correct.

Steaming

This means cooking food slowly by moist heat in the steam rising from boiling water. It is a nutritious method of cooking as many of the minerals and vitamins which would normally dissolve into the cooking water are retained. It is ideal for invalids as the long, slow cooking makes the food digestible.

Steaming may be carried out by one of three methods:

a. In a steamer over boiling water.

b. Between two plates over a saucepan of boiling water. This is useful for thin pieces of food such as fish fillets.

c. In a covered container in a saucepan. Boiling water should come half way up the side of the container, and the covering must be heat-resistant and waterproof to prevent the steam from getting into the food.

Food may also be placed directly onto a well greased clean cloth or aluminium foil, tied tightly and cooked in the water. Puddings such as steak and kidney pudding or plum pudding are suitable for steaming.

Steamed foods are often insipid. To improve the flavour it may be possible to

a. Season well
b. Grease containers with butter and dot a few pieces of butter on steamed foods before or after cooking. (This may not be permitted in invalid cooking.)
c. Marinate meat or fish before steaming.
d. Serve with a well-flavoured sauce.

Conservative Cooking

This method is a combination of boiling and steaming, which conserves vitamins and minerals and is used for cooking vegetables. To every 500 g of sliced vegetables allow approximately 150 ml boiling water, 25 g margarine and ½ teaspoon salt. Simmer gently for 20–30 minutes in a saucepan or casserole in a slow oven for 40–50 minutes.

PRESSURE COOKING

Principle

The boiling point of water is 100°C at normal atmospheric pressure, but the temperature at which water boils can be altered by lowering or raising the pressure. An increase in pressure will make water boil at a higher temperature. This is what happens in pressure cooking.

Pressure cooking is really steaming under pressure. Because of the high temperature within the cooker, steam is forced through the food so that it cooks very quickly – often in a quarter or a third of the normal cooking time (e.g. a stew cooks in 20 minutes instead of 1½ hours).

Water is put in the pressure cooker and brought to the boil. Once the lid and weights have been placed in position neither air nor steam can escape easily, and so the pressure builds up until a certain level is reached.

Steaming: vitamins and minerals are retained and the food is digestible – but it may taste insipid (John Topham Picture Library)

At atmospheric pressure water boils at 100°C (228°F). When *steam* pressure is increased by

5 lb per sq. in. (0.35 kg per sq. cm) water boils at *108°C (228°F)*

10 lb per sq. in. (0.7 kg per sq. cm) water boils at *115°C (240°F)*

15 lb per sq. in. (1.05 kg per sq. cm) water boils at *122°C (252°F)*

Air pressure at 5 lb, 10 lb and 15 lb pressure boils water at temperatures of only 73°, 90° and 100°C. For this reason a pressure cooker will not function effectively until the steam has built up and driven all the air out.

Most foods can be cooked in a pressure cooker by steaming, braising, boiling or stewing. It is useful for making soups and can also be used for making jams and bottling fruit. (It is no longer considered wise to preserve meat or vegetables at home.)

The Pressure Cooker

The pressure cooker is a large saucepan made of heavy-gauge aluminium. It has a special tightly-fitting, locking lid with a rubber gasket or ring which makes an airtight seal. The cover has a central vent which must be kept clear at all times to allow excess steam to escape. There is a safety valve on the lid, consisting of a rubber plug with a metal centre. If the pressure inside the cooker builds up to 20 lb or over because the central vent is blocked, the metal centre of the safety plug pops out, releasing the steam. If the cooker boils dry and overheats, the metal centre will melt, also releasing the steam.

Valve weights. In most pressure cookers the pressure is controlled by weights; a 5 lb weight is used for dishes containing a raising agent and also for bottling fruit; a 10 lb weight is used when making preserves and a 15 lb weight is used for most other pressure cooking. (These are now called low, medium and high.) Some cookers have weights which cannot be removed and which exert 15 lb pressure only. Others have an indicator dial showing when each pressure has been reached. This is adjusted by raising or lowering the heat under the pressure cooker.

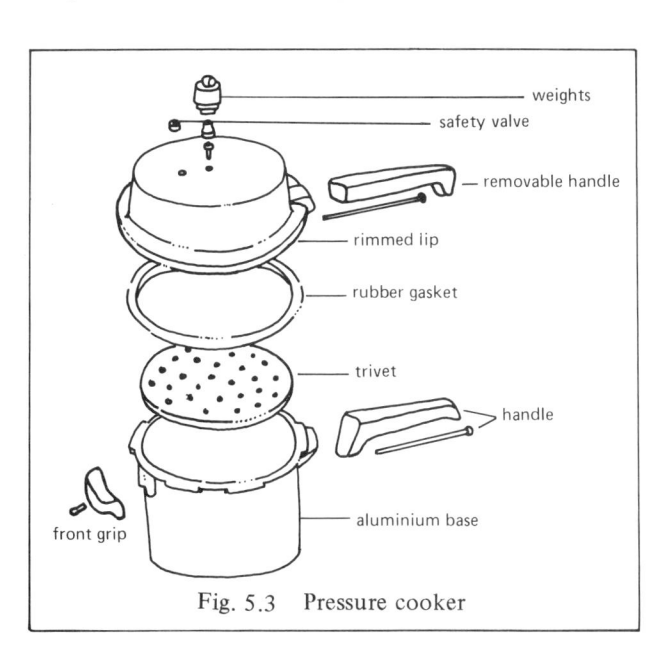

Trivet. A perforated metal tray which is used to keep some foods (e.g. joints, chickens) out of the water so that they are steamed rather than boiled.

Basket, separators. Many pressure cookers are equipped with a light metal basket or perforated separators so that different items can be cooked at the same time. A whole meal can be cooked at once provided care is taken to time each item correctly.

Automatic pressure cookers are now available. These can be set for the required time, and when that time is up, they release the pressure automatically so that the food is not overcooked.

Timing. A list of cooking times for different types of food will be included in the manufacturer's booklet of instructions supplied with each pressure cooker. The short list below will give some idea of how much quicker pressure cooking is than conventional cooking:

stew	20 minutes
potatoes	5 minutes
carrots	4 minutes
boiled fowl	30 minutes
stock	30 minutes

Rules for Using Pressure Cookers

1. Follow the manufacturer's instructions *implicitly*. Remember that otherwise, pressure cooking can be dangerous.
2. Add the correct minimum amount of water recommended in the instructions — and remember that the cooker should never be filled more than half full of liquid foods or two-thirds full of solids.
3. Take care to place the lid and weights in position correctly.
4. When cooking time is reached, reduce pressure by leaving cooker to stand at room temperature or by running cold water over the side of the cooker. *Never* open a pressure cooker while there is still pressure within.
5. After use wash the cooker very thoroughly, especially the rim, gasket and steam valve. If any particles of food remain on these, the cooker may not function properly. Dry well and store without the lid on.

Advantages
1. It saves time
2. It saves fuel

Fig. 5.3 Pressure cooker

3. Food value and flavour are retained.
4. A whole meal can be cooked in one saucepan, which saves on washing-up.

Disadvantages
1. Unless cooking is timed precisely, food will be overcooked.
2. Food cannot be tested to see if it is done without reducing pressure and opening the cooker.
3. As foods cannot be stirred there is a possibility of food sticking to the base of the cooker. To avoid this, grease the base before use and avoid thickening stews etc. until end of cooking time.

FUEL SAVING

As the price of most fuels is continuing to rise dramatically, there is more of an emphasis on saving fuel now than ever before.

Oven Economy
1. Make full use of the oven when it is on. When the main course is to be cooked in the oven choose vegetable dishes and puddings which can be cooked at the same time and temperature, and consider doing some baking as well.
2. Use rising heat for casseroles and retained heat for items such as meringues.
3. Electric ovens can be turned off 15 minutes before the end of the cooking time as they retain the heat well. This does not apply to gas ovens.

Other Useful Hints
1. Use steamers, pressure cookers, electric slow cookers or three-sectioned saucepans to cook complete meals.
2. Plan menus so that double amounts can be cooked on one day and the remainder utilised the following day, e.g. *Day 1:* boiled bacon, cabbage, boiled potatoes (double), stewed apples (double), custard (double). *Day 2:* Pork chops, peas, sauté potatoes (remainder), apple sauce (remainder), trifle using custard (remainder). This saves time and fuel.
3. Serve foods which need no cooking – salads, cheese, cold sweets.

Haybox Cooking
This is a very old method of using retained heat to cook food, which can be used for stews, curries, porridge, soup, milk puddings or any dish which requires long, slow cooking. The 'haybox' can be a large square box packed with an insulating material; originally hay was used but now polystyrene granules or packing are more easily obtainable. A hollow is made in the centre for the saucepan or casserole and a pillow filled with insulating material is used to cover it.

The food is prepared, brought to the boil, simmered for 5–15 minutes and placed in the haybox. The insulating material must fit snugly around the cooking pot. It is left for 6–8 hours, then brought to the boil and simmered for 5–10 minutes to heat through.

Pressure Cookers – see pp. 104–6.

Electric Slow Cookers – see p. 103.

SOUPS

The basis of good soups and sauces is good stock, which gives them a rich flavour. Although stocks and soups generally have little nutrient value, their ultimate food value will depend on the ingredients used and the method of cooking. Sauces, on the other hand, are usually nourishing due to the use of milk, eggs and other ingredients.

Stock
This is the liquid into which the flavour and soluble nutrients of a food have passed during cooking. Bones, lean meat, root vegetables and seasonings are usually used to make stock, and the basic principle is to extract the maximum amount of flavouring and nutrients from these. The extraction depends on the solvent action of water and prolonged slow cooking. Ingredients are put into cold water which is heated gradually to boiling point. The stock is then simmered for 2–6 hours, strained and cooled quickly.

Nutritive Value. The nutritive value varies according to the ingredients used, but is generally rather low in view of the high water content. A little protein is present, mostly in the form of gelatine. There are also traces of some minerals such as iron and potassium.

Varieties of Stock
1. *Brown stock.* Browned bones and chopped raw lean beef simmered with onions, celery, carrot, herbs, seasoning.

2. *White stock.* Chicken or veal bones simmered with water and vegetables.

3. *Fish stock.* Bones, skin, head and trimmings of white or shell fish simmered with seasoning, onion and lemon slices. Unlike most other stock, fish stock requires a short cooking time – about 20 mins.

4. *Vegetable stock.* The water in which vegetables have been cooked.

Note. Stock cubes are concentrated dried stock containing meat extractives and flavourings and may be used if real stock is not available, but they contain even fewer nutrients and do not taste as good.

Soup Types

There are three main types of soup:
1. Thin soups
2. Thickened soups
3. Purées

1. Thin Soups are of two types: (a) *Clear soups* are made with a rich bone stock clarified with egg whites. The name of the soup is often taken from the garnish used, e.g. *Consommé à la Julienne* contains thin julienne strips of root vegetable, while *Consommé à l'Italienne* is garnished with pasta. (b) *Broths* are made from meat and vegetables, thickened with a cereal such as pearl barley or rice, or purely from a mixture of diced vegetables cooked in a well flavoured stock. *Examples:* Scotch broth, chicken broth, vegetable broth.

2. Thickened soups are made from meat, fish or vegetables and are thickened by the addition of a liaison (see below). *Examples:* chicken, oxtail, fish chowder, lobster bisque.

3. Purées are soups thickened with a purée of their own ingredients. They are almost always made from vegetables which are sieved or blended after cooking. A liaison is sometimes used to keep the purée in suspension. *Examples:* asparagus, carrot, pea, mushroom, potato. Many of these soups are also referred to as cream soups, e.g. cream of tomato.

Liaisons

These are the binding or thickening agents used in soups to hold the ingredients in suspension or to give a thickened consistency:

1. Starch. Flour, cornflour, arrowroot, semolina, ground rice. These must be blended with a little cold liquid before they are added to the soup.

2. Roux. Equal amounts by weight of flour and fat cooked as for a white sauce (see p. 108). The sieved soup is added gradually, then simmered for 5 minutes to cook the flour.

3. Beurre manié. Equal quantities of butter and flour are blended together well, then dropped into the soup, a little at a time. It should be stirred briskly to avoid lumping, and simmered for 5 minutes to cook the flour. This can also be used to thicken stews and sauces.

4. Egg yolk and cream. These are used to thicken and enrich pale soups. The yolks are blended with milk or cream and a little of the soup. This mixture is then added to the remainder of the soup which is stirred briskly. Heat through but *do not boil* as the eggs would curdle.

Nutritive Value of Soups

Soups are over 90 per cent water and have little value in the diet as a source of nutrients. Clear soups have only 1 per cent gelatine, a little protein and small amounts of mineral elements Thickened soups and purées contain some starch and possibly a little more protein, fat and mineral matter depending on the ingredients – for example meat soups will have slightly more protein. The vitamin C content is negligible and there is some loss of B group vitamins due to prolonged cooking.

Dietetic Value

Soup has an important role to play in the diet. It stimulates the appetite, encourages the secretion of digestive juices and introduces variety and warmth to the diet. Broths are given to invalids and convalescents as a dilute and easily digested form of food.

Disadvantage. When served before a meal, many soups give a feeling of fullness, thus reducing the capacity to eat a more nourishing main course. This applies particularly to children. For this reason soup should only be served in small portions.

Rules for Making Soup

1. Use good quality fresh ingredients.
2. Use a well-made stock rather than water.
3. The soup should taste of its main ingredient and ingredients should be carefully blended to give a good flavour.
4. Season well using salt, pepper, herbs and other flavourings.
5. Skim grease off if necessary by drawing kitchen paper over the surface of the soup.

Note. A well-made soup should have a good flavour, colour and consistency. It should provide a contrast in colour, taste and texture with the course that follows it. Serving portion: 150 ml per person.

Garnishes. Chopped parsley or chives, softly whipped cream, grated cheese, croûtons.

Accompaniments. Dinner rolls, Melba toast, brown or white bread, croûtons.

Making Soups Quickly

The use of a blender/liquidiser and a pressure cooker save much of the time and labour otherwise spent in making stocks and soups. A pressure cooker reduces the cooking time of stock and soup by about one-third, and a blender is ideal for puréeing the mixture at the end of the cooking time. In fact, it is possible to purée vegetables before they have fully softened; this way they will retain more nutrients and flavour.

Commercially Prepared Soups

Dehydrated soups. The sales of these soups have increased enormously since they were first introduced in the early 1950s. As a result of strict quality control and much scientific research, the standard of dehydrated soups has improved a great deal since their introduction, but their flavour and texture are still inferior to those of home-made soups.

Manufacture

1. Meat, vegetables and other ingredients are sampled and tested to ensure high quality.
2. They are then prepared. Meat is steam cooked and minced, and vegetables are blanched and diced.
3. The fat and gravy from the meat are concentrated and dried with the meat by hot air. The vegetables are dried separately.
4. Cereals, flours and other thickeners have their moisture content reduced further.
5. Batches of soup are now made by mixing different blends of dehydrated ingredients together with the addition of seasonings, flavourings and additives.
6. Some samples are tested for quality while others are cooked in the factory test kitchen.
7. The mixture passes into filling machines which weigh the required amount into the foil envelopes. These are then sealed and packed.

As the food value of most soups is quite low, it follows that dehydrated soups, when reconstituted, are of even less importance as a source of nutrients, as any vitamin B or C will have been greatly reduced by drying. The calorific value of packet soups is relatively high. Dehydrated soups are quick to prepare and cook and easy to store, but home-made soups taste better, contain no additives and are generally more nourishing.

Canned soups are easier to reconstitute than packet soups, taste better and have fewer additives, but they are bulkier to store.

SAUCES

Sauce-making is an important part of a cook's repertoire. A good sauce will improve any food, adding flavour, food value, moisture and often colour.

A sauce is a liquid food which is usually flavoured and thickened. Although most sauces are smooth, in some cases the ingredients are left unsieved to provide texture (e.g. curry sauce). Many sauces provide a contrast in flavour to the dish with which they are served – e.g. apple sauce with pork provides a piquancy lacking in the meat and the acid present makes the meat more digestible.

A good sauce will be:
a. the correct consistency
b. well seasoned and flavoured
c. smooth and glossy (except in the case of a textured sauce)
d. either piping hot or very cold.

Uses of Sauces
a. It may be *part of a dish*, e.g. beef stew, braised celery.
b. It may be used to *coat food*, e.g. cauliflower au gratin.
c. It may be used to *bind food*, e.g. rissoles.
d. It may be served with food as an *accompaniment*, e.g. gravy, custard

Nutritive Value

As the ingredients used in sauces vary considerably, it is impossible to give a general picture of their food value. Obviously the inclusion of milk, cheese or eggs will do much to increase their nutritive value.

Dietetic Value

Apart from their food value, sauces are important in the diet as they stimulate the appetite and aid digestion. They give flavour to insipid foods and help to counteract the richness of indigestible foods (e.g. roast duck and orange sauce).

Classification

1. Simple sauces, blended (e.g. cornflour or custard powder), infused (e.g. jam and syrup sauces thickened with cornflour) or made from purées (e.g. apple, cranberry, tomato).

2. Roux sauces, including white sauce (Béchamel), brown sauce (Espagnole) and their variations, and chaudfroid sauce.

3. Egg-based sauces, including egg custards, Hollandaise sauce (a hot egg/vinegar/butter emulsion) and mayonnaise (a cold egg/oil/vinegar emulsion).

4. Cold sauces and dressings such as mint sauce, horseradish sauce, French and other salad dressings.

5. Butter sauces such as brandy butter and maître d'hôtel butter.

Dehydrated Sauces

Like dehydrated soups, these sauces are made from dried, powdered ingredients such as vegetables, skimmed milk, fat, emulsifiers, thickeners, seasonings, and additives such as monosodium glutamate. Casserole sauce mixes are also available. Like the soups, they save time and are handy in emergencies, but home-made sauces are tastier and more nourishing.

HOME BAKING

Home baked breads, cakes and pastries are more nutri-

Basic ingredients for home baking (Bill Doyle)

tious and economical than bought varieties. To ensure successful results good ingredients should be used and they should all be accurately weighed or measured — especially the raising agent. The oven temperature should suit the food being baked; bread and pastry usually require a hotter oven than cakes.

Ingredients

Wheat flour forms the structural framework of most bread, cakes and pastries, because of its gluten content. Strong flour should be used for yeast cookery and puff pastry in order to provide sufficient elasticity in the dough.

Shortening should suit the product being baked. Margarine is a good all-round shortening but harder vegetable fats will give better results in pastry making. Butter, although difficult to cream, has a superior flavour for cakes. Cooking oils may be used for all-in-one pastry and cakes — that is, recipes where all the ingredients are mixed together at the same time. However, the more conventional step-by-step method produces better results.

Eggs, which should be at room temperature for baking, act as emulsifiers holding the creamed fat in suspension. As they can entrap air when beaten, eggs may also help to lighten baked products.

Raising Agents

In order to achieve a light, palatable texture in bread, cakes and many other foods it is necessary to *aerate* the ingredients by introducing a gas into the mixture. This is based on the principle that gases expand and rise upon heating. If sufficient gas is enclosed in the mixture, it will push the dough upwards as it rises until the heat of cooking sets the dough. If the dough is to stretch as the air expands, it must have an elastic texture. This elasticity will depend on the gluten present in the flour. The introduction of gas can take place:

a. *Mechanically* – by trapping air in the mixture, e.g. beating.

b. *Chemically* – by using breadsoda or baking powder to form carbon dioxide.

c. *Biologically* – by using yeast to form carbon dioxide (see p. 110).

a. Mechanical Method

Many cakes and batters are aerated by beating. Sponges and soufflés are whisked to entrap air and rich pastries such as puff or flaky pastry incorporate air by repeated rolling and folding. After air has been introduced, the mixture must be cooked in order to set the gluten framework and hold the shape. Gelatine is used to 'set' air in cold mixtures.

As *water vapour* also expands at high temperatures, some batters and fairly wet mixtures such as choux pastry are further aerated because of their water content.

b. Chemical Method

The most usual form of chemical raising agent is *sodium bicarbonate* or *bread soda*, which may be used on its own or with an acid.

In the latter case, the carbonates in bread soda react with the acid to produce carbon dioxide.

$$\underset{\text{acid}}{H^+} + \underset{\text{carbonate}}{HCO_3} \rightarrow \underset{\text{salt}}{NaCl} + \underset{\text{water}}{H_2O} + \underset{\text{carbon dioxide}}{CO_2}$$

This reaction will only take place when moisture is present to dissolve the acid and release the hydrogen ions. Bread soda is often used with acids such as sour milk in order to form carbon dioxide but as the exact acidic con-

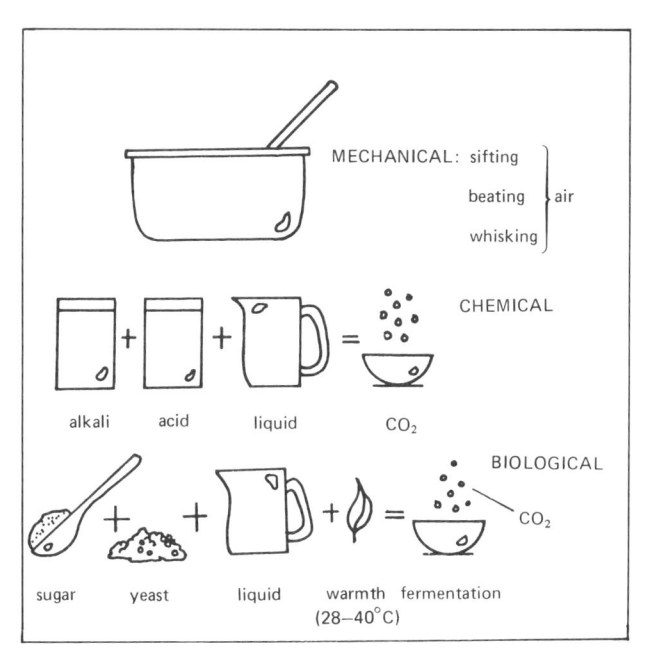

Fig. 5.4 Raising agents

tent is unknown, the results are not as accurate as when an exact acid such as potassium hydrogen tartarate (cream of tartar) is used.

Baking powder is a mixture of sodium bicarbonate and potassium hydrogen tartarate, with a third ingredient known as a filler. The most usual fillers are cornflour or rice flour and their presence prevents the premature reaction of the carbonate and acid by keeping them dry. All baking powders are of a standard strength and most have a delayed action which prevents immediate release of carbon dioxide upon moistening. In spite of this, it is wise to put mixtures containing baking powder into the oven within 15 minutes of moistening.

When bread soda is used alone, the dry heat of the oven decomposes the sodium bicarbonate and produces carbon dioxide.

$$\underset{\text{sodium bicarbonate}}{2\,NaHCO_3} + \text{heat} \xrightarrow{\text{heat}} \underset{\text{sodium carbonate}}{Na_2CO_3} + \underset{\text{water}}{H_2O} + \underset{\text{carbon dioxide}}{CO_2}$$

The resulting sodium carbonate (washing soda) has a dark colour and a bitter taste which is disguised by the treacle and spices used in this type of recipe, e.g. gingerbread.

Nutritional Value of Raising Agents:

As yeast (see below) is rich in the B group vitamins the food value of baked goods is improved by its use. On the other hand, the alkaline residue from chemical raising agents causes a reduction in the thiamine content of the foods in which they are used.

c. Biological Method (yeast)

Yeast is a living micro organism — it is a single-celled plant of the fungus group. As it cannot manufacture its own food, it depends on the sugar in other foodstuffs for its nourishment. Given suitable conditions — food, warmth and moisture — yeast reproduces itself by budding, at the same time giving off carbon dioxide and alcohol.

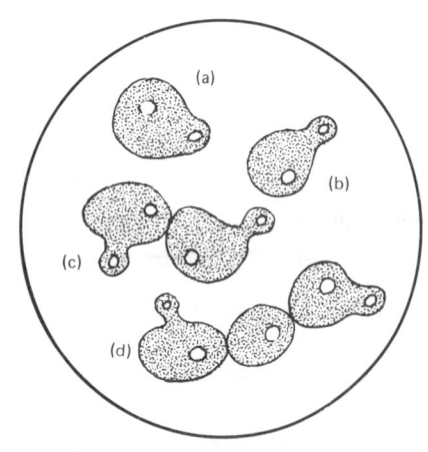

Fig. 5.5 Yeast cells budding

The yeasts used for baking and alcohol manufacture are from the genus *saccharomyces*. Although the yeast used in baking is cultivated, wild yeasts are plentiful in the air and on the skins of fruit.

Yeast is affected by extremes of heat and cold. It will multiply rapidly in warm, moist conditions — 25 to 28°C — but is inactivated by cold or freezing conditions and is destroyed by heat, that is, a temperature over 52°C.

Fermentation.
1. Flour contains an enzyme *diastase* which converts some of the starch present in flour to maltose. As most types of flour already contain 1–2 per cent maltose, it is not necessary to add sugar to yeast bread, although this is usually done in order to speed up fermentation. Both maltose and sucrose, if they are used, are disaccharides, which must be broken down before they can be absorbed by the yeast cells. Enzymes present in the yeast bring about this change, converting maltose and sucrose to glucose by hydrolysis.
2. The yeast cells absorb the glucose, converting it to alcohol and carbon dioxide. The alcohol is evaporated and the carbon dioxide, when heated, causes the bread to rise.

$$\underset{\text{glucose}}{C_6H_{12}O_6} + \underset{\text{fermentation}}{\text{yeast}} \rightarrow \underset{\text{alcohol}}{2C_2H_5OH} + \underset{\text{carbon dioxide}}{2CO_2}$$

The group of enzymes which bring about fermentation of yeast are known as the *zymase* group

Forms of Yeast Used in Baking

Fresh yeast (baker's or compressed yeast). This is an active form of yeast with a putty colour, and a 'beery' smell. It should crumble easily. It can be kept for two to

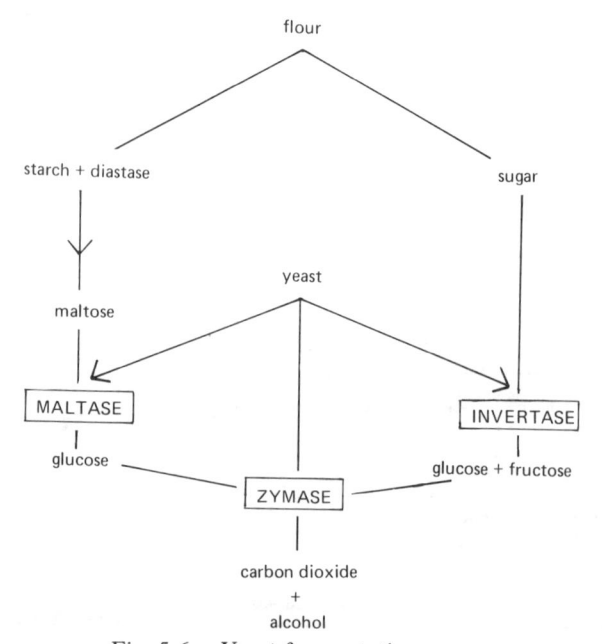

Fig. 5.6 Yeast fermentation process

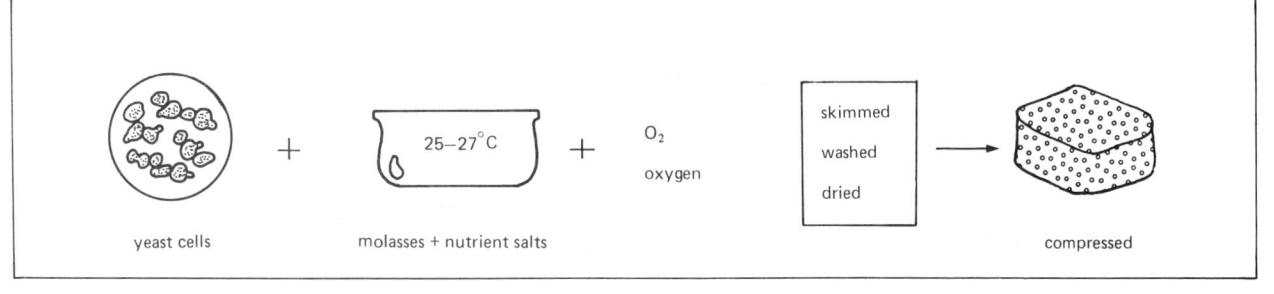

Fig. 5.7 Cultivation of yeast

seven days in a refrigerator or for up to three months in a deep freeze. Fresh yeast can be bought in some bakeries, delicatessens and health food shops.

Dried baker's yeast is dehydrated active yeast. As it is more concentrated than fresh yeast, only half the weight of fresh yeast should be used. It can be stored for up to six months in an airtight container.

Note. Brewer's yeast and *inactivated yeast* (dried brewer's yeast) are unsuitable for baking.

Cultivation of Yeast

Yeast is produced by placing healthy yeast cells, *Saccharomyces cerevisiae,* in a growing medium such as molasses + nutrient salts. The medium is kept at a temperature of 25–27°C and oxygen is introduced to encourage growth. The yeast is skimmed off, washed, dried in a vacuum and finally compressed into blocks.

Yeast extract ('Marmite') is obtained by heat-treating yeast until it forms a thick liquid. This is rich in B vitamins and ideal in vegetarian diets as a substitute for meat extract.

Using Yeast in Baking

1. Yeast must be fresh. Stale yeast will result in insufficiently risen bread and a strong yeasty taste.
2. To encourage fermentation, all ingredients and utensils should be very slightly warm (about 25°C).
3. Fresh yeast should be blended with about half of the measured, warm liquid from the recipe being used. Dried yeast should be reconstituted in half of the measured liquid with a little sugar, and left to stand for 15 minutes.
4. Remember that too much salt or sugar may retard the growth of yeast. It is now accepted that the idea of creaming the yeast with sugar destroys many yeast cells, and reduces the amount of carbon dioxide produced. It is preferable to blend the yeast with the warm liquid.
5. Yeast and liquid are poured into a well in the centre of the flour, a little flour is blended in and the mixture is allowed to stand until frothy. This is an indication that fermentation has begun.
6. When other ingredients have been added and the mixture thoroughly beaten, the dough should be left to rise or prove in a warm place until double its size. It should be covered with a damp cloth or placed in a greased polythene bag, to keep in moisture and prevent the formation of a skin.
7. The mixture may be kneaded at this stage, shaped and proved again in the tin for a short time. Thorough kneading is essential to strengthen the gluten and counteract the softening effects of the yeast.
8. Yeast dough should be baked in a hot oven in order to kill the yeast. The temperature is then reduced to allow the dough to cook through without burning. Rising continues in the oven for the first 10–20 minutes until the carbon dioxide stops expanding. The yeast cells are destroyed when the loaf temperature reaches 55°C and the gluten sets at 70°C.

Use of ascorbic acid in fermentation. It has been discovered that the addition of ascorbic acid or vitamin C will speed up the fermentation of yeast. Vitamin C tablets (25 mg) may be added to the liquid before mixing and then a very thorough beating or kneading is necessary to distribute the ascorbic acid. This method, which is now used extensively in bakeries and is known as the *Chorley Wood Process,* reduces proving time by one third, but it requires extra yeast. Usually one and a half times the normal quantity is used, together with a little fat.

Baking with Yeast

Proportions

15 g (1 level tbsp) yeast to 500 g flour.
30 g (2 level tbsp) yeast to 1.5 kg flour.
45 g (3 level tbsp) yeast to 3 kg flour.

Liquid: 300 ml milk or water to each 500 g flour. Extra ingredients such as sugar, fat, eggs, can affect fermentation. Too much sugar or fat will reduce fermentation. Eggs help to aerate the mixture. Strong flour gives best results in yeast cookery.

Basic Recipe for Yeast Bread

500 g strong flour or wholemeal
3 level tsp. salt
1 level tsp. sugar

Yeast breads, plain and fancy (Bill Doyle)

15–20 g yeast
300 ml water

Grease tins and leave in a warm (but not hot) place. Slightly warm flour and sieve with salt into warm bowl. Add a little measured warm water to yeast and cream. Blend with the remainder of the water. Pour into the flour and mix to a loose dough. Beat for 5 minutes by hand. Shape and turn into two warmed loaf tins. Allow to rise until double the volume, approximately 45 minutes.

Bake in a hot oven for 30–35 minutes. Test by knocking the base of the bread. If it is cooked there will be a hollow sound. Cool on a wire tray.

Various shapes may be used instead of a loaf:

Coburg – shape into a ball and cut a cross on top.
Bloomer – shape into an oval and score the top four or five times with a sharp knife.
Plaits – divide the dough into three, roll into long sausage shapes and plait evenly.

Richer buns and breads made with yeast are: Chelsea buns, tea breads and barm brack. Croissants and Danish pastries are made from a yeast dough which is rolled several times to form layers like puff pastry. Savarins and babas are rich yeast puddings.

Pastry

Pastry is a mixture of flour and fat moistened with water. Other ingredients such as salt, egg yolk, sugar, lemon juice, may be added.

Plain flour should be used for all pastry except suet pastry. As air is the raising agent used in pastry making, it should be enclosed where appropriate.

Many forms of shortening may be used, and the method of adding the shortening will depend on the type of pastry: for example, it can be rubbed in, flaked or folded. The minimum amount of water should be added to give a stiff rather than a soft dough.

Note. In recipes using pastry, the weight refers to the weight of flour, e.g. 100 g short pastry means the amount of pastry made from 100 g flour.

Classification

Suet	Short	Rich (layered)	Hot
suet crust	plain shortcrust	rough puff	choux
	rich shortcrust	flaky	raised pie
	biscuit pastry	puff	crust
	cheese pastry.		

Note. Most types of pastry freeze well either raw or cooked.

Rules for Pastry Making (excluding suet and hot pastry)

1. Ingredients and utensils should be cold. Work in a cool room. If a large amount of baking is being done start making the pastry before the room becomes warm from the heat of the oven. Water should be ice-cold. If possible use a marble slab for rolling.
2. Weigh ingredients accurately. Incorrect proportions can change the character of the pastry completely.
3. Aerate the mixture as much as possible. Sieve the flour, lift hands when rubbing in the fat, enclose air during folding.
4. Mix the pastry with a knife. As well as being colder it helps to make a stiffer dough.
5. Handle the pastry as little as possible. Use a knife to turn the pastry as the heat of the hands melts the fat.
6. Avoid using too much flour when rolling as this makes the pastry hard.
7. Roll lightly with a to and fro movement. Rolling from side to side makes the pastry uneven.
8. Do not stretch the pastry or it will shrink on cooking. Allow it to relax in a cool place between rolling and before cooking.

Baking pastry. Bake in a hot oven, preheated to 220°C (425–450°F)/Gas 7. Intense heat is needed to burst the starch cells so that the fat will be absorbed as it melts. The air which is trapped in the bubbles or layers of pastry expands, pushing up the dough until the gluten sets in its risen state. Too low a temperature will make the fat melt before the starch cells burst, resulting in a heavy, greasy, pastry. After 10–15 minutes, the temperature should be reduced to prevent burning and allow the fillings in the pastry to cook.

Cake Making

Most cakes belong to one of the following groups:

Plain – rubbed in method, e.g. scones and tea breads

Rich – creamed method, e.g. madeira and fruit cakes

Whisked – eggs and sugar whisked, e.g. sponges and meringues

Melted fat – e.g. gingerbread

All-in-one – fork or mixer method, variations on the above

Cake Mixes

These consist of a mixture of ingredients such as flour, raising agent, fat, dried milk, sugar and sometimes dried egg, which are blended together in the correct proportion with various additives to make cakes, puddings, scones, pastry, pancakes, icing and many other products simply by adding liquid and mixing.

Mixes contain a specially blended cooking fat which emulsifies easily. Most manufacturers use additives such as emulsifiers, non-staling agents, improvers, artificial colour and flavourings in cake mixes. The moisture content of the mix must be sufficiently low to prevent the action of micro-organisms and enzymes. After weighing, the ingredients are sieved, blended, and aerated before being packed automatically into envelopes or cartons. The material used in packaging must be grease and vapour proof to prevent contamination and dehydration.

To use a mix
1. Follow the directions on the packet very carefully.
2. It is usually necessary to moisten the mix with egg, milk or water or a mixture of liquids.
3. Remember that mixes can be used as a base for many dishes. Sponge mix can be used as a topping for puddings such as Eve's pudding or as a steamed pudding. Scone mix can be used as a topping for a casserole such as Beef Cobbler or for a quick pizza.

Mixes are quick and easy to use, and may be cheaper than the home-made or bought ready-made equivalent, but the taste, texture and appearance are usually inferior and the volume is often smaller than anticipated. Most mixes contain a number of artifical additives.

CONVENIENCE FOODS

Food may be classified as a convenience food when it has been prepared and processed by the manufacturer in such a way that it saves work and/or is easily stored.

Convenience foods include canned and bottled foods; frozen foods; dehydrated foods; mixes; 'instant' foods and synthetic foods.

Although many cookery experts tend to dismiss them, convenience foods are a fact of modern life. Modern technology and food processing ensure that plenty of food, in good condition, is available all year round. The standard of prepared foods nowadays is far superior to that of twenty or thirty years ago. Urbanisation, industrialisation and population growth have created the necessity for processed foods on a large scale.

The changing patterns of life – mainly the fact that many women go out to work as well as running a home – have meant that meals must be quick and easy to make. Even many women who do not work are not prepared to spend hours in the kitchen each day preparing family meals; nor are they prepared to shop each day. Single people who have been out at work all day and who live in flats or bedsitters, often with limited cooking facilities, also want meals which are quick to prepare.

Just as the modern way of living has led to an increased demand for convenience foods, so foreign travel has brought about increased interest in continental and exotic foods which are often conveniently packaged or tinned.

Advantages of Convenience Foods
1. Time-saving.
2. Labour-saving.
3. No waste.
4. Handy for beginner cooks, elderly and handicapped people.
5. Useful in emergencies, when time is limited or when cooking facilities are poor.
6. Require little fuel in cooking.
7. Easy to store and with a long shelf-life.
8. Provide a wide variety of foods, including foreign foods and native foods out of season.

Disadvantages
1. Usually more expensive than fresh or homemade equivalent.
2. Most have an inferior flavour to that of the fresh or homemade variety.
3. Many have reduced amounts of vitamins and minerals, particularly vitamins B and C.
4. Many are overpackaged and packaging is often deceptive – i.e. a small amount is packed into a large carton.
5. The large amount of money spent on advertising must be built into the retail price.

6. Many contain additives such as colouring and preservatives.

Nutritive Value

Although convenience foods are often accused of having little or no food value this is not true. Many are equal in value to fresh foods and some, such as frozen vegetables, may have a higher vitamin C content than ordinary fresh vegetables. Protein, fat and carbohydrate content generally remains unchanged. Frozen foods tend to retain their nutritive value, flavour and colour better than canned and dehydrated foods. These used to lose substantial amounts of vitamins B and C during the canning and drying process, but the development of aseptic canning and freeze-drying has reduced these losses to the amounts normally lost in cooking. Vitamins and minerals may leach into the canning liquid, which should therefore be used.

Convenience foods – dried, tinned and frozen (John Topham Picture Library)

NB: To obtain correct results it is essential to follow the manufacturer's directions regarding the storage, making-up and reheating of the food.

---

1. Canned and bottled foods. Canned foods have been around for 150 years. Most canned foods, especially fruit and vegetables, are soft in texture as a result of the high temperatures used in processing. They need no further cooking, just reheating and serving. Vitamin B and C content is reduced by traditional canning processes. Colourings and additives are often used. Canned foods are relatively cheap and keep for a minimum of a year, and often for much longer.

2. Frozen foods. These are probably the most acceptable form of convenience food as they are superior in taste, texture, appearance and food value to most other pre-packaged foods. They are generally the most expensive form to buy and they also require storage in a deep freeze or star-marked refrigerator. The popularity of home freezing has, if anything, increased the sales of commercially frozen foods because once a housewife owns a deep freeze, she tends to buy frozen food as well as to freeze fresh food.
Commercial freezing process – see p. 140.

3. Dehydrated foods. These have been in use for thousands of years but modern research has greatly improved their quality. Most dried foods appear to have a high calorie content but once reconstituted they revert to the proportions of the fresh variety. Some dried fruit is used without rehydrating, e.g. raisins.
Dehydration – see p. 138.
Accelerated freeze drying – see p. 139.

4. Mixes. These are quick to use but most contain little more than salt, raising agent and sugar; the more expensive ingredients such as fat, eggs and milk must be added by the user. See *Cake mixes,* page 113. Types include cake mixes, scone, batter, pastry mixes, stuffings, pudding mixes, icings, fillings, meringue powder.

5. Instant foods. These are the most convenient of all convenience foods as they can be eaten as they are or with the addition of liquid. Instant foods include processed cheese, ice cream, yoghurt, processed meats, instant desserts, coffee, teabags, instant porridge, breakfast cereals, snacks e.g. potato crisps, peanuts, concentrated fruit juices.

6. Synthetic foods. These foods have been developed as a substitute for other foods. Margarine has been on the market for about a hundred years. Synthetic meat (TVP) is a relatively new food made from soya beans (see p. 61). Other synthetic foods include synthetic cream and coffee creamers.

Other Forms of Convenience Foods

Complete meals. These are expensive, portions are small and meat content is low, but they are useful for those living alone or with bad cooking facilities. Available in frozen, dried and canned form.

Take-away foods. The growth of the take-away business has increased greatly in the past few years for many of the reasons already mentioned. Take-away foods include fish and chips, hamburgers, hot dogs, Chinese meals, pizzas, meat pies, cold meats and salads from delicatessens.

The sales of convenience foods are increasing rapidly, helped by tempting supermarket displays and clever packaging and advertising. A growing range of boxes and bags, cans and packets, is being absorbed into our way of life. Sliced pans, frozen peas, gravy powder and custard are an everyday part of our eating pattern, and it is possible that before long instant potato, instant desserts and TVP will be found in almost every store cupboard.

It is important to keep these foods in perspective. Many are convenient but not all are. Some packages are difficult to open; some are fiddly to make up; some need to be thawed out long in advance. Keep convenience foods in their place: use them in emergencies and when time is limited, but do not become too dependent on them. It is a good idea to avoid using more than one type of convenience food in a meal, adding fresh ingredients and garnishes to improve its flavour, food value and appearance.

Clever use of convenience foods can give greater variety in the minimum of time, but where there is a choice it is preferable to eat wholesome natural foods instead.

MODERN FOOD PRODUCTION

As the world population rises, so too must the amount of food available to feed the human race. In order to increase the supply of food produced and to reduce waste, certain methods of production with undesirable side-effects have been pioneered; in addition, various substances which can be added to food during processing to improve it in some way have been developed. Substances which are present in food unintentionally are called *contaminants*, while those which are added intentionally are called *additives*.

CONTAMINANTS

1. *Antibiotics* are used when rearing animals and poultry in order to prevent disease and also to fatten the animals. Many animal foodstuffs contain antibiotics and other chemicals. If farmers defy instructions and give animals too much or have animals slaughtered too soon after the use of antibiotics, traces will remain in the meat or milk. This could cause humans to become insensitive or allergic to antibiotics which may be important in the treatment of serious human diseases. Legislation on the use of antibiotics in Irish farms is not strict enough to prevent accidents due to carelessness and ignorance.

2. *Hormones* such as oestrogen are also used in animal feeds in order to speed up the fattening of animals.

3. *Insecticides.* Millions of tons of food are lost annually because of damage from insects and fungi. Chemicals have been developed to reduce these losses but many people believe that these chemicals can be dangerous, either directly or in a cumulative way – and not without reason. Certain chemical substances are particularly persistent and toxic and have destroyed the wild life in the areas in which they are used. Because of this fresh fruit and vegetables should always be washed before eating.

4. *Factory farming.* By this method, chickens, turkeys, pigs and calves are deliberately housed closely together

Factory farming: battery hens (John Topham Picture Library)

Crop-spraying with chemicals reduces damage from insects and fungi – but the chemicals can also be harmful (Camera Press)

Cattle and sheep in double-decker building (John Topham Picture Library)

so they do not lose weight by moving about and in order that more animals can be accommodated in a limited space. In this way hens produce many more eggs and animals fatten up more quickly. Although this method gives good results, diseases which may cause food poisoning are easily spread from one animal to another. Factory farming is considered by many to be unnecessarily cruel.

Other food contaminants. These include mercury in fish, lubricants, dissolved metals and, occasionally, foreign bodies which find their way into food during manufacture.

ADDITIVES

These are natural or artificial substances which are added to food to improve its appearance, taste or keeping qualities. Of the 1,420 lbs of food consumed by the average Irish person each year about one pound consists of food additives. They may be classified as follows:

a. Colourings
b. Flavourings
c. Preservatives
d. Nutritional additives
e. Improvers or physical conditioning additives

a. Colourings

A good colour improves the appearance of food and stimulates the appetite. The colour of food is generally a good indication of its maturity, quality and freshness, but as colour is often lost in processing, colourings, either natural or artificial, may be added to replace it. Although colourings are not essential additives, it has been proved that if they are removed from processed foods, consumers refuse to buy them as they look most unappetising.

Natural food colours occur naturally in food. They include:
1. Carotene or pro-vitamin A, a colouring pigment found in carrots, tomatoes, peppers, peaches and some shellfish.
2. Chlorophyll, a green colouring present in green vegetables.
3. Tannin, a brown pigment found in tea, coffee, cocoa, beer.
4. Cochineal, a red colouring made from dried insects.
5. Saffron, a bright yellow colouring from the dried stigmas of a type of crocus.
6. Caramel, a brown colour obtained by prolonged boiling of sugar.
7. Turmeric, a yellow spice used in curry and Eastern dishes.
8. Annatto, a yellow colouring used to give butter and cheese a consistent yellow colour.

Synthetic food colours are mostly derived from mineral substances such as coal tar. These are cheaper, more permanent and available in a wider range of colours than the natural variety. Many are positively safe and freely used in foods such as jam, ice cream, soft drinks, sweets and jelly. Others are still being tested as it is feared they may be harmful in large quantities. Only 48 out of over 100 colourings are permitted in Britain. Most synthetic colourings are known by a name followed by a letter or number, e.g. Green S; Fast Red E; Ponceau 6 R.

Colourings are not permitted in: raw or processed meat, poultry, fish, fruit, milk or cream, baby foods.

Bleaches are sometimes used in food production to remove colour, e.g. during flour manufacture chlorine dioxide is used.

b. Flavourings

Flavourings have been used in food for thousands of years. They improve the flavour of the food and may indirectly improve its digestibility. Some flavourings such as salt and spices have a secondary effect as preservatives or colourants.

Natural flavourings are obtained from natural sources such as the seeds, bark, leaves or roots of plants, the origin of many herbs and spices. Other examples are meat extractives; oils from the peel of citrus fruits; salt; sugar; citric and acetic acid; concentrated fruit juices; vegetable flavourings; strongly flavoured extracts from plants, called essential oils.

When the plants are dissolved in alcohol or another suitable solvent they result in an *essence,* e.g. vanilla pods steeped in alcohol produce vanilla essence.

Synthetic food flavourings. Some of these are obtained from natural products by chemical processes, while others are wholly chemical.

1. Chemical flavours include chemical compounds such as esters or aldehydes which are blended together to produce new flavours, e.g. pear flavour – amyl acetate; strawberry flavour – benzyl acetate; rum flavour – ethyl acetate; cherry and almond flavour – benzaldehyde. Many of these chemicals are also used in the manufacture of paint thinners and nail polish remover!

2. Artificial sweeteners are frequently used in diet foods and diabetic foods where a low energy value is required (see p. 93), e.g. sorbitol, saccharine. Another sweetener, sodium cyclamate, is now banned in food production.

3. Monosodium glutamate is a flavour intensifier, which brings out the natural flavour in foods. It is the salt of the amino acid, glutamic acid, and has been used in Chinese cooking for thousands of years. It is present in soy sauce, seasoning powders and many convenience foods such as stock cubes, packet soups, potato crisps and so on. It is known that over-use may induce headaches, nausea and other similar side effects. Further tests are being carried out.

c. Preservatives

It is necessary to use preservatives in many modern foods in order to facilitate distribution, prevent waste and prolong shelf-life. Preservatives may
a. destroy micro-organisms
b. inactivate enzymes
c. prevent oxidation

Natural preservatives include:
1. sodium chloride (salt) used for preserving meat, fish, vegetables;
2. acetic acid (vinegar), also known as ethanoic acid;
3. sugar;
4. herbs and spices (see p. 208)
5. alcohol;
6. wood smoke, which produces a coating of tar that has a preservative effect on meat and fish.

Most forms of mould and bacteria cannot survive strong concentrations of these substances.

Chemical preservatives are only permitted in certain foods in limited amounts.

The commonest permissable preservatives are:
1. sulphur dioxide which is used to preserve wine, fruit drinks, sausages etc;
2. benzoic acid used in the production of coffee;
3. phenyls, impregnated into fruit wrappers to prevent mould growth;
4. potassium salts or salt petre, used in restricted amounts for curing bacon;
5. scorbic acid, used to preserve flour, confectionery and margarine;
6. nisin, an antibiotic used in cheese and canned milk;
7. nitrates and nitrites, antibacterial preservatives used in canned meat products. They have the added effect of inhibiting the growth of the deadly *Clostridium botulinum* (see p. 133).

Antioxidants are added to cooking fats and oils and to many foods which have a high percentage of fat in order to slow down oxidation of fats and oils and thus to delay rancidity. Some, such as ascorbic acid and tocopherols, occur naturally in food. Others such as BHA (butylated hydroxyanisole), BHT (butylated hydroxytoluene) and propyl gallate are chemically produced to withstand the high temperatures used in baking and frying. Foods containing antioxidants include cooking oils and fats, butter, margarine, suet, confectionery, crisps and other snack foods and many dehydrated foods such as soups. Antioxidants are also used to prevent discolouration of fruits.

d. Nutritional Additives

Nutrients (usually vitamins and minerals) may be added to food to replace those lost during processing or to increase the food value of a product. Some foods such as margarine and flour must, by law, be fortified — margarine with vitamins A and D and flour with iron, calcium and (in some countries) vitamin B. Breakfast cereals may be enriched with B vitamins; fruit drinks and instant mashed potatoes with vitamin C. Manufacturers add these nutrients voluntarily in order to increase sales. Many baby foods and formula milks also contain added vitamins and minerals.

In order to reduce dental caries, the water supply in many counties in Ireland contains sodium fluoride.

e. Improvers (Physical Conditioning Additives)

The functions of this group of additives vary considerably. In broad terms, they create and maintain the correct consistency of food.

1. Emulsifiers help to bring about the correct consistency of a food, e.g. to distribute droplets of oil in water in order to produce an emulsion. Lecithin, which is found in egg yolk and gelatine, is a well known natural emulsifier. Some natural gums also have an emulsifying effect. Glyceryl monostearate (GMS) is a synthetic emulsifier.

2. Stabilisers help to retain an emulsion for its normal shelf life, and many also increase viscosity — that is, they have a thickening effect and help to keep food smooth and uniform in texture. They are used in sauces, syrups and custards. Lecithin, pectin and gums have a stabilising effect. Methylcellulose and propylene glycol are synthetic stabilisers.

3. Thickeners often double as stabilisers. Their use is restricted to certain foods such as ice cream, imitation cream and instant desserts, and they are forbidden in some foods such as real cream. Alginates are synthetic thickeners.

4. Humectants are used to prevent foods from drying out and losing their palatability, e.g. in desiccated coconut. They also act as an anti-staling agent. *Examples:* glycerine, sorbitol.

5. Flour improvers. Chemicals such as bromate are used to speed up the natural maturing of flour.

6. Acids are used to give a sour taste to food, and many also have a preservative effect. *Alkalis* may be used to control acidity. *Examples:* citric, tartaric and acetic acids; sodium bicarbonate (alkali).

7. *Anti-caking agents* are employed to prevent powdered substances such as salt, icing sugar and cake mixes from lumping or caking.

8. *Inert gases* such as hydrogen and nitrogen are blown into packages before sealing to eliminate air as otherwise the oxygen present would produce off-flavours or rancidity during the product's life. Instant mash and instant coffee are packaged in this way. Other conditioning agents include sequestrants (which prevent mineral elements in food from affecting it adversely), anti-spattering agents, anti-foaming agents and firming agents.

Advantages of Additives

Food additives have many beneficial and even necessary effects.
1. They preserve food, thus reducing waste.
2. They improve colour and flavour.
3. Preservatives help prevent food poisoning.
4. Additives enable many bland and unpalatable foods to be used. Synthetic foods can be made tasty by additives.
5. They facilitate distribution of food and thus make a wider choice of foods available.

Disadvantages
1. Large amounts may be toxic.
2. It is impossible to test all the permutations of chemicals which mixtures of the various additives may produce.
3. Some may have a cumulative effect in the long term if they are stored in the body until toxic levels are reached.

Testing Chemicals

Each country has legislation which controls the use of chemicals in food. Most countries have food and drugs organisations which investigate new drugs and test their toxicity, usually on animals. The FAO (Food and Agricultural Organisation of the United Nations), the United States Food and Drugs Administration, the WHO (World Health Organisation) and the Scientific Committee for Food of the EEC are all organisations involved in this type of research. In spite of these precautions, several chemicals were used for years before it was discovered that they had toxic or carcinogenic effects on the human body. The insecticide DDT and the artificial sweetener cyclamate are examples of these.

Most chemicals are tested individually but there is always the possibility that, as with mixed drinks, two or three chemically combined in the body could have a much more serious effect than the effect of each when taken individually.

People who are worried by this possibility usually eat organically grown produce and avoid convenience foods, most of which contain chemical additives.

In Ireland, government Acts, such as the Health Regulations Act 1973, forbid the use of certain additives and restrict the use of others. A Food Advisory Committee of doctors, dietitians and scientists has been set up under the Department of Health to advise on further legislation.

The EEC is also introducing legislation to have standard controls on additives in each member country.

QUESTIONS

1. Taking one method of cooking, e.g. roasting, stewing, state briefly
   a. its basic principle
   b. the foods suited to this method
   c. its advantages and disadvantages
   d. its effect on the nutrients in food.
   Discuss any modern advances in cookery processes or equipment which improve or speed up the method chosen.

2. Describe the basic structure of a pressure cooker, mentioning any advances in design made in the last few years.
   State the principle of pressure cooking.
   Why is it necessary to exhaust air from the pressure cooker before bottling food?
   What is the temperature within the pressure cooker at high, medium and low respectively?

3. Differentiate between a chemical and a biological raising agent, and describe in detail how each works.
   Give ingredients and the method for making homemade bread using a biological raising agent.

4. Compare homemade pastry with a readymix packet under the following headings:
   a. convenience
   b. quality and results
   c. economic value
   d. nutritive value
   What are the basic principles of pastry-making?

5. Discuss soups *or* sauces using the following headings:
   a. classification
   b. nutritive and dietetic value
   c. economic value
   d. importance in meal planning
   Describe how a dehydrated soup is prepared and packaged.

6. What factors do you think have contributed to the spiralling sales of convenience foods?
   Classify the various convenience foods available to the consumer. In the case of two foods, describe briefly how they are processed and state how the processing has affected their food value.

7. To what extent do you feel the addition of chemicals to our food should be condoned?

Classify the main groups of additives used in modern food and write a brief description of two, mentioning (a) sources; (b) functions; (c) advantages and (d) foods in which these additives are permitted.
8. Assuming that you have taken part in a project relating to convenience foods, describe how you set about it.

What was your aim?
Discuss
a. the practical work undertaken
b. research work
c. any scientific experiments
d. your findings

6 Microbiology and Food Spoilage

Many living organisms are so tiny that they cannot be seen with the naked eye. These *micro-organisms* such as bacteria, fungi and viruses can only be studied, therefore, with a powerful microscope, and *microbiology* is the term used to describe such study.

Bacteria and viruses cause many diseases, while yeasts and moulds are chiefly concerned with food decay and plant diseases.

Louis Pasteur (1822–95) was the great pioneer of microbiology. It was he who proved that micro-organisms are present in the air and that warmth, pH and oxygen have an effect on the growth of bacteria. He coined the terms aerobic and anaerobic to describe the bacteria which need oxygen and those which multiply without it. He showed that fermentation was brought about by micro-organisms.

In his studies of infectious diseases, Pasteur isolated many bacteria and discovered that immunity to certain diseases was possible. He also developed artificial immunisation, a breakthrough which has saved the lives of many millions of adults and children in this century. As a result of painstaking research, he developed the process of heat-treating milk in order to destroy pathogenic bacteria, among them those causing diptheria and tuberculosis. To this process he gave his name – pasteurisation. After his death the Pasteur Institute in Paris continued his research on micro-organisms and at the end of the last century organisms even smaller than bacteria, which were known to cause many diseases, were isolated. These eventually became known as viruses.

Classification of Living Matter

All the plants, creatures and organisms in the world can be classified under three main divisions or kingdoms.

a. *Protista* containing the simplest organisms, generally unicellular micro-organisms such as algae and fungi, e.g. moulds;
b. *the plant kingdom;*
c. *the animal kingdom.*

Bacteria, originally classified as protista, have an entirely different cell strucutre from organisms in the three main kingdoms and for this reason they are now placed in a separate group – *procaryotes.*

Within each of the kingdoms, there are thousands, and in some cases millions, of species. For the purposes of detailed classifications, the species are further subdivided and placed in groups with similar characteristics. The order of classification is as follows: kingdom; phylum; class; order; family; genus; species.

As far as micro-organisms are concerned, it is sufficient for us to know the genus, which is the equivalent of the family name, e.g. saccharomyces, and the species which can be compared with the first name of an individual organism. Both are usually written in italics, e.g. *Saccharomyces cerevisiae* – yeast, the fungus used in brewing and breadmaking.

FUNGI

Fungi belong to the protista kingdom. As they do not contain chlorophyll, they cannot make their own food by photosynthesis. They depend on energy-rich organic matter for the nutrients they require. Fungi which live in

Louis Pasteur: pioneer of microbiology (John Topham Picture Library)

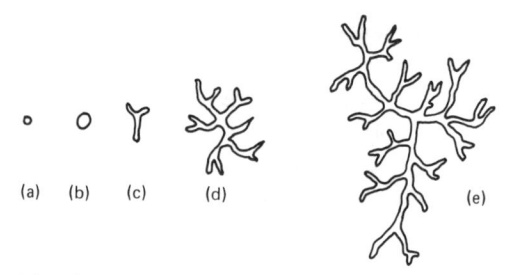

Fig. 6.1 Stages in the development of a fungal mycelium

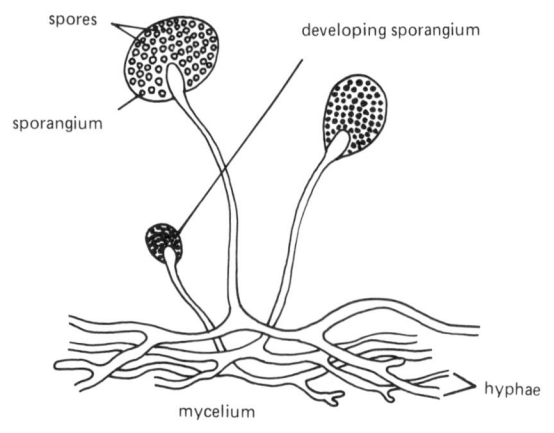

Fig. 6.2 Mucor

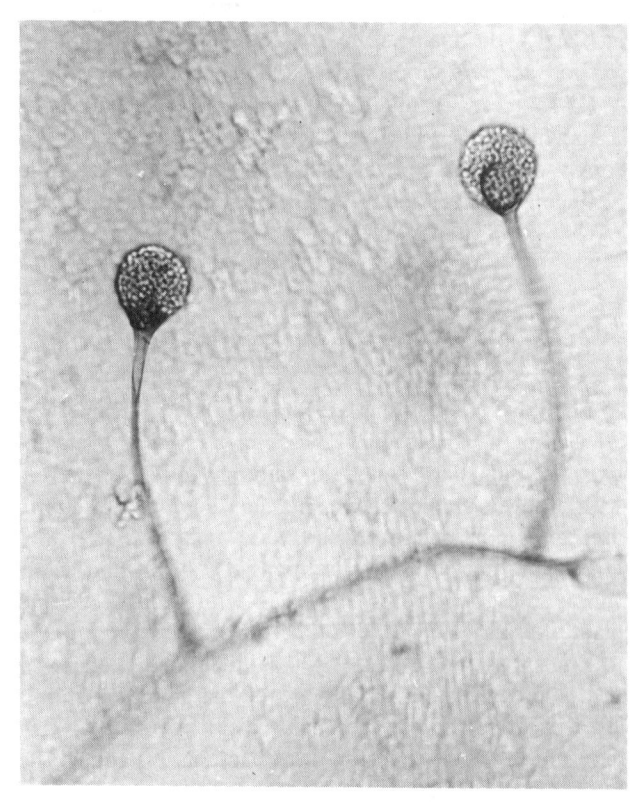

Growth of mucor on bread (C. James Webb)

or on the tissues of living organisms are *parasites*. Those which feed on dead or decaying matter such as soil or sewage are called *saprophytes*. Although the structure of many fungi consists of a microscopic single cell, large, easily-identifiable structures such as mushrooms come under this heading.

Basic Structure

Most fungi begin as a single cell or spore which, when it lands on a suitable growth medium, sends out a single tube-like thread called a hypha. This branches repeatedly to form a large spread of intertwining filaments (hyphae). Unlike true cells, these have few dividing walls. The mass of hyphae, called a mycelium, embeds itself in a suitable food source and absorbs nutrients from it.

Reproduction may be asexual or sexual. Many fungi can reproduce by either method.

Asexual reproduction. Vertical branches grow upward from the mycelium enlarging at the top to form a swelling (sporangium) which contains the spores. Moulds often vary in colour according to the colour of the sporangia. The sporangium breaks off or bursts, releasing the spores into the air, where they are dispersed by air currents.

Sexual reproduction. This is less common and involves the growing together of two hyphae. At the point of contact a swelling (zygospore) forms within which the spores develop. When fully formed, the zygospore breaks away and after a resting period, the spores are released. Spores will germinate into hyphae in a suitable environment, although they can survive for long periods in adverse conditions.

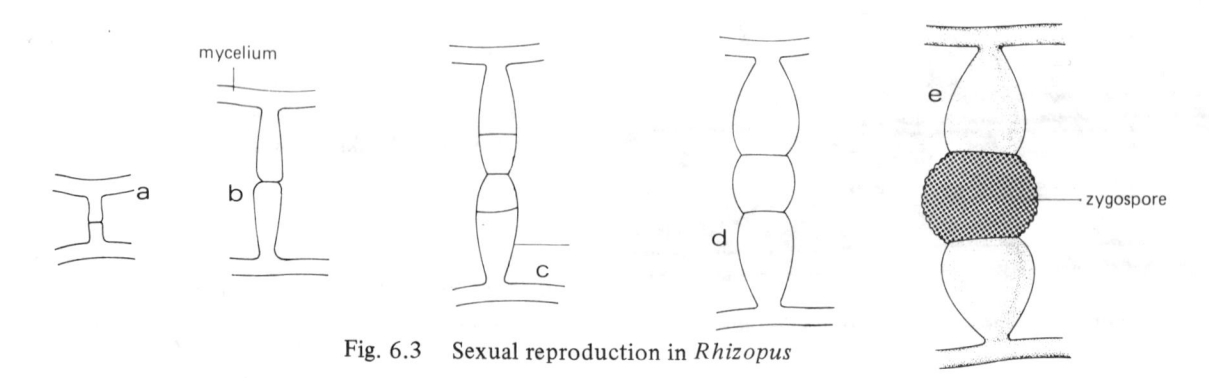

Fig. 6.3 Sexual reproduction in *Rhizopus*

Fungi, which belong to the phylum 'Mycota', are usually divided into three classes:

Phycomycetes	*Basidiomycetes*	*Ascomycetes*
Mucor	*Agaricus* (mushrooms)	*Penicillium*
Rhizopus	Puff balls	*Aspergillis*
Phytophthora	Rusts and smuts	*Saccharomycetes* (yeasts)

Some fungi do not fit into this classification and are known as *fungi imperfecti*. Included in this group is a common mould, *Cladosporium*.

Many fungi are harmful, causing both plant and animal diseases. Rusts and smuts are responsible for large-scale destruction of cereal crops, while potato blight, caused by the parasitic fungus *Phytophtora infestans,* resulted in the Great Famine (1845–7) in which almost a million people died from hunger and tens of thousands fled the country.

Other harmful fungi attack valuable plants and trees. Dutch elm disease, chestnut blight and many plant mildews are all caused by fungi. Athlete's foot, ringworm and thrush are fungal diseases which affect man. Moulds cause food spoilage and certain fungi, e.g. such toadstools as *Amanita* sp. (p. 123) are directly poisonous to man.

On the other hand, some members of this phylum are beneficial. As well as producing food (mushrooms, morels and truffles), fungi are used in the production of many cheeses. Yeasts are important in the production of vinegar, wine and other alcoholic beverages as well as aerating bread by fermentation. They are also used in the synthesis of vitamins. Penicillin and other antibiotics are produced from moulds. Decomposition of dead plants and animals is hastened by fungi present in the air and soil which help to recycle valuable nutrients. The water mould *Saprolegnium* assists in the breakdown of sewage.

The following moulds cause spoilage of food:

Phycomycetes

a. **Mucor,** a group of moulds, which grow on decaying fruit, bread and other similar matter.

Sporangia of mucor (C. James Webb)

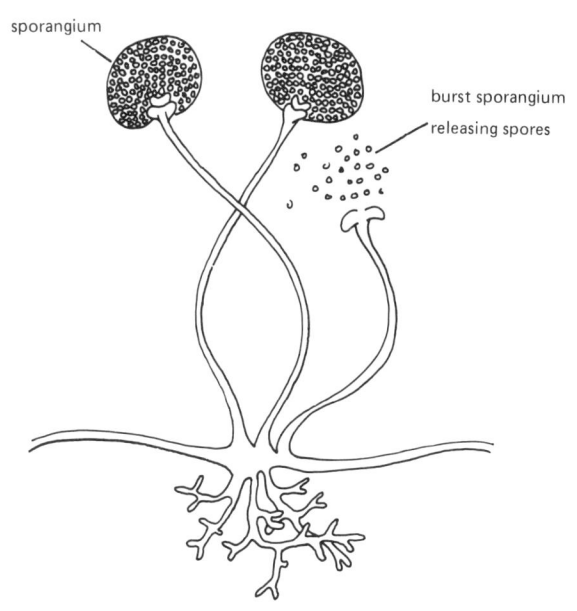

Fig. 6.4 *Rhizopus*

b. **Rhizopus,** a saprophytic fungus similar to mucor. *Rhizopus nigrans* (black bread mould) forms readily on bread, fruit and vegetables and produces distinctive black sporangiae.

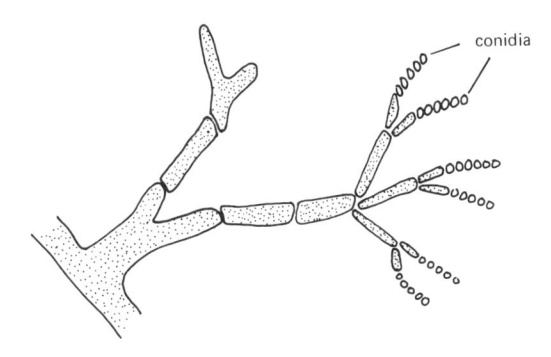

Fig. 6.5 *Penicillium*

Ascomycetes

a. **Penicillium.** Like most ascomycetes, penicillium reproduces by developing chains of spores (condidia) at the end of a hyphal branch. It forms a greenish blue mould on many foods, particularly fruit. *Penicillium roquerforti* is the culture used in the ripening of Roquefort and Stilton cheeses to give them their characteristic colour and taste. Another penicillium mould is used in the production of the antibiotic penicillin.

b. **Aspergillis.** This mould, which looks very like penicillium, grows in plants and causes destruction of stored cereals. If ingested it can be toxic. *Aspergillis niger*, a black mould, is used commercially in drug manufacture.

c. **Saccharomycetes** (yeasts). See below.

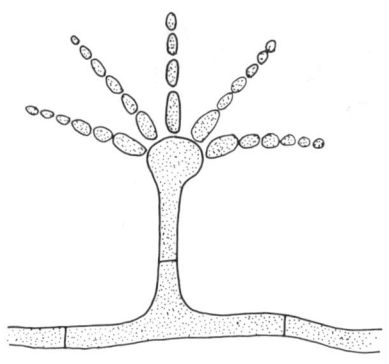

Fig. 6.6 *Aspergillis*

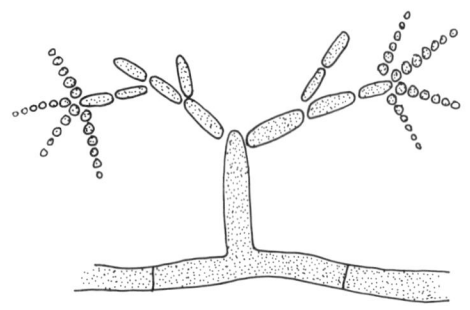

Fig. 6.7 *Cladosporium*

Fungi Imperfecti

Cladosporium is a green mould very similar to the ascomycetes which is usually present in the air. It attacks fruit, vegetables and meat.

Conditions Required for Growth of Fungi

Fungi require food, oxygen, water, a warm temperature and a slightly acid medium if they are to reproduce successfully.

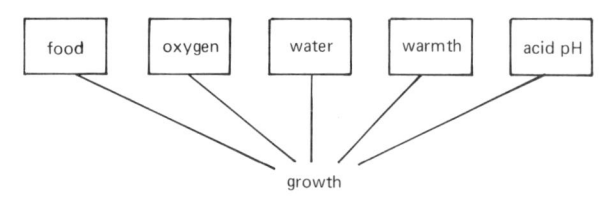

Fig. 6.8 Conditions required for growth of fungi

1. *Food* requirements include carbon which is obtained from the host cells in the form of polysaccharides and amino acids which provide nitrogen. Some types of fungi obtain their nitrogen from the soil. Feeding: fungi secrete an enzyme which softens the food so that the nutrients can be absorbed through the walls of the hyphae.

2. *Oxygen* is required for respiration. For this reason fungi tend to grow on the surface of most foods.

3. *Water* is necessary for normal metabolism. This is usually provided by the food source.

4. *Temperature*. Most fungi grow and reproduce at an optimum temperature of 30°–45°C. Certain *thermophillic fungi* prefer a higher temperature of about 50°C while others, called *psychrophillic fungi*, grow at low temperatures of 0°–30°C. High temperatures destroy most fungi and their spores (i.e. 70°C and over).

5. *pH Value*. Most members of the fungi family favour a slightly acid medium (pH 5–7). For this reason citrus fruits and other acidic foods tend to attract them.

Basidiomycetes

This group of multicellular fungi produces large, fruiting bodies which are clearly visible to the naked eye and include mushrooms, puffballs and many woodland fungi. Tasty morels and truffles, although similar in appearance to mushrooms, belong to the *ascomycete* group.

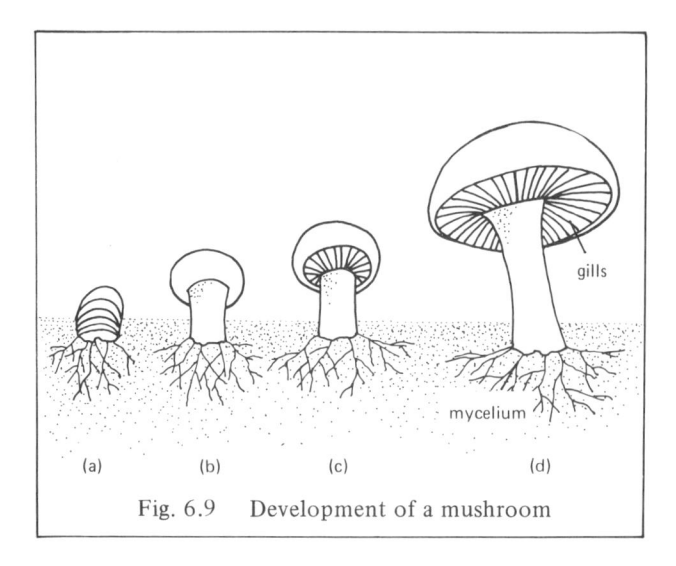

Fig. 6.9 Development of a mushroom

Basic structure of a mushroom. Before the mushroom appears, a mycelium forms from a germinating spore beneath the surface of the soil. Many of these may spread over a wide area. When conditions are right, usually during a spell of warm, wet weather, a stalk grows upwards surmounted by a cap. This structure consists of tightly packed hyphae. Beneath the cap radiating from the centre are rows of gills to which are attached thousands of spore-bearing bodies called basidia. As the mushroom ripens, the cup flattens, the gills become darker in colour and eventually millions of spores (basidiospores) are released. These are carried by the wind to new sites where, under suitable conditions, the spores germinate and new mycelia are formed.

Spore print. If the ripe cap of a mushroom is placed gills facing down on a sheet of white paper and left overnight, the spores, as they fall between the gills, will form a pattern on the paper.

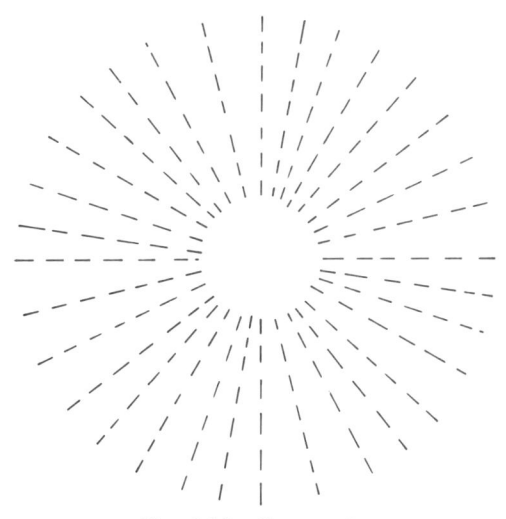

Fig. 6.10 Spore print

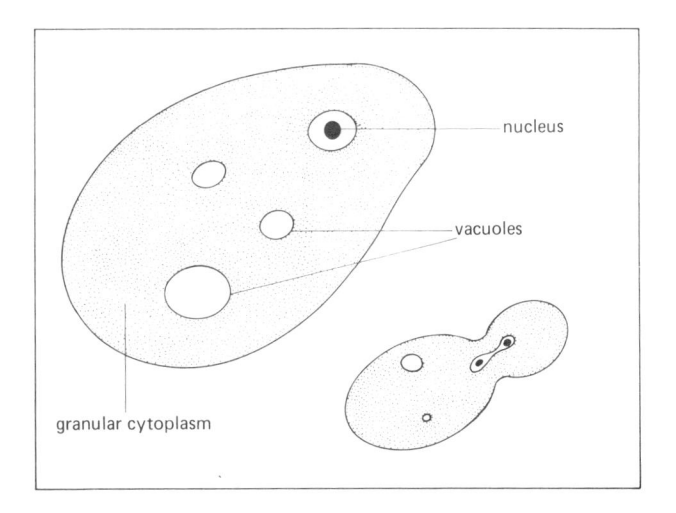

Fig. 6.11 Structure of a yeast cell

Identification of mushrooms. The cultivated mushroom *Agaricus bisporus* and the field mushroom *A. campestris* are the most widely eaten fungi. Other lesser known fungi may be eaten but it is vital that they should first be identified correctly as several mushrooms are deadly poisonous, including *Amanita phalloides* (death cap) and *Amanita muscaria*. It is possible to identify the toxic varieties by using a reliable reference book but it is essential to check and double check each species with the illustration in the book. If the mushroom is immature it may be difficult to identify it. Never eat a fungus which has a sac-like structure around the stem and avoid fungi with white gills; these should be pink or brownish in colour.

YEASTS

Yeasts are parasitic fungi. They are generally unicellular, are present in the air and are plentiful on the skins of fruit.

Structure

Most yeast cells are small oval structures with a thin outer wall. The cell is filled with cytoplasm which contains one or more vacuoles and a nucleus. Granules of glycogen and other food reserves are also present in the cytoplasm.

Reproduction

Normal reproduction of yeast is unusual as unlike other fungi, no spores are produced. Cells reproduce by 'budding'. At first a small projection appears at the edge of the parent cell, and from this cytoplasm and nutrients flow. As the bud grows, the nucleus moves towards it and divides so that a new nucleus enters the bud. When the bud is almost as large as the parent cell, a wall forms, separating it from the parent cell, and it then breaks away. When yeasts are reproducing rapidly, the buds do not

Deadly mushrooms: *Amanita virosa* (Destroying Angel) (left) and *Amanita muscaria* (Scarlet Flycap) (above) (John Topham Picture Library)

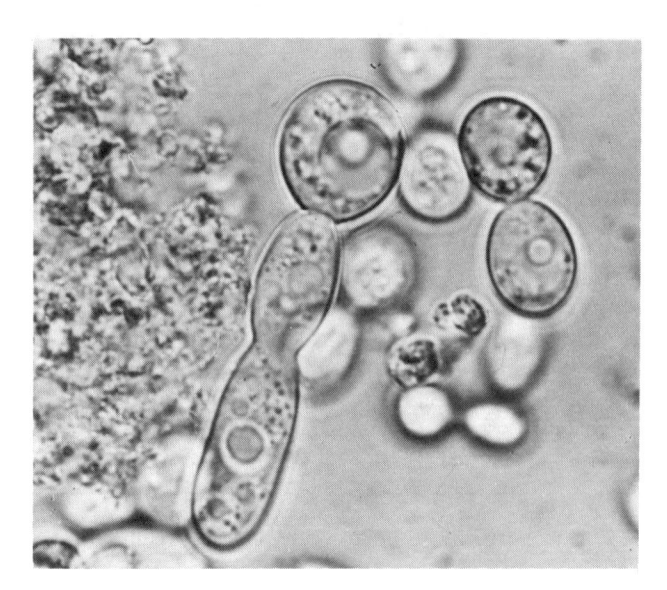

Yeast cells budding (C. James Webb)

break away but continue to reproduce until long chains of yeast cells are formed.

Conditions Required for Growth of Yeast

Yeast cells grow and reproduce in conditions similar to those required by other fungi. They need oxygen, warmth, food and moisture in order to grow successfully. Yeasts grow best at temperatures between 25°C and 30°C. Extreme heat destroys all yeasts and most are destroyed at temperatures above 60°C.

Yeast cells live mainly on foods containing sugar, e.g. the surface of fruit and jams. The 'bloom' on the skin of a grape consists of yeast cells.

Fermentation

The power of breaking down sugar into carbon dioxide and alcohol is called fermentation. This chemical change is caused by a group of enzymes known as zymase, which are present in the yeasts themselves. Although the process of fermentation is employed by the brewer and the baker alike, its actual purpose is to produce energy for the metabolism of the yeast cell.

$$\underset{\text{glucose}}{C_6H_{12}O_6} + \text{yeast} \rightarrow \underset{\text{carbon dioxide}}{2CO_2} + \underset{\text{alcohol}}{2C_2H_5OH} + \underset{\text{energy}}{118\text{kj}}$$

The yeasts used in brewing and breadmaking are *Saccharomyces cerevisiae*. Fermentation is a vital part of beer and wine making. Yeasts are added to the fermenting barley or grape solution and alcohol and carbon dioxide are produced. If the liquid is bottled while carbon dioxide is still present, a fizzy drink such as beer or champagne will be produced, whereas if fermentation is allowed continue until no carbon dioxide remains, a still drink such as table wine will result.

Fermentation in breadmaking – see p. 110.

Nutritive Value of Yeast

Yeasts contain many essential amino acids and B group vitamins. Yeast culture is a quick and easy way to produce protein but so far it has only been used in this way to provide food supplements.

BACTERIA

Bacteria are single-celled microscopic organisms, varying considerably in size and structure. Most are between 1/60,000 cm and 1/6000 cm in diameter. In order to study the bacterial cell in detail an electron microscope must be used. Species of bacteria can be identified if colonies are grown on agar jelly.

Bacteria are present in the air, soil and water, particularly polluted water. They multiply readily on plants, animals and man and are the cause of many diseases. Human food is also a source of nutrients for bacteria which contaminate it, making it a potential source of food poisoning.

Structure

The structure of bacterial cells is much simpler than that of normal plant or animal cells. There is no nuclear membrane; nuclear material may be found in one or more areas within the cell. Many of the other structures found in the normal cell are not present. Ribosomes (concerned with protein synthesis) are suspended in the cytoplasm.

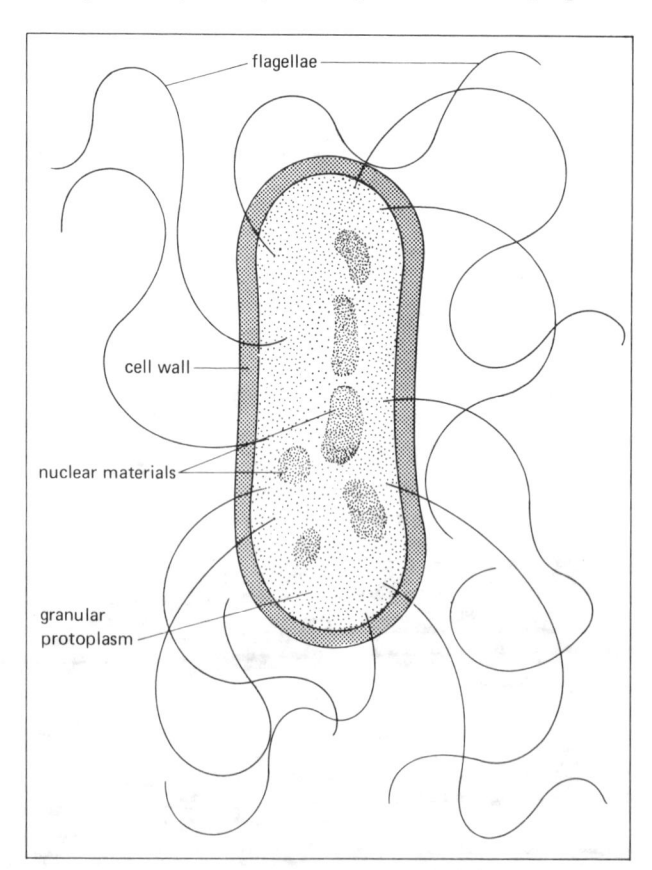

Fig. 6.12 Structure of a typical bacterial cell

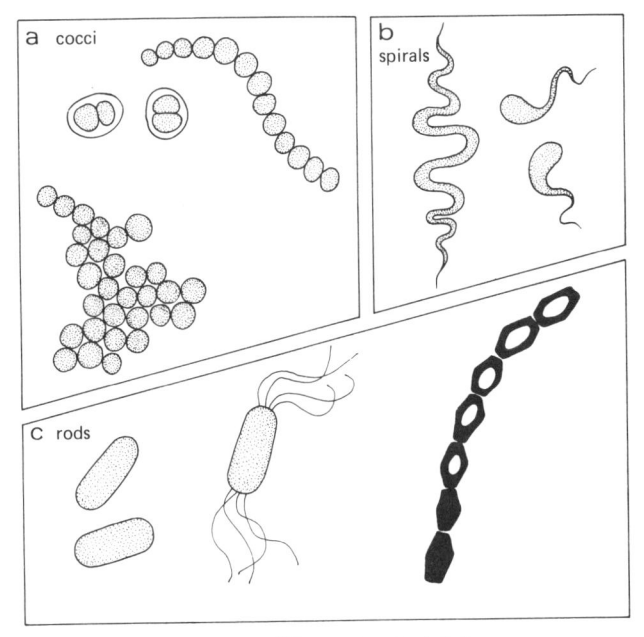

Fig. 6.13 Different bacterial shapes

Most bacteria have a rigid cell wall composed mainly of proteins and lipids. Some have hair-like protrusions called flagellae, which help to propel them about in liquids.

Shapes

Bacteria may be grouped according to their shapes:

1. Spherical bacteria (cocci). These may occur singly or in pairs, called *diplococci*. Pneumonococcal bacteria which cause pneumonia are typical diplococci. Some form chains and are known as *streptococci*. These cause throat infections, tonsilitis and scarlet fever. Others form clusters; these, called *staphylococci*, are often found on the skin and cause skin infections such as boils, and food poisoning.

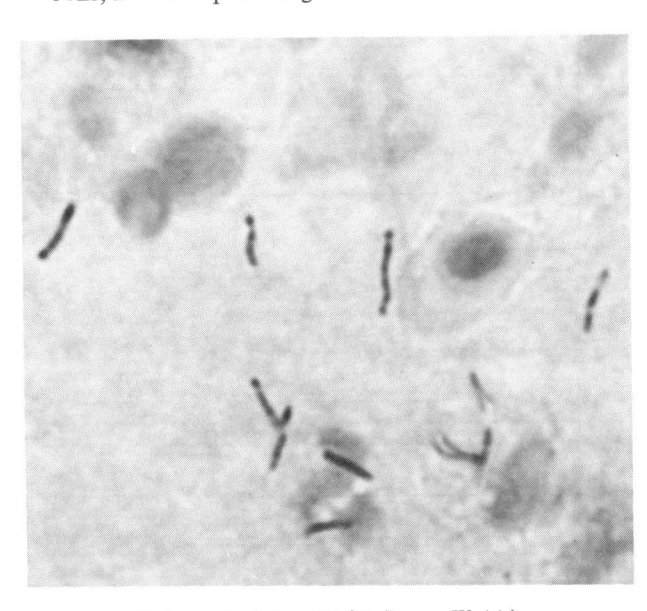

Tuberculosis bacilli (C. James Webb)

2. Rod shaped bacteria (bacilli and clostridia). These include many spore-forming, food-poisoning bacteria and the tuberculosis bacterium. *Escherichia coli, Salmonellae* and *Shigellae* belong to this group.

3. Spiral bacteria (spirilla). These have a spiral structure and include the spirochete which causes syphilis. The curved bacterium *Vibrio cholerae,* which causes cholera, is also included in this group. Spiral and curved bacteria are not very common. They are usually found in water or other liquid media.

Classification

The kingdom *procaryotae*, to which all bacteria belong, contains several sub-divisions. Because of the great diversity of structure, function, metabolism and genetic types in the kingdom (there are over 230 genera of bacteria identified), it is difficult to sort bacteria into similar groups. So far, however, microbiologists have organised bacteria into twenty sub-divisions, most of which can be more simply classified into three groups:

a. *Mycoplasmas,* which have no outer cell wall, just a thin membrane. These are the smallest bacteria. They are plentiful in mucous membranes where they live as harmless parasites, but some mycoplasmas which cause disease have been identified.

b. *Gram-positive bacteria* have a thick single-layered cell wall. Most are aerobic and many produce spores (see p. 126). Gram-positive bacteria are generally immobile and those which possess flagellae have them distributed all over the cell rather than at one end. Gram-positive bacteria include *Streptococcus, Clostridium* and *Streptomyces*.

c. *Gram-negative bacteria.* The cell wall of this group consists of two thin layers. Almost all Gram-negative bacteria are mobile; some have flagellae at one or both ends, while others have a gliding movement. They do not produce spores but many are resistant to antibiotics such as penicillin. Bacteria in this group include *Escherichia coli* which is present in normal intestinal flora; *Salmonellae* and *Shigellae* which cause enteric (food-poisoning) diseases; *Rickettsia typhus,* which causes typhoid and *Treponema pallidum* which causes syphilis.

The Gram stain. Called after its inventor, Christian Gram, this is a method of staining bacteria in order to identify them. It involves staining a smear of growing cells with crystal violet dye, then with a dilute solution of iodine. The smear is then subjected to an organic solvent such as alcohol or acetone. Gram-positive bacteria resist decolourisation and remain a blue-black colour, while Gram-negative bacteria are completely decolourised.

Reproduction

Most bacteria reproduce by binary fission. One cell divides in two, these divide again to form four and so on. Growth and reproduction in bacteria are essentially the same process. Under ideal conditions (see below) many bacteria can reproduce every twenty minutes: in fact, one bacterium could become several million in a few hours. Luckily for us, bacteria cannot multiply at this optimum rate for long. They quickly exhaust the available nutrients and in certain cases produce toxins during metabolism, which cause a decline in growth and eventually arrest it completely. The bacteria die when cellular reserves of energy are depleted.

Spores

Many Gram-positive bacteria have the ability to form a dormant cell known as an *endospore* (spore). Spores are not formed during active growth, but when conditions are unsuitable for growth. This could occur, for example, when nutrients are not available or when conditions are dry. Spore formation could be compared with hibernation: metabolism ceases and the cell enters a period of dormancy while retaining its capacity to germinate and develop into normal vegetative (growing) cells. Spores may remain inactive for several years and during this time they are highly resistant to heat, radiation and many toxic chemicals. Bacteria and their spores also survive freezing temperatures.

Endospore formation. Under adverse conditions, the nucleus divides and one end of the cell is separated by a dividing wall. The larger part of the cell wraps itself around the new cell or spore. The spore wall thickens and when ripe is released from the mother cell. After a period of dormancy, the spore, when subjected to suitable conditions, will germinate and grow into a typical bacterial cell.

Spores are present in the air, soil, dust and dirt. Dried out sputum and faeces contain millions of spores and must therefore be disposed of hygienically.

Conditions for Growth of Bacteria

As bacteria lack chlorophyll they exist as parasites or saprophytes, taking nutrients from food or man.

1. Food. Bacteria require an adequate supply of food in a suitable form. Liquids or semi-liquids such as broths, milk, and jelly are an ideal medium for the growth of bacteria. For this reason agar or nutrient jelly is used in the cultivation of bacteria. Feeding: enzymes synthesised by the bacterial cell are secreted into the food. These break down the large food molecules outside the cell and convert the food to soluble organic compounds which can be absorbed through the cell walls.

2. Temperature. The range of temperatures over which growth is possible varies considerably among bacteria. Most thrive between 30°C and 40°C. As this range includes the normal body temperature of humans, they are suitable targets for bacterial infection.

Thermophiles grow at temperatures above 55°C.

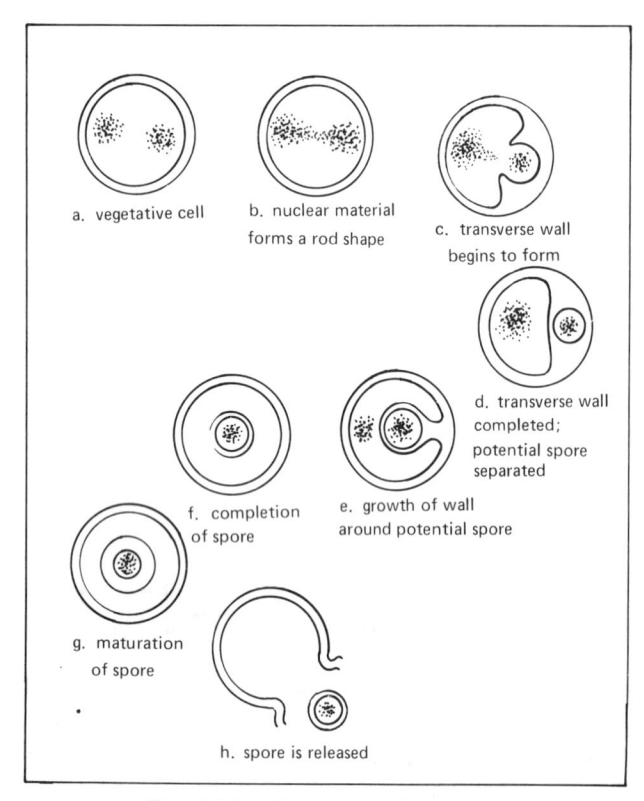

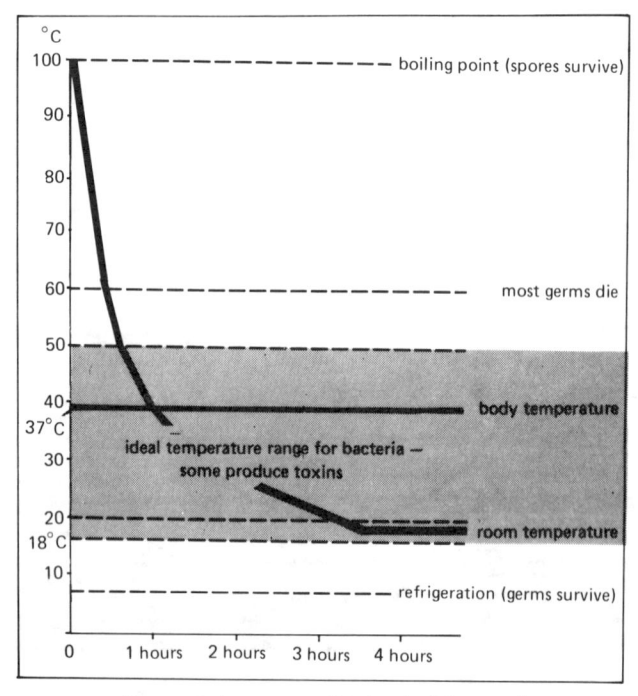

a. vegetative cell
b. nuclear material forms a rod shape
c. transverse wall begins to form
d. transverse wall completed; potential spore separated
e. growth of wall around potential spore
f. completion of spore
g. maturation of spore
h. spore is released

Fig. 6.14 Formation of a spore

Temperature range for bacterial growth

Mesophiles grow well in the middle temperature range — 20° to 45°C.

Psychrophiles like cold temperatures. These include many marine bacteria which grow well at 0°C.

Most bacteria are destroyed at temperatures over 60°C, and all are destroyed by boiling (100°C). Many spores, however, survive boiling and may require temperatures over 140°C to destroy them.

Cold temperatures (0°C to 10°C) retard growth of bacteria and freezing (−18°C) prevents growth, but *does not* destroy bacteria. Bacteria in frozen food will become active upon thawing.

3. *Oxygen requirements* are variable. Aerobic bacteria require oxygen for metabolism but anaerobic bacteria cannot utilize O_2. Some bacteria use oxygen when it is available but can also grow without oxygen. *Clostridium botulinum,* the bacterium which causes botulism, the frequently fatal form of food poisoning, is an anaerobic bacterium which multiplies in dark, airless places, e.g. cans of food.

4. *Moisture* is necessary for normal metabolism. Bacteria thrive in moist, badly-ventilated kitchens.

5. *Darkness.* Bacteria multiply well in damp, dark places. Fresh air and sunlight destroy germs and have a sterilising effect.

6. *pH* varies according to the species of bacteria — generally neutral or very slightly acid or alkaline, 6.7–7.5.

Toxins are highly poisonous substances mainly composed of protein, which damage the cells of the infected host, causing inflammation, illness and sometimes death. They are produced during rapid division of bacterial cells.

Importance of Bacteria

Most people associate bacteria with decay and disease, but many bacteria are beneficial to man, as already mentioned. Bacteria are used in the production of vinegar, cheese and synthetic vitamins. Many bacteria live in the human intestine where they assist in the decomposition of faeces and produce some B vitamins. (Overuse of antibiotics and similar drugs may interfere with intestinal flora.) The most important function of bacteria is their ability to break down or decay organic materials so that simpler compounds may be released for use by other living things. This occurs in the soil and in sewage treatment works. Without this recycling action, life would cease to exist.

Many bacteria are capable of causing disease. Bacterial diseases include diphtheria, typhoid fever, scarlet fever, whooping cough, pneumonia, meningitis, dysentry, cholera and enteric fevers such as gastroenteritis and salmonella poisoning.

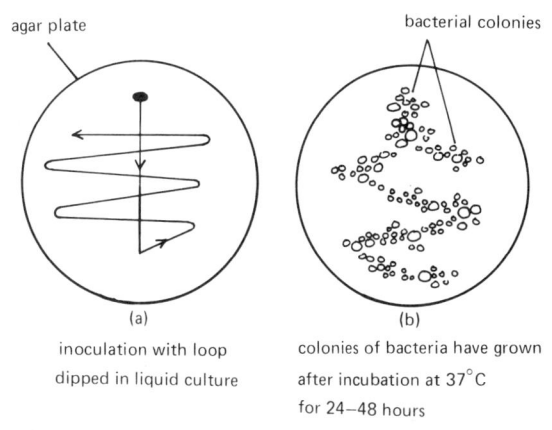

Fig. 6.15 Cultivating bacteria from liquid culture

Cultivating Bacteria

Note. Absolutely sterile techniques must be used in order to prevent contamination of cultures. All equipment must be sterilised by heat and all cultures protected from the air. Avoid cultivating pathogenic bacteria. Further details on the techniques of agar cultures should be studied from a reliable biology textbook.

Procedure. Wash and dry a number of petri dishes thoroughly. Sterilise these and all other implements before use. Dissolve nutrient agar according to the directions. Pour it into petri dishes, cover quickly, allow to set. Keep one dish as a control.

Inoculate one dish with liquid culture by streaking an 'S' shape on the surface of the jelly using a special metal loop (which has been passed over a bunsen burner flame to sterilise it). Cover quickly.

'Contaminate' other agar plates in some of the following ways:

1. Leave one exposed to the air.
2. Cough on another.
3. Put some dust on another.
4. Touch one with an unwashed finger.
5. Slide a single hair over one.
6. Touch one with a dish cloth.
7. Place a drop of water on one.

Incubate plates for 48 hours at 37°C. In a few days colonies of bacteria will develop.

VIRUSES

Viruses are on the borderline between living and non-living things. They are minute organisms, even smaller than bacteria, which grow as parasites within the living cell. Viruses are transmitted from cell to cell in the form of tiny infectious particles called *virions*. The core of the virion contains nucleic acid surrounded by a protein membrane.

Viruses cause many diseases such as polio, smallpox, chickenpox, measles, mumps, influenza and colds. Viral

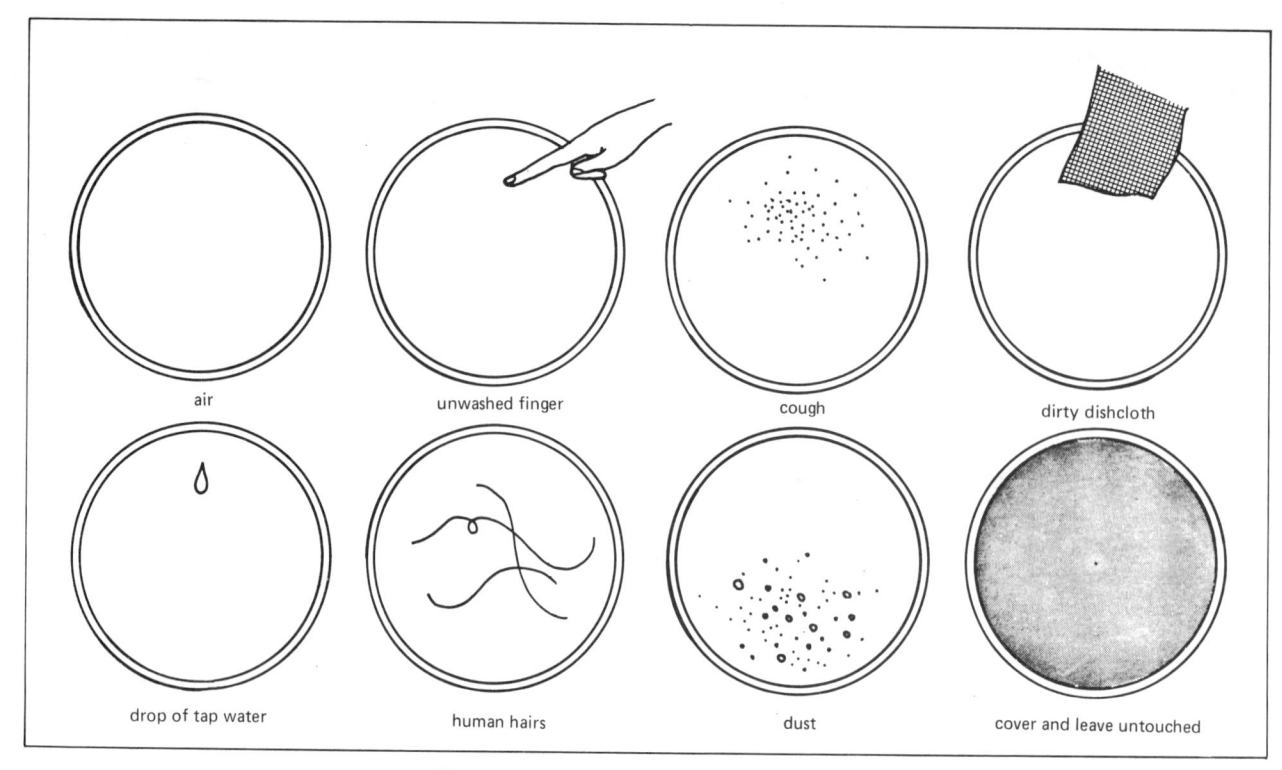

Fig. 6.16 Cultivating micro-organisms

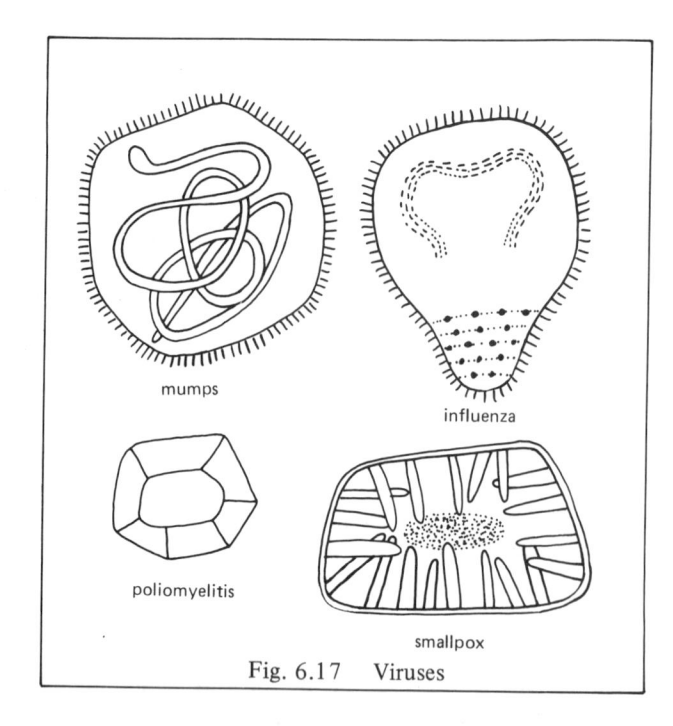

Fig. 6.17 Viruses

diseases are usually spread by close personal contact or by droplet infection – the spreading of germs by spraying people or objects during coughing or sneezing.

Viruses multiply in the body producing toxins which destroy the cell and cause disease within the body. They then leave the host to continue their life cycle in another body.

FOOD SPOILAGE

Most natural foods have a limited life. Perishable foods such as fish, meat, milk and bread have a short lifespan. Other foods keep for a considerably longer time but decompose eventually. Fats become rancid; flour becomes stale and infested with flour mites. Even preserved foods will decompose unless they are stored and sealed correctly.

There are many causes of food spoilage. Enzymes within some foods bring about their destruction, while chemical reactions such as oxidation and rancidity decompose others – but the main single cause of food spoilage is invasion by micro-organisms such as moulds, yeasts and bacteria.

Micro-organisms are found everywhere, since the conditions which bring about their growth are readily available. Like human beings, they prefer a warm moist environment, a supply of oxygen and food – and because their nutritional requirements are similar to ours, they readily contaminate our food supplies. Tons of food are wasted each year through contamination by micro-organisms. When food spoilage is caused by the growth of yeasts and moulds it is self-evident; a furry growth covers the food and it becomes soft and often smells bad. Bacterial contamination is more dangerous because very often the food does not look bad: even though severely infected, it may appear quite normal. The presence of highly dangerous toxins and bacterial spores is often not detected until after an outbreak of food poisoning, when laboratory examination and experiment uncovers the infecting agent.

Carriers of bacteria: flies on a rabbit carcase (John Topham Picture Library)

All living things carry bacteria. Many are harmless and some are even beneficial, but food may readily become contaminated with *pathogenic* (disease-bearing) bacteria. These are transmitted by flies, vermin, animals and especially by humans with careless and unhygienic habits.

Bacteria Carriers

Flies feed indiscriminately on animal faeces and on food. When they leave dung or rubbish heaps, they carry pathogenic bacteria on their legs and deposit them on their next landing site – possibly a clean kitchen surface or an uncovered cream bun. They soften food by using their saliva and often as they feed they vomit and excrete on the food. It goes without saying that they should be banned from every kitchen and food preparation centre. To prevent flies from breeding near the house keep dustbins clean and as far from the house as practicable.

Household vermin include rats and mice, both of which are prolific breeders. They are attracted both to decaying refuse and to fresh food and have dirty habits. They excrete on and near foodstuffs and are attracted to a warm, dirty environment. It is essential to store food in clean, cold storage areas in vermin-proof containers. As rats and mice can chew through paper, polythene, cardboard and even wood, if they are hungry enough, glass and metal containers are the best way of keeping food clean and safe from vermin. Other disease-producing vermin include lice, fleas, bed bugs and cockroaches; the latter carry salmonella bacteria.

More vermin-proof containers needed in this grocer's shop! (Rentokil Ltd.)

Animals. Household pets are unhygienic and attract flies and fleas. Cats and dogs should not be allowed in the kitchen at any time – especially during meal preparation. It is important to wash the hands after handling animals.

Humans. The most basic principles of hygiene are ignored by many people. It is essential to wash hands thoroughly after using the lavatory as many food-poisoning diseases are spread through the faeces. Hands should also be washed before food is handled or prepared, and before meal times. Bacteria normally safe in the nose or bowel may cause disease if they are spread by unwashed hands. Human 'carriers' may be immune to a disease themselves but pass it on to more susceptible victims.

Where Bacteria Spread

Bacteria may infect food in one of the following ways:

1. *On the farm.* Strict hygiene must be observed, particularly in areas directly concerned with food production. Milking parlours, hen batteries and so on must be kept scrupulously clean. Utensils must be sterilised and sheds hosed out daily. Feeding stuffs which have been carelessly handled or stored are often a source of salmonella infection, and rats and mice also spread disease on farms.

2. *Animal transport.* Trucks used for carrying animals are often covered with dung and urine, both of which contain pathogens. They must be hosed down after use.

3. *Close contact.* When animals are packed together, as they are in some forms of intensive farming, disease spreads easily from one contaminated animal to another. In addition, close contact often causes stress which lowers the animals' resistance to infection.

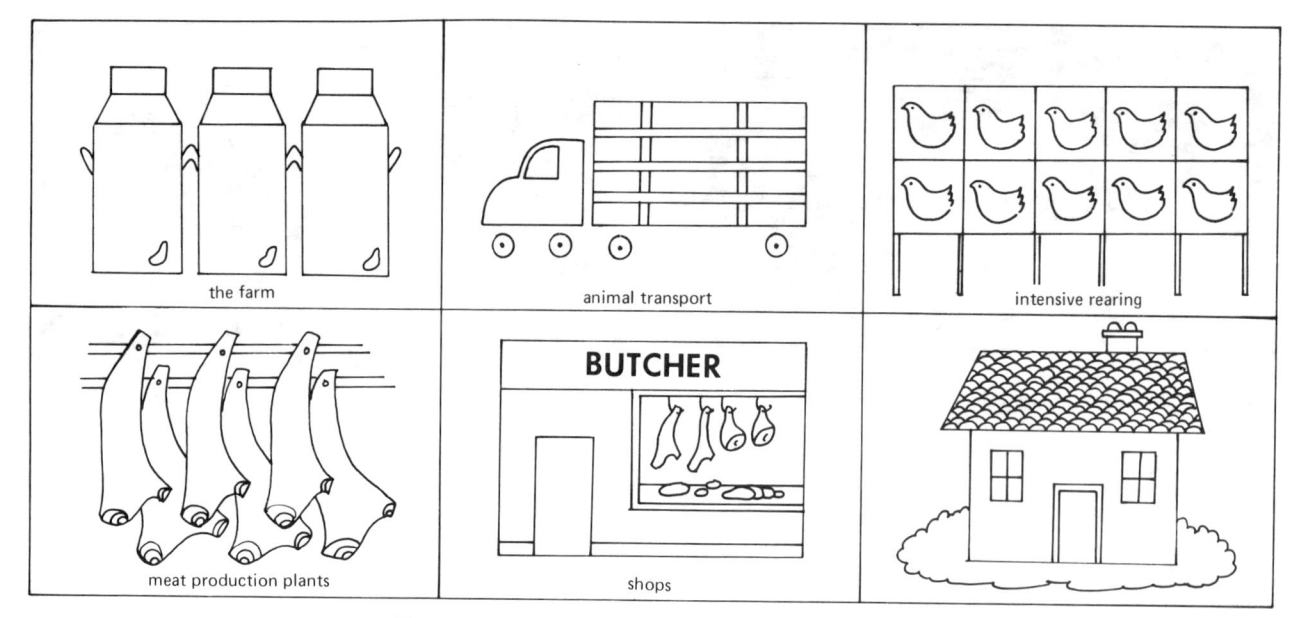

Fig. 6.18 Sources of bacterial infection in food

4. Abattoirs/poultry plants. Disease is easily spread in live animals or carcases through the handling of one contaminated animal. Veterinary checks are essential before slaughter and during production.

5. Meat production. Mincers, conveyor belts and utensils may become contaminated from one diseased animal. Chilling meat to 5°C lessens the risk of spreading infection as most bacteria cannot multiply at this temperature. EEC legislation ensures a high standard of hygiene in meat plants.

6. Shops, supermarkets and catering establishments. A high standard of general hygiene is essential.

Cooked and raw meat must never be prepared on the same surfaces. If raw meat is contaminated with pathogenic bacteria, normal cooking will almost certainly destroy them — but if cooked meat is placed on surfaces contaminated by infected raw meat the germs will remain active and, as there is no further cooking, the germs will not be killed. Strict application of this principle must be observed in catering establishments, shops and household kitchens.

Rules for Avoiding Food-Poisoning
1. Every person working with food should maintain a high standard of hygiene. Hands should be washed (with soap and hot water) *after* using the lavatory or changing a baby's nappy and *before* handling food or feeding a baby.
2. Avoid coughing and sneezing over food.
3. Cover cuts with elastoplast and change it frequently.
4. Keep all food covered and handle it as little as possible: use tongs if you can.
5. Cook food thoroughly — especially frozen meat and poultry. It should first be thawed out completely and then cooked at a high enough temperature for long enough to destroy all bacteria.
6. Eat food soon after it is cooked. Avoid keeping food warm, as germs multiply in warm foods. Leftovers should be cooled quickly, covered, stored in the refrigerator and used soon. They should then be reheated thoroughly: a temperature of 100°C (boiling point) is necessary to destroy pathogens.
7. As bacteria do not 'crawl' they are usually found on the surface of meat and normal cooking will quickly kill them. But when meat is 'boned and rolled' or minced or chopped, surface bacteria may reach the centre. For this reason rolled roasts, hamburgers, meat pies and sausages must be thoroughly cooked.
8. A hygienic kitchen should be well ventilated to prevent humidity and should have easy-to-clean surfaces. Floors, walls and work surfaces should have the minimum number of joins as they may harbour bacteria. Stainless steel sheeting is preferable to tiles — as is laminated plastic and vinyl: tiles may harbour germs in the grouting between them.
9. Adequate dry, well ventilated storage is required, together with refrigeration for perishables.
10. Good washing-up facilities are essential. A dishwasher is ideal as the high temperature of the water leaves the dishes sterilised; if this is not available plenty of hot water must be used. Cloths should be washed daily and dried out of doors; weekly boiling or bleaching will keep them germ-free. Never leave wet cloths lying about, as bacteria thrive in damp, dirty cloths. Paper towels are more hygienic.
11. Use disinfectants regularly in sinks, drains and lavatories.

INFECTIOUS DISEASES

Many diseases occur when the body is invaded by bacteria or viruses. When a disease is spread from one person to another either directly or indirectly through a 'carrier' or infected article, for example, it is called an infectious or communicable disease. Many infectious diseases are caused by contaminated food or water.

Resistance to Disease

Man has several defences which protect him from invasion by pathogens.
1. The skin protects the body from the entry of disease-bearing germs.
2. Openings in the skin, such as the nose, ears, eyes, are protected by mucous, wax and enzyme secretions which form a barrier against any foreign bodies.
3. Many of the pathogens which reach the stomach are destroyed by the hydrochloric acid secreted by the gastric lining.
4. White blood cells (leucocytes) engulf and destroy bacteria.
5. Antitoxins present in the blood neutralise the poisons made by bacteria.
6. The lymph system is also involved in the fight against pathogens.

In spite of such strong defences, the body can become infected and succumb to disease if it is invaded by large numbers of bacteria or if the general health of the infected person is not good. Obviously if the body is kept fit with plenty of fresh air and exercise, a good diet and sufficient rest, it has a better chance of fighting most diseases.

Disease Carriers
Diseases may be spread by:

1. *Humans*, by a. *droplet infection*. When an infected person coughs or sneezes he sprays a large area with germs. Bacteria or spores from dried droplets may be breathed in by other people, particularly in crowded places or in badly ventilated rooms or dwellings. Tuberculosis, diphtheria, pneumonia and many respiratory infections can be spread in this way. Humans also spread infection by b. *direct contact*. Many skin diseases, respiratory complaints and venereal diseases such as syphilis and gonorrhoea are spread by close personal contact.

2. *Animals.* Rats, mice and insects such as flies and mosquitoes carry various diseases such as rabies, malaria and anthrax. These may be transmitted to humans by biting or by puncturing the skin with a sting.

3. *Inanimate objects.* Certain diseases may be spread through the handling of objects infected with pathogenic bacteria. Infections which produce food poisoning symptoms such as vomiting and diarrhoea are often contracted by using infected lavatories, hand towels, kitchen cloths, chipped cups, dirty cutlery and so on.

4. Food and water

FOOD POISONING

Food poisoning occurs when food or water which has been contaminated by poisons or infected by pathogenic bacteria is ingested. The symptoms are abdominal cramps, vomiting, diarrhoea and sometimes fever. These generally occur within 24 hours of ingestion of the contaminated food.

Chemical Food Poisoning

Food may be contaminated by toxic chemicals present in the food. This may occur in one of several ways.

1. Natural toxins. Some toxins may be naturally present in the food. Many fungi and berries are highly poisonous and must not be eaten; rhubarb leaves contain a high concentration of oxalic acid, which is toxic to man, but the stems, containing smaller amounts, are quite safe to eat. Some harmless foods contain substances which are potentially toxic if taken in excess — e.g. green potatoes contain solanine, a substance which can be dangerous if large amounts are eaten.

2. Metallic poisoning. This occurs when metal from a container contaminates the food it contains. Vitreous enamel cooking pots may release cadmium, particularly if the inside is coloured. All new cooking pots of this type are now lined with white enamel. Zinc poisoning has been caused by the use of zinc ice-lolly moulds or by preserving acid foods in galvanized iron vessels.

3. Chemical additives. Substances which are toxic to body cells may be added to food either deliberately or by accident. Many synthetic additives used in modern food production may, in spite of stringent testing, prove to be harmful (see p. 116). Examples of chemicals which were used extensively in food processing but which are now banned in several countries are cyclamates (synthetic sweeteners) and DDT (an insecticide). Many chemicals such as pesticides and weed killers leave residues in animals and plants which may prove to be toxic, especially if the quantity of chemicals has built up gradually over a long period. Some metals also have a cumulative effect if they are absorbed into food, and radioactive materials and chemicals from severely polluted air may affect our food, too. Drinking water may become contaminated with chemicals or effluent from sewers either accidentally or through carelessness, even though local authority by-laws are intended to prevent this. Household utensils can become contaminated with chemicals through careless use of cleaning agents and insufficient rinsing.

There is a growing awareness of the dangers of toxicity in food. To avoid such dangers, only fresh foods should be eaten as far as possible, and any foods which could be contaminated – e.g. fruit, vegetables – should be washed before eating.

Bacterial Food Poisoning

Most food poisoning results from the ingestion of pathogenic bacteria. In most cases this type of poisoning can be prevented by high standards of hygiene, especially by those involved in the preparation of food. It is essential that any person who handles food should be aware of the very real dangers of food contamination and food poisoning. The standard of hygiene in a home is generally the responsibility of the mother, and it is essential that she should set a good example and instill in young children sensible habits for healthy living. A casual attitude to hygiene can lead to disease.

Bacterial food poisoning can occur in two ways:

Infectious food poisoning, in which pathogens are ingested and multiply within the body of the infected person. Most food poisoning bacteria settle in the intestine and pass from the body in the faeces. Unwashed hands can quickly spread the infection. *Salmonellae, Shigellae* and *Clostridium welchii* are the main causes of infectious food poisoning.

Toxic food poisoning is caused by poisons produced by bacteria on food or in the body. These damage or kill the cells of the host, causing unpleasant symptoms such as inflammation of the gastro-intestinal tract and diarrhoea, and may in severe cases kill the patient. Toxic food poisoning is generally caused by *Staphylococci* but occasionally results from *Clostridia,* e.g. *Clostridium welchii,* and rarely *Clostridium botulinum,* which may be fatal.

1. Salmonellae. There are several hundred species of salmonella, the majority of which cause food poisoning. They are short, rod-like germs which move by means of many flagellae which are attached to them. They live as parasites in the intestines of animals and man and are usually found in human faeces. *S. typhimurium* is among the most common food poisoning bacteria. *S. typhi* causes the serious disease typhoid.

Meat and poultry are easily infected with salmonella as it can easily pass from the intestines to the carcase during slaughter and butchering. Sausages and cooked meats are another common source of this infection and egg dishes in which the eggs are uncooked (mayonnaise) or cooked at a very low temperature (custard) can also be a source. Fish, especially shellfish, is a common carrier of salmonella.

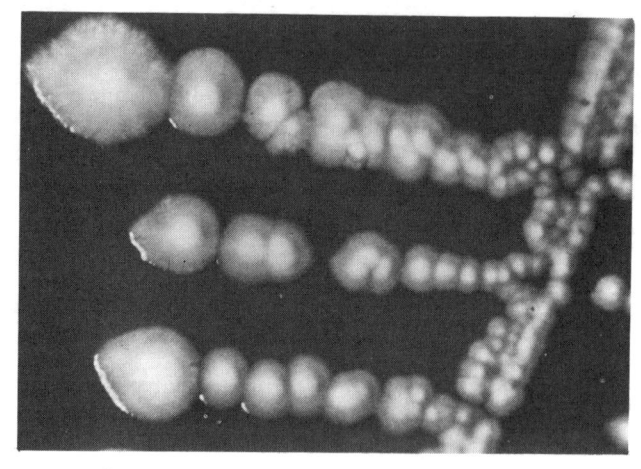

Salmonella typhi bacteria (C. James Webb)

Water contaminated by sewage and food infected by flies and vermin will carry countless numbers of salmonella bacteria. The incubation period is 12–36 hours and the symptoms of the infection nausea, abdominal pain and diarrhoea. Food poisoning may continue for up to five days, and patients are highly infectious during and immediately after the disease.

2. Shigellae. These are short, rod-like germs which live in the intestines of man only. They can survive outside the body and are often found on dust, dirty dishes, cloths and especially around lavatories. They are spread by careless hygiene and passed from unwashed hands to food and utensils, eventually entering the mouth of another victim. *S. dysenteria* causes dysentry, a serious form of food poisoning which produces a high fever together with the usual symptoms of food poisoning. Other more common examples are *S. sonnei* and *S. shigae.*

3. Staphylococci are round germs which occur in clusters. (The name comes from Greek words meaning bunches of grapes or berries, which they resemble.) They are easily visible under the microscope and when grown on agar they resemble little pieces of butter. These parasites of

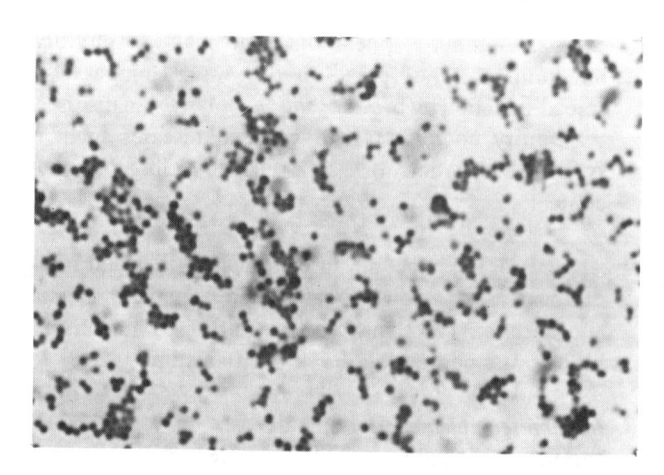

Staphylococci (C. James Webb)

animals and man are normally present on the skin, nose and throat. *Staphylococci* are faculative bacteria – which means they can grow with or without oxygen, on animate or inanimate objects (e.g. towels and dirty sinks). They are often found in cowsheds and readily infect milk which provides an ideal medium for their growth, but pasteurisation destroys them. Flies spread staphylococcal bacteria and cooked meat is a common source of infection. Infection is spread frequently by human skin – dirty hands, cuts and sores – and also by coughing over food. Onset of illness is very rapid as the toxin has an immediate effect on the gastrointestinal tract. Symptoms may show within half an hour but usually within 1–3 hours and include nausea, vomiting, diarrhoea and abdominal cramp. The illness rarely lasts longer than 24–36 hours but leaves the patient weak and dehydrated.

4. Clostridia. *Clostridium welchii* is a rod-shaped bacterium which lives in the soil and thereby infects fruit and vegetables. As its spores can survive in a dry atmosphere it is found in dust on floors and on kitchen surfaces. It may also be found in water, milk and particularly in sewage. Because it is anaerobic, it multiplies readily in the centre of rolled joints of meat, meat pies and stews which have been allowed to cool slowly. Illness may take the form of infectious food poisoning caused by active bacteria multiplying in the intestine or it may be toxic. Rapidly multiplying *Clostridia* produce a toxin which results in the characteristic food-poisoning symptoms of cramps and diarrhoea. It is milder than most other forms of food poisoning, and fever and vomiting are rare. Incubation takes 8–12 hours and the illness rarely lasts longer than a day or two.

Clostridium botulinum, although a very rare cause of food poisoning in this country, is important as it is often fatal. *C. botulinum* is a rod-shaped bacterium which produces a very powerful toxin. It lives in soil and decaying matter and is commonly found on vegetables – but as it is anaerobic, (i.e. will not multiply in air) vegetables may be safely cooked and eaten. If, however, these vegetables are canned and, through some fault in the canning process, the temperature reached is not sufficiently high, the bacteria and their spores will survive and multiply – and produce deadly toxins. Any can or bottle which appears to be 'blown', leaking, rusted or in any way inferior should not be used. The contents may look normal but one teaspoonful could be sufficient to kill.

Canned fruit will not cause botulism because of its acidity. However, home canning and bottling of vegetables and other non-acid foods must be discouraged because temperatures of home pressure cookers will rarely be sufficient to destroy bacterial spores (150°C).

Home canned produce, canned fish such as salmon and tuna, canned vegetables, meat and fish pastes are all possible sources of botulism. The symptoms of the disease, which occur within 1–3 days, at first appear similar to normal food-poisoning symptoms, but are followed by a gradual paralysis. Slurred speech, double vision and breathing difficulties ensue and death occurs when the respiratory organs become paralysed. Those who are lucky enough to survive have a long, slow recovery.

Other Food-borne Diseases

Brucellosis. A bovine disease picked up by those in frequent contact with cows. Veterinary surgeons and farmers are the most usual sufferers.

Scarlet fever, caused by a streptococcal bacterium.

Cholera, a serious disease usually caused by contaminated water. *Vibrio cholerae* is the infective agent.

Immunity

Lymphocytes produced by the lymph nodes produce antibodies which interfere with bacteria by causing them to clump together or by destroying them, thus lessening the risk of disease. They also produce antitoxins which neutralise the toxins produced by bacteria. Antibodies remain in the blood for some time after infection in order to prevent a recurrence of the disease while the individual is still weak. If he comes in contact with the disease after manufacture of antibodies has ceased, they will be stimulated into production again in order to fight the new infection. In this way, antibodies in the blood bestow immunity on the individual.

Vaccines can be produced which contain dead or less virulent strains of bacteria. Inoculation with these harmless substances stimulates antibody production and gives the individual an acquired immunity to a specific disease.

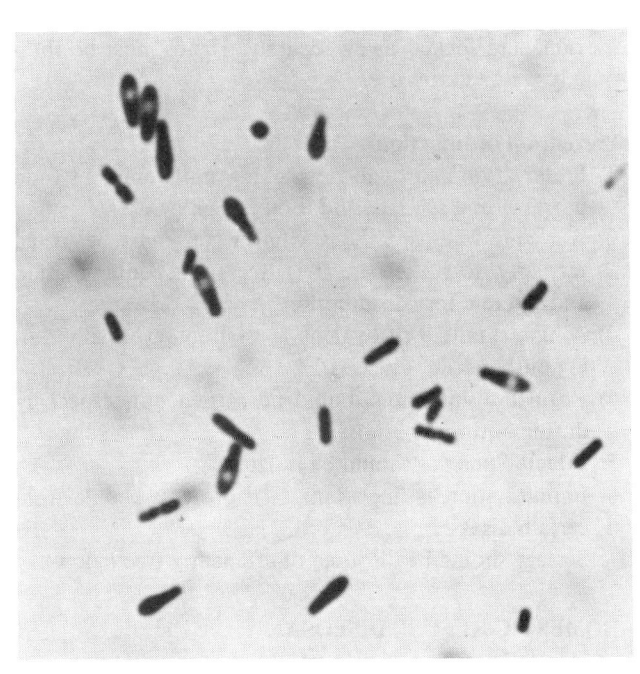

Clostridium botulinum bacilli (C. James Webb)

Vaccines are available which control the following diseases:

Smallpox	Whooping cough
Diphtheria	Tuberculosis
Polio	Typhoid
Tetanus	Cholera.

Many of these are available free at health centres. Mothers should make sure that their children are immunised against all dangerous diseases where possible.

Antibiotics. These are chemicals which destroy many pathogenic bacteria. They are produced from fungi such as moulds and yeasts, from bacteria and other protists. They can be taken orally or administered directly into the bloodstream by injection. Penicillin, the first antibiotic, was derived from the *Penicillium* mould and is still used for many bacterial infections. Streptomycin, obtained from bacteria, is another antibiotic in general use both for animals and man.

Antibiotics work by killing bacteria without harming the body cells. However, many bacteria have developed resistant strains which are unaffected by certain antibiotics. Furthermore, there is the danger that man may become resistant to certain antibiotics by overuse or by ingesting large amounts in meat from recently treated animals or poultry.

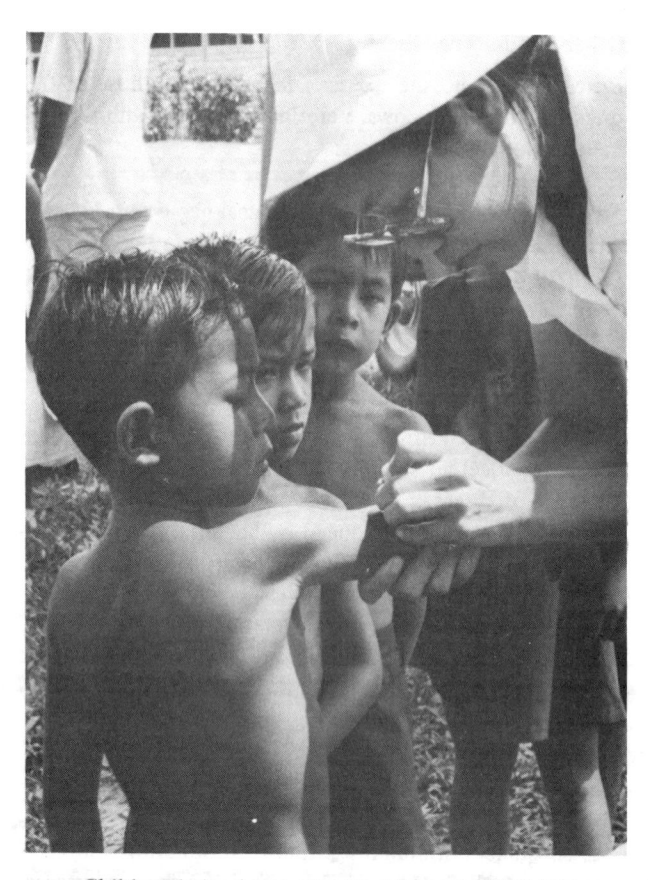

Children being immunised in Malaysia (W.H.O.)

Food Poisoning: Summary
Foods readily contaminated:
Meat, fish, milk, eggs.

Confectionery. Cream-filled cakes, sticky icing, especially when left uncovered in warm shops.

Cooked meats and reheats

Shellfish – a particularly common source as they feed by filtering seawater through their bodies. Bacteria from polluted water are absorbed during this process.

Basic rules
1. Cool dry storage is essential.
2. Meat and fish should be kept under refrigeration and should be thoroughly cooked.
3. Reheated dishes should be brought to boiling point.
4. Scrupulous personal hygiene is essential. Wash hands after using the lavatory. Hand towels must be changed regularly: there is little point in washing hands if they are dried on a dirty towel.
5. Choose food shops and restaurants which have a high standard of hygiene.
6. Remember that heat and disinfectants destroy most germs and toxins. Use disinfectants liberally in bathrooms, lavatories, kitchens and drains.

Rural Households
It is important to pay particular attention to hygiene in farms. Hands must be washed after handling animals. Avoid having pigstyes, hen houses etc. too near the dwelling house. Insecticides and other chemicals should be used with care, stored in the original containers, marked 'poison' and locked away securely. Hands must be thoroughly washed after use.

Prevention of Infection
1. Preservation and hygienic packing and storage of food prevent food spoilage and food poisoning.
2. A clean water supply is vital: water must be tested regularly to make sure it is free from contamination and suitable for consumption.
3. A high standard of hygiene is essential, especially during outbreaks of disease.
4. Regular disinfection should be carried out, especially during outbreaks of disease.
5. Infected persons should be isolated.
6. Immunisation is important as it protects people from certain diseases.
7. Sewage should be disposed of efficiently (see below).

DOMESTIC REFUSE DISPOSAL
Domestic refuse consists mainly of
a. organic waste such as food, plants, peelings etc;

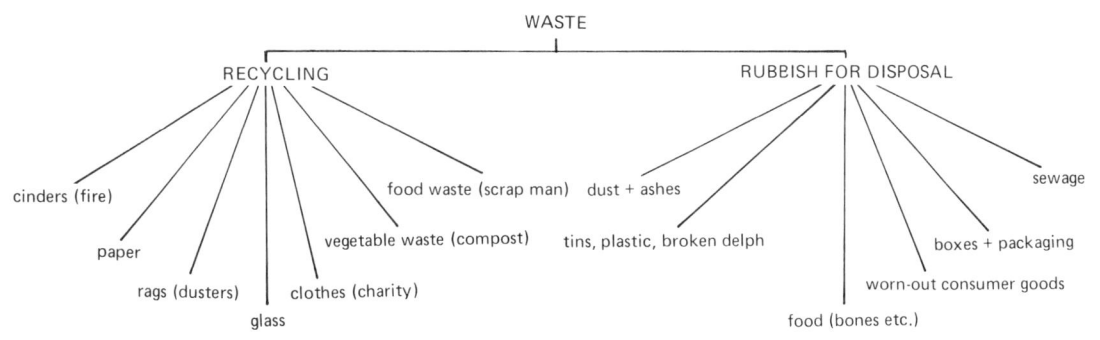

Fig. 6.19 Domestic refuse disposal

b. *inorganic waste* such as cans, polythene and plastic containers, and delph, glass, metal, fibre or wooden household articles.

Recycling

There is increasing emphasis on conservation and recycling because of a growing awareness that the world's resources are dwindling. For this reason any items which may have some recycling value should be set aside. Much organic waste can be used on a *compost heap* – although protein foods such as fish and meat trimmings should not be put on a compost heap as they attract flies and vermin. All vegetable waste such as tea leaves, fruit and vegetable trimmings can join the garden waste on the compost heap. They will, in time, help to produce valuable compost.

Other food waste, apart from bones, may be sought after by *someone rearing animals,* who may collect it if it is kept in a separate covered bucket. Newspapers and magazines should be tied in bundles and given to *waste-paper collections* for local charities. In some areas there are collection centres for *bottles.* These are sorted out and sold to glass manufacturers for recycling.

As yet, no local authorities in Ireland are taking separate collections of specific types of refuse such as paper, glass and metal – but this has been done in some parts of Britain and it has been found that the extra trouble and expense has been offset by the money earned.

Disposal

Most cities and large towns in Ireland have a weekly refuse collection for householders, and a daily service for busy shopping areas. The waste is collected by large mechanical trucks which compress and eventually empty it into a local authority tiphead. These are usually situated in low-lying areas suitable for reclamation. They may be temporary eyesores, may smell unpleasant and attract rats, but eventually they are covered with soil and used for public amenities such as sportsfields and parks.

Some smaller towns provide a dumping area which is used by local people for disposing of refuse. In rural areas each householder must make provision for his own waste disposal. The usual system is a pit which is filled with refuse over a period of time and then covered over with soil. Any refuse pit or dump should be as far as possible from the house in order to avoid infestation of the home by rats, mice and flies. Households with certain types of solid fuel cooker may find it more satisfactory to burn refuse.

Waste disposal units. Electric waste disposal units which fit into the sink outlet and grind up scraps of food are useful and hygienic (see Part II).

Kitchen bins. These should be easy to clean and ideally should have a removable inside. They should be made of hard-wearing, rustproof material such as stainless steel or plastic. They should have a tight fitting lid and a pedal, so that they need not be handled during cooking. Line the bin before use with paper or polythene bags and clean it regularly.

Dust bins. These should be made of hard wearing zinc, galvanised iron or heavy duty polythene. They should have a lid which fits tightly, to keep out animals and vermin. Empty, clean and disinfect weekly. Keep bins at a convenient distance from the house, but not too near as they attract flies. Avoid using polythene bags alone for refuse disposal as scavenging animals can easily tear holes in them. If used, they should be tied securely with twine. Wrap food and other moist waste in newspaper before disposing of it.

Incinerators. Disposal chutes are sometimes used in blocks of flats. The rubbish is tipped into a chute in the kitchen which carries it into an incinerator in the basement or into large bins which are collected regularly.

Sewage disposal. As faeces and urine contain countless germs, they must be disposed of quickly and hygienically in order to prevent the spread of disease. Modern flush toilets, piped water supplies and sewage disposal pipes do this effectively. This subject is dealt with in more detail in Part II.

In the sewage works anaerobic bacteria break down solid matter, releasing soluble compounds. These are sprayed or aerated in other ways to encourage further breakdown, this time by aerobic bacteria which convert the compounds into harmless salts to be released into the soil, river or sea nearby. Many chemicals interfere with this important process, hence the drive towards the use of *biodegradable* detergents which can also be broken down into harmless compounds. As some factory effluent can destroy the valuable bacteria which cause decay, local authorities now restrict the amount and type of effluent which is released from industries and processing plants.

QUESTIONS
1. Write a note on fungi under the following headings:
    a. classification
    b. conditions for growth
    c. reproduction
    Describe in detail the structure of two members of the fungi phylum. Use diagrams to illustrate the answer.
    List the beneficial uses of fungi.
2. Describe the structure of a yeast cell.
    How does reproduction of yeasts differ from normal fungal reproduction?
    What is the optimum temperature range for the growth and reproduction of yeast?
    Describe in detail the chemical changes which take place during fermentation.
3. Sketch and describe the structure of a bacterial cell. Classify the different types of bacteria and write a detailed note on two bacteria included in the classification. Write a brief note on the Gram stain or Gram reaction.
4. Discuss the effects of temperature and pH on the growth of bacteria.
    Explain the terms (a) thermophillic; (b) aerobic and (c) pathogenic.
    Describe a laboratory test which shows the presence of bacteria in food.
5. Discuss food hygiene under the following headings:
    a. food production
    b. food processing
    c. kitchen practice
    d. storage
    What advice would you give to a young housewife in relation to the planning of her kitchen, so that a high standard of hygiene would be maintained?
    List the methods by which refuse may be disposed of in (a) urban and (b) rural areas.
6. List the main causes in infectious disease.
    What inbuilt defence mechanisms has the human body to protect us from disease?
    List the steps which should be taken at home to prevent the spread of infection.
7. What is food poisoning?
    Write a note on *two* of the following bacteria: Salmonella; Clostridium; Staphylococci. Use these headings: (a) source; (b) structure; (c) method of infection; (d) symptoms.
    Explain the term immunity.

7 Food Preservation

Because some of the factors which control growth — such as the weather — are still beyond our control, it follows that there are times when many foods are plentiful and cheap and other times when they are scarce. It makes sense to preserve such foods so that they are available when out of season.

Preserving also facilitates the storage, transport and distribution of food on a very wide scale, with the result that a varied selection of food is available in most countries. Food may be preserved for a short time by correct storage (e.g. in a refrigerator), or for long periods by methods such as drying, bottling or deep freezing.

Reasons for Preserving Food
1. To avoid waste.
2. To use up fruit and vegetables when they are cheap and plentiful, and make them available when they are not in season.
3. To make full use of garden produce.
4. To introduce flavour and variety to the menu (e.g. with jams, pickles, chutneys).
5. To save money through economies of scale, e.g. in jam-making.
6. On a commercial scale to facilitate distribution and export of food.

Principles of Preserving
1. To inhibit or destroy the action of the *enzymes* which cause decay in food.
2. To destroy the *micro-organisms* (i.e. moulds, yeasts and bacteria) which decay food and/or to remove the conditions (warmth, moisture) which allow the growth of such micro-organisms.
3. To prevent *oxidation and recontamination* of the preserved food by sealing it quickly and completely.
4. To keep the *colour, flavour, appearance and nutrient value* of the food as near as possible to those of the unprocessed variety.

THE CAUSES OF DECAY

a. Enzymes

Fruit and vegetables are subject to a natural cycle of growth, maturity and decay. In food preservation this cycle is interrupted by delaying or preventing decomposition, thus prolonging the edible life of the food. Decomposition or decay is speeded up by *enzymes* present in the food which continue to bring about reactions (such as discolouration) long after the food has been harvested.

Enzymes are chemicals present in all living cells which act on the carbon compounds within the cells (e.g. the proteins and carbohydrates) causing them to break down or decompose.

Enzymes are destroyed by temperatures above 70°C and low temperatures retard enzyme activity. Enzymes are inactive at −18°C but above this they can cause slight deterioration of food. They are most active in warm, moist conditions (between 20° and 40°C).

b. Micro-organisms

Micro-organisms are also responsible for the contamination of food. The micro-organisms which destroy food are moulds, yeasts and bacteria (see p. 119).

1. Moulds. Mould spores are present in the air and multiply on suitable food in a warm, moist environment. Temperatures above 75°C will destroy most moulds and their growth is inhibited by cold or dryness and also by vinegar and sulphur dioxide.

These dried beans have no moisture in which micro-organisms can flourish. In the bean pickle, salt, sugar and vinegar inhibit their growth (Syndication International)

2. Yeasts. These organisms flourish in a warm, moist atmosphere and especially in acid conditions and in the presence of sugar. They are destroyed by extreme heat — i.e. temperatures over 60°C — and inactivated by cold storage (temperatures below 16°C). Deep freezing destroys some types of yeast. Although yeasts multiply rapidly on foods which contain sugar, high concentrations of sugar (above 65 per cent) will prevent growth, as will alcohol and vinegar.

3. Bacteria. These are found almost everywhere. They are destroyed by temperatures in excess of 100°C, although some bacterial spores can survive temperatures of over 140°C. Most spores are destroyed by pressure cooking (120°C) for one hour. Growth of bacteria is inhibited by deep freezing, but the cells remain dormant and become active again upon thawing.

Because enzymes and micro-organisms need warmth, moisture and usually air in order to multiply, it is obvious that if any one of these factors is removed food may be preserved:

Freezing — removal of warmth
Drying — removal of moisture
Canning/Bottling — removal of air

Often it is necessary to apply heat first in order to destroy micro-organisms or enzymes. This occurs in the blanching of vegetables before freezing and during the sterilisation processes of bottling and canning. Substances such as sugar, salt, vinegar and sulphur dioxide, all of which inhibit the growth of micro-organisms, are often used alone or in conjunction with other preserving processes.

METHODS OF PRESERVATION

1. Heat Treatment

Bottling, canning and jam-making employ temperatures which are high enough, and maintained for long enough, to destroy micro-organisms. Non-acid foods such as vegetables and protein substances require higher than normal temperatures (125–140°C) to destroy spore-forming bacteria. As it is not possible to achieve these temperatures with domestic equipment, such foods should not be preserved by home bottling or canning.

Pasteurisation and sterilisation of milk (see p. 75) are also forms of heat processing.

Canning and bottling principle. Food is placed in the container and both are brought to a high temperature. When the contents are sterilised, the containers are removed from the heat and cooling creates a vacuum which holds the lid on tightly, sealing the contents. As long as the bottle or can remains sealed, the contents are safe from bacterial contamination.

Commercial canning and bottling. These processes are carried out commercially in much the same way as in the

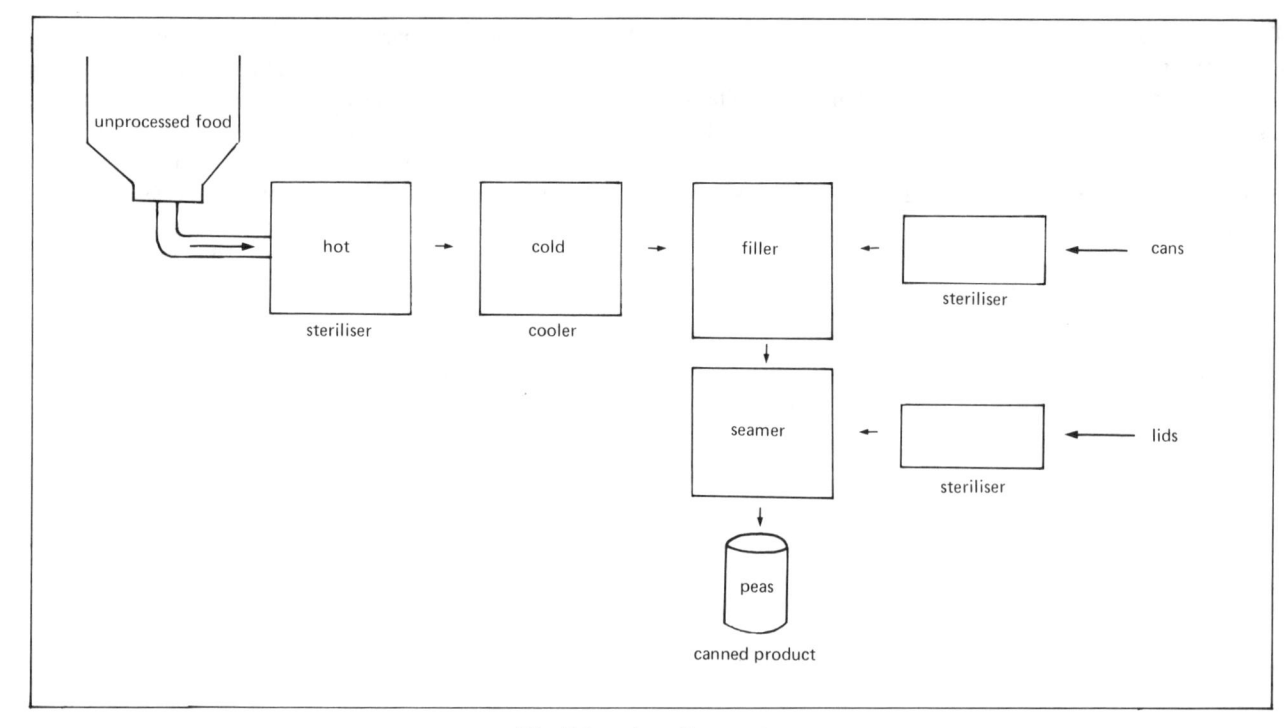

Fig. 7.1 Aseptic canning

home, but on a larger scale. Special machinery prepares the food, blanches it or cooks it where necessary and fills it into lacquered tin cans or bottles. Syrup, brine or sauce is poured in. The containers are 'exhausted' to exclude air and then hermetically sealed. Cans or bottles are sterilised for the required time in huge autoclaves (pressure-cooking ovens), then sealed and cooled quickly.

Over-processing will produce inferior canned foods which are too soft, while under-processing may allow bacterial spores to survive in the can, and possibly cause food poisoning.

Additives are often used to preserve flavour, colour or nutritive value. Nisin is a chemical which is often added to non-acid canned foods as its presence inhibits the growth of the deadly *Clostridium botulinum* (see p. 133).

Meat and fish are available canned, as are stews, soups, luxury sauces, baby foods and many other products. The bottling process is used for fruit, vegetables, baby foods and sauces.

Aseptic canning process. This is a new method of canning which involves sterilising the food at ultra-high temperatures (UHT) for a very short time. It is then quickly filled into sterilised containers and sealed. As this method is quicker than normal processing it has less effect on the colour, flavour and food value of the product and enables additives such as vitamins to be included in containers before they are sealed.

Spoilage. Leaking, rusty or blown cans should never be used as there is a danger of spoilage by food-poisoning bacteria. Most spoilage is self-evident: as soon as the cans are opened the contents give off an unpleasant smell and generally they look unwholesome. However, canned food infected with *Clostridium botulinum* usually appears normal so there is nothing to warn the person preparing the food that the contents are infected.

Storage. Store canned and bottled foods in a cool, dry place. Canned foods may be expected to last at least two years, although they will almost certainly last for several years without deterioration. Canned food abandoned at a camp site in the Antarctic used by Shackleton in 1908 was found to be in perfect condition fifty years later!

2. Dehydration

This method of preservation has been used for centuries but is less common now as the popularity of more successful methods such as freezing has increased. It is still used commercially for large-scale drying of fruits (e.g. raisins, prunes) and pulses (e.g. peas). At home herbs can be dried successfully. Apple rings, onions, mushrooms and a few other foods can also be dried.

Principle of dehydration. Both enzymes and micro-organisms need moisture to survive. If water is evaporated from the cells in the food, the salt and sugar concentration is increased and through osmosis, water is taken from the cells of the micro-organisms. This destroys the moulds, yeasts and bacteria as they cannot survive without moisture. In some commercially dehydrated foods such as dried vegetables, the amount of liquid present in the food is low enough to prevent the growth of micro-organisms but not sufficiently low to prevent the action of enzymes. For this reason it is necessary to blanch these foods before drying them.

Domestic dehydration. Home drying can be carried out in an oven at a very low temperature, over a solid fuel cooker or even in a hot press. The food (e.g. apple rings) is placed on latticed wooden trays or strung on pieces of string and dried very slowly. Slow drying is essential to prevent the outside from hardening and perhaps bursting before the centre has dried out. All dried food should be cooled for 12 hours at room temperature and then packed in layers into airtight boxes. Greaseproof paper can be used to separate the layers.

Note. Before drying all food should be in peak condition, dry and freshly picked.

Dried herbs. Small-leaved herbs can be tied in bundles and hung up in a warm place such as a hot press for a few days. Other herbs should be blanched for 1 minute, shaken and left to dry at 120°F (48°C) approximately until crisp; this usually takes one hour. Crush the herbs and store in an opaque, airtight jar. Sunlight and air will cause deterioration of colour and flavour.

Note. Most dried foods are soaked in water (rehydrated) before cooking. Exceptions are raisins, sultanas etc.

Commercial Dehydration: Originally fruit and vegetables were dried in the sun in countries where the climate was suitable such as Mediterranean areas, California and Australia. It is more usual in large food processing plants to dry food in heated chambers or on a conveyor belt passing through a tunnel oven.

The food is prepared and chopped. Most vegetables and fruit are blanched to preserve colour, flavour and vitamin C content. Meat and fish are steamed before chopping or mincing. Foods are spread thinly on perforated trays and are then dried in special ovens which blow hot air over them until the moisture content is reduced to 5–10 per cent. There is considerable loss of vitamins A and C by this method but other nutrients remain virtually unchanged.

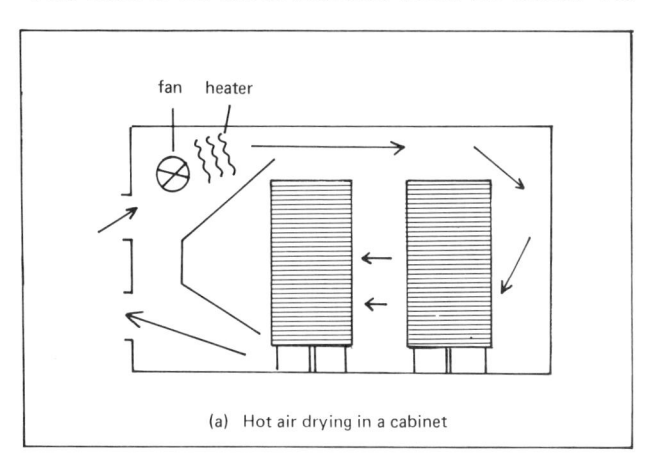

Accelerated freeze drying (AFD). This combines freezing and drying and utilises the principle of *sublimation*. The moisture is removed in the form of ice crystals which are changed from solid (ice) to vapour (steam) without passing through the liquid (water) state. The same thing happens on a frosty day when frozen washing on a line dries quickly without becoming wet again. Foods dried by this method (now more commonly known as freeze drying) are superior in flavour and nutritive value to those dried by the conventional heat drying process.

The food is prepared, thinly sliced or chopped, and blanched where necessary. It is spread on trays and frozen very quickly so that tiny ice crystals form in the food cells. The frozen food now enters a vacuum chamber where the ice crystals are quickly sublimated by using moderate heat. (Water in a vacuum evaporates below its normal boiling point.) Because the drying temperature is so low, the structure of the food remains unchanged, resulting in a honeycomb of cells with their moisture removed. The low moisture content (2–3 per cent) results in longer shelf life and a lightweight food which is quickly and easily reconstituted.

Freeze-dried foods are so light that they are easy to pack, transport and store. On the other hand, foods processed in this way are very fragile: being porous, they are readily affected by moisture and oxidation which can cause discolouration, off-flavours and rancidity. For this reason they must be carefully packed, often with nitrogen gas injected before sealing. Freeze drying is used to preserve some vegetables (e.g. peas); expensive fish (e.g. prawns); coffee; complete meals (e.g. curries, chow mein). Vitamins B and C are reduced by this process but to a lesser extent than with normal drying.

Puff drying. This is done by vacuum-drying the food until the moisture is reduced to 30 per cent. It is then subjected to heat under reduced pressure, which causes the moisture left in the food to expand. The expanded food is further dried until only 5 per cent of the original moisture remains. This results in a light, puffed product which reconstitutes very quickly. Many breakfast cereals are processed by this method.

Fig. 7.2 Drying
(a) Hot air drying in a cabinet
(b) Hot air drying in a tunnel oven

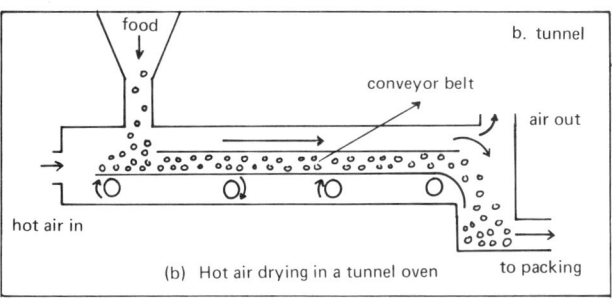

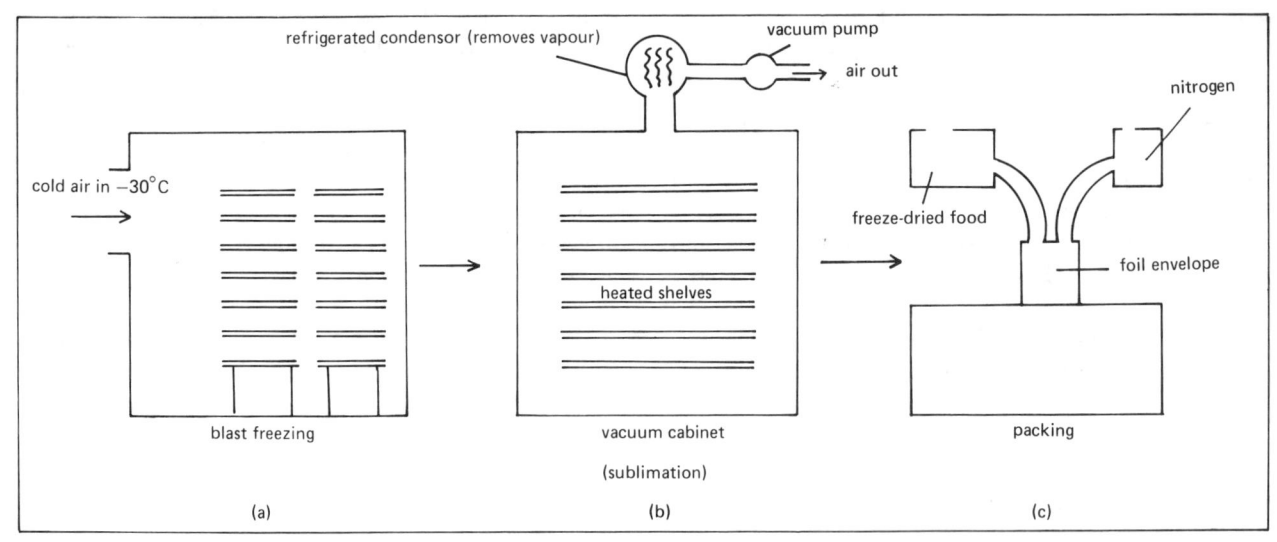

Fig. 7.3 Freeze drying

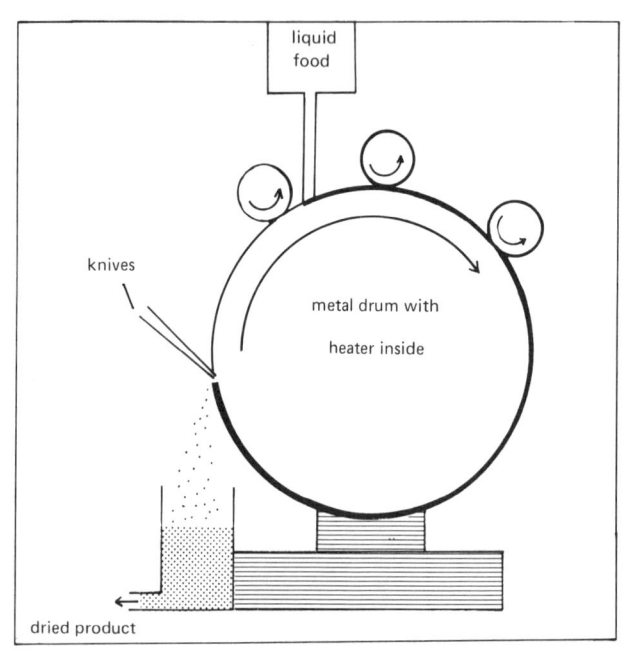

Fig. 7.4 Roller drying

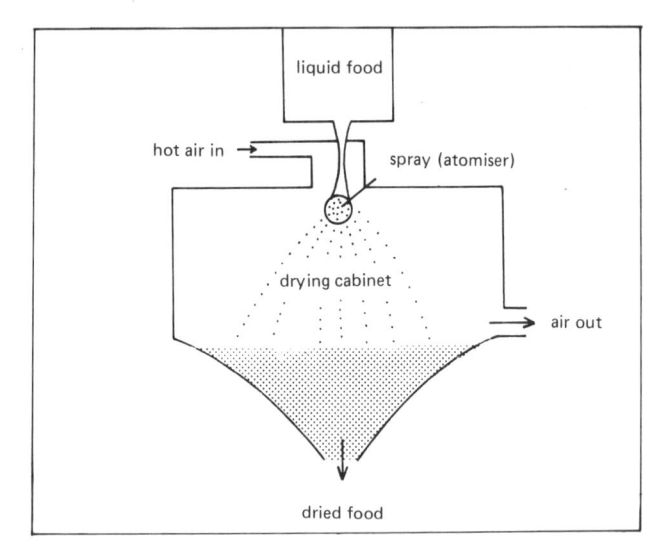

Fig. 7.5 Spray drying

Roller drying is one method of dehydrating liquids such as milk. Milk is run over a heated revolving drum, where it dries quickly and is scraped off as a powder.

Spray drying involves spraying milk into the top of a heated chamber. As the drops fall, they are quickly dried and fall to the bottom as a powder. Eggs can also be preserved by this method for use in bakeries, but because of the greater risk of bacterial contamination, they must first be pasteurised.

Both methods involve some loss of vitamin A, B and C, although roller drying causes a greater loss of all vitamins. Spray-dried milk is easier to reconstitute and superior in flavour.

3. Freezing

By this method the water present in the food is converted into ice crystals. Quick freezing at $-25°C$ to $-34°C$ produces very small ice crystals which cause less damage when the food is thawing. Slow freezing produces large crystals which distort and damage the cell walls, resulting in loss of flavour, appearance and nutritive value. Fluctuating temperatures also cause large crystal formation as the ice crystals melt and reform. Home freezing is carried out at temperatures of approximately $-25°C$. The frozen food is then stored at $-18°C$. Commercial freezing is carried out at $-30°C$ or lower using blast freezers. Enzyme activity is reduced to a minimum and micro-organisms are inactivated by freezing.

Commercial quick freezing. Most commercial freezing plants are built close to their source of supply – for example many factories which specialise in freezing fresh foods are surrounded by acres of fields growing fruit and vegetables, and factory ships catch and process huge quantities of fish on board. This ensures the absolute freshness of the product, which is frozen in the minimum time, and also means that the maximum amount of nutrients is retained.

Blast freezing. This method of freezing is used in factories and meat plants. The food is placed on trays on refrigerated shelves and cold air is blown over them at temperatures of −30°C. Food may also be quickly frozen by *immersion* in freezing brine; poultry is often frozen in this way. Some food is *spray frozen:* ice-cold brine or syrup is sprayed over unpackaged foods.

Cyrogenic freezing. This is the quickest method of freezing. Small pieces of food are sprayed with liquid nitrogen, an extremely cold substance. The food is frozen instantly and it retains its shape and appearance well. It is used to preserve expensive and out of season luxuries such as alpine strawberries and prawns.

Flo-freezing. Small vegetables and fruits such as peas and berries are moved along on a cushion of cold air. This prevents foods from sticking together.

4. Chemical Preservation

Chemicals have been used to preserve food for thousands of years. Most of these chemicals are found in an average kitchen cupboard. Salt, sugar and vinegar are traditionally used to preserve food as their presence inhibits the growth of bacteria.

Basic principle. Many chemicals work by dissolving in the water which is present in food, forming a concentrated solution. Water from the bacterial cell passes out by the process of osmosis in an attempt to equalise the concentration. This dehydrates the cell and as bacteria cannot survive without water they are eventually destroyed. Some preservatives which create conditions unfavourable to microbial growth (e.g. a low pH) are used; others are toxic to bacteria (e.g. antibiotics such as nisin).

Traditional preservatives. Because these have been used for centuries they are considered to be above suspicion and are not subjected to the stringent tests which modern preservatives must undergo.

Salt. Salted or cured meat was more common before the advent of refrigeration. Up to the early part of this century it was common to slaughter large numbers of animals before the winter and salt them to provide meat for the few months when food stocks were low. Bacon, ham and corned beef are still preserved by salting (see p. 54), although with modern refrigeration the salt is used less as a preservative than a method of flavouring. Sodium chloride (salt) and potassium nitrate (saltpetre) are used in curing bacon and corned beef.

Sugar. Bacteria cannot survive in the high level of sugar concentration in jam and other preserves. A 50 per cent sugar solution prevents bacterial action and a 65–70 per cent concentration inhibits mould and yeast formation. If too little sugar is used, fermentation will occur, while too much causes crystallisation. Crystallised fruit is also preserved by saturation with sugar.

Vinegar. Vinegar contains acetic acid which lowers the pH

Millions of peas move along a freezing tunnel (Bird's Eye Foods Ltd.)

Natural preservatives

| salt | sugar | vinegar | alcohol | smoking |

Permitted artificial preservatives

sulphur dioxide   phenyls

nisin   potassium salts

benzoic acid   scorbic acid

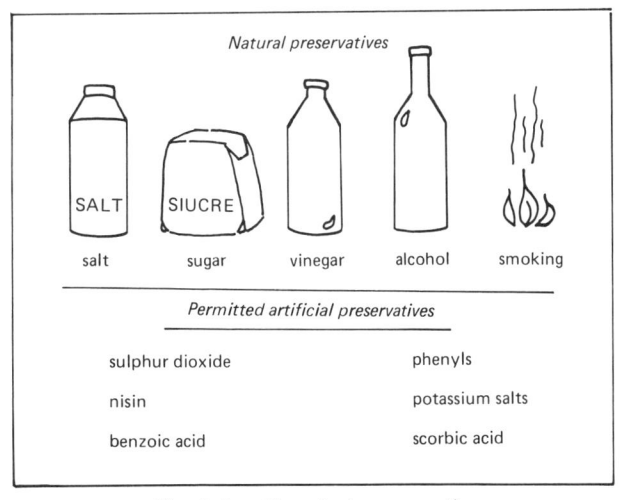

Fig. 7.6 Chemical preservatives

of the food to 2.7, a level in which bacteria cannot survive. Pickles, chutneys and ketchup contain high concentrations of salt, sugar and vinegar.

Alcohol. Alcohol contains ethanol, a substance which denatures proteins, preventing the enzymes in the bacterial cell from carrying out their function – in effect destroying it.

Smoking. When food, usually salt meat or fish, is subjected to the fumes from smoking wood, the outer layer of the food becomes covered with a layer of concentrated tars, phenols and aldehydes, all of which prevent bacterial growth (see p. 67).

Permitted preservatives. As many preservatives have toxic effects, most countries have drawn up a list of permitted preservatives. Health regulations in Ireland permit the use of certain substances which 'inhibit, retard or arrest the deterioration of food caused by micro-organisms' (see also pp. 117).

The most commonly used preservative is *sulphur dioxide* which is used in sausages, dried fruit and vegetables, pickles and fruit juice. It inhibits the growth of moulds, yeasts and aerobic bacteria. Sulphur dioxide is available in tablet form for the home preservation of stone fruits and apples.

Sulphur dioxide is also used commercially to preserve fruits for a limited period until they can be made into jam – for if jam factories only produced jam when fruits were in season, they would be closed for six months of the year. After picking, fruit is packed into plastic barrels which are filled with a sulphur dioxide solution. During subsequent processing, the sulphur dioxide evaporates until only traces remain in the jam.

Other permitted preservatives include *benzoic acid* and *scorbic acid,* which lower the pH of the food to a level unfavourable to the growth of bacteria. Phenyls, antibiotics and antioxidants may also be used in certain foods.

Nisin, an antibiotic used as a preservative in milk and cheese, is one of the few preservatives whose use is permitted in cheese and milk products.

Radiation. Research on radiation as a form of preserving food is still being carried out. When X-rays penetrate food, they are absorbed by the cells and destroy micro-organisms. This method is sometimes used on cereals and vegetables during storage as it prevents sprouting and destroys insects, but as there is considerable risk attached to its use, radiation is restricted at present.

HOME PRESERVING

JAM MAKING

Making jam is a useful and economical way of using up fruit when it is plentiful. A good jam should have a clear, bright colour, a good flavour of fruit, be well set and should keep for one year. It is essential that the jam should have a 65 per cent sugar content if it is to keep well; less than this will allow the growth of yeasts and moulds.

Pectin. The setting of jam depends on this gum-like polysaccharide which is found in the cell walls of fruit. It is present in unripe fruit in an insoluble form called pectose. On ripening the pectose changes to pectin, which is water-soluble. Pectin is released from the cell walls when fruit is stewed. If the correct amounts of sugar and acid are added the jam will set. Pectin changes to *pectic acid* as the fruit becomes over-ripe. As pectic acid has no setting properties over-ripe fruit is unsuitable for jam making.

Acid. Acid is necessary to assist in the extraction of pectin and the formation of a gel. It also improves the flavour of the jam and prevents crystallisation. The amount of acid used in a recipe will depend on the amount present in the fruit. Strawberries, blackberries and some sweet apples are deficient in acid, so it may be added to these fruits as follows:

To each 2 kg fruit   lemon juice – 2 tablespoons
(1 small lemon)
or
citric or tartaric acid – ¼ teaspoon

Note. Acid should be added during the initial cooking of the fruit to assist in the extraction of pectin.

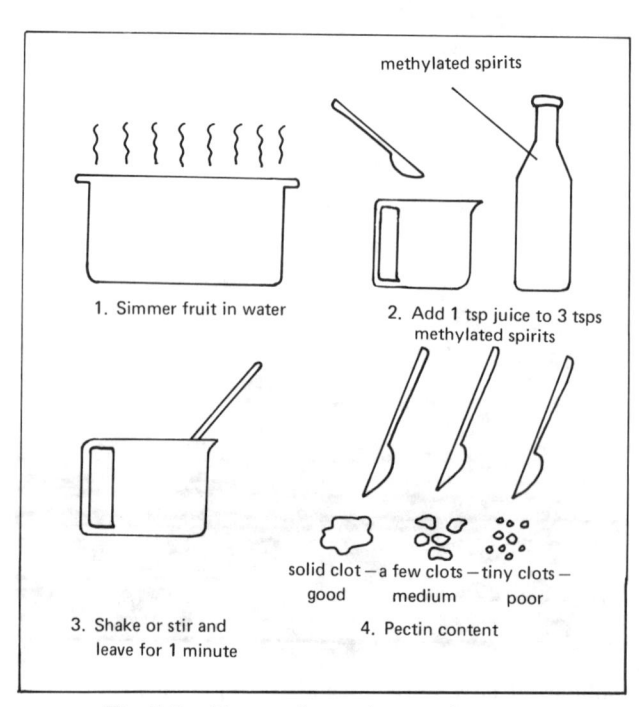

Fig. 7.7 Test to determine pectin content

Fruits rich in pectin and acid: apples, blackcurrants, green gooseberries.

Fruits with a medium amount of pectin and acid: raspberries, plums, apricots, early blackberries.

Fruits with little pectin and acid: cherries, pears, strawberries, late blackberries.

Acid and commercial pectin (available in powder and liquid form) may be added to these fruits to assist in the formation of a gel.

A mixture of fruits rich in pectin with those lacking in pectin will make a successful jam. So, too, will a mixture of ripe and unripe fruit. *Never use over-ripe fruit.*

To test for pectin. Stew a little of the fruit to be tested in a small amount of water until the skins are soft. Cool slightly. Put 1 teaspoon of the fruit juice into a screwtop jar; add 3 teaspoons of *methylated spirits*. Shake and leave for 1 minute.

Results

Fruits rich in pectin – a single firm clot.

Fruits with average pectin – a soft clot or two or three firm clots.

Fruits poor in pectin – many small pieces of jelly.

Water. The amount of water required depends on how juicy the fruit is. Soft fruits need no water, but otherwise enough water must be used to soften the fruit and extract the pectin without burning the jam. A wide topped preserving pan allows evaporation to take place more quickly than an average sized saucepan. It also means that the jam can be boiled quickly after the sugar is added.

Sugar. Many errors can occur in jam making due to carelessness in the weighing and treatment of the sugar in the recipe.
1. It is essential to use a reliable recipe which has at least 60–65 per cent sugar content. Too little sugar will cause fermentation and mould formation, while too much will cause crystallisation.
2. The fruit must be completely softened before the sugar is added; sugar has a hardening effect on the skins of fruit.
3. The jam must be stirred until the sugar is dissolved.
4. The sugar should be completely dissolved before the jam is reboiled. Failure to do this causes crystallisation.

Rules for Jam/Jelly Making
1. *Fruit.* Sound, dry fruit must be used. Do not use over-ripe fruit, as the jam will not set. Avoid diseased or bruised fruit, as it will not keep. Dried fruit (e.g. apricots) or tinned or frozen fruit (e.g. oranges) may be used.
2. *Preparation.* Food is prepared as for cooking. Wash, peel, remove stones and chop if necessary. Pick over soft fruits. Skins and cores may be used for jellies; pips and pith for marmalade.
3. *Containers.* Use a large aluminium or stainless steel preserving pan or saucepan. It should have a thick machine-ground base to make full use of heat and prevent burning. It should be large enough to allow the fruit and sugar boil vigorously. Grease the base to prevent the jam from sticking. Avoid chipped enamel pans or copper or brass containers as these reduce the vitamin C content of the jam.
4. *Cooking.* Simmer fruit gently to soften the skins and extract the pectin before adding sugar.
5. *Add warmed sugar.* Stir until dissolved and dissolve thoroughly before boiling. Boil quickly until setting point is reached.
6. *Test for setting* (see below). Jams with large pieces of fruit (e.g. marmalade, strawberry) should be cooled for 10 minutes to prevent the fruit from rising in the jars.
7. *Jars.* These should be thoroughly washed and checked for flaws. The jars should then be sterilised in a warm oven for 15 minutes to destroy micro-organisms and prevent the jars from cracking when filled with hot jam.
8. *Finishing.* Stir the jam and pour into heated pots, filling them to within 1.5 cm of the top. Wipe rims and cover jam at once with waxed discs and cellophane covers, which should be stretched tightly to prevent entry of air. Label and date.
9. *Storing.* Store in a cool, dark, dry place to preserve colour, flavour and quality.

Tests for Setting
a. *Cold plate:* Put 1 teaspoon of jam on a cold plate and cool. The surface of the jam will wrinkle when pushed with the finger if setting point has been reached.
b. *Temperature:* A warmed sugar thermometer placed in the jam should register 104°C when the correct sugar concentration has been reached.
c. *Flake test:* When the jam falls from the edge of a clean wooden spoon in a broad flake instead of a continuous stream, the jam will set.

Faults in Jam Making
Fermentation
1. Insufficient boiling.
2. Not enough sugar used.

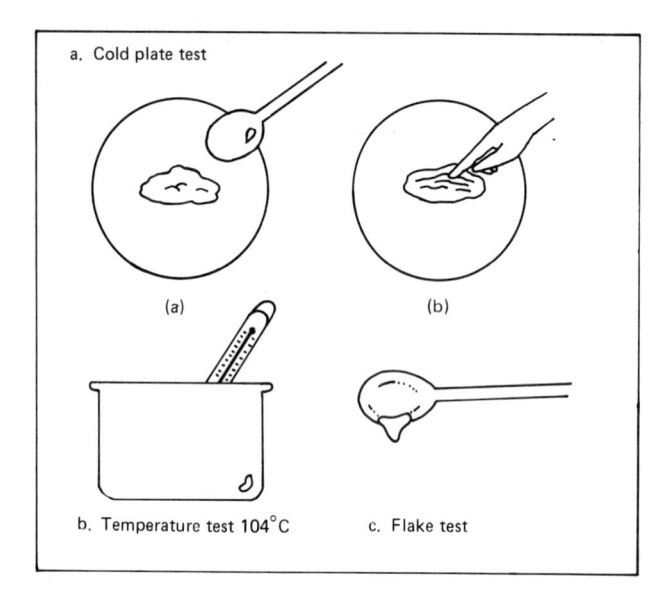

Fig. 7.8 Tests for setting

Crystallisation
1. Too much sugar.
2. Not enough acid.
3. Prolonged cooking resulting in over-evaporation.

Jam too thick and sticky
1. Over-cooking.
2. Too little water used.

Jam not set
1. Not boiled for long enough.
2. Insufficient pectin or acid. Unset jam may be corrected by (a) reboiling or (b) adding pectin or lemon juice and reboiling for a short time.

Mould Growth
1. Incorrectly sealed.
2. Incorrectly stored.
3. Too little sugar.
4. Bruised, wet or damaged fruit used.

Shrinkage
Incorrectly sealed or stored.

Strawberry Jam

    1.75 kg strawberries
    juice of 2 lemons
    1.5 kg sugar

1. Prepare fruit, halve strawberries.
2. Heat gently in preserving pan with the lemon juice.
3. Add warmed sugar and stir until it is dissolved.
4. Boil rapidly for approximately 20 minutes. Test.
5. Cool for 10 minutes. Stir, pot, cover and label.
Yield: 5 × 1 lb jars.

Gooseberry Jam

    2 kg green gooseberries
    1 litre water
    2.75 kg sugar

1. Wash, top and tail gooseberries. Halve large fruit.
2. Simmer with water for approximately 40 minutes.
3. Add warmed sugar and stir over gentle heat until dissolved.
4. Boil rapidly for approximately 20 minutes. Test.
5. Cool for 10 minutes. Stir, pot, cover and label.
Yield: 10 lbs

Dried Apricot Jam

    900 g dried apricots
    3.4 litres water
    juice of 2 large lemons
    2.5 kg sugar

1. Wash fruit thoroughly. Soak for 24 hours in measured water.
2. Add lemon juice and simmer for approximately 40 minutes.
3. Add warmed sugar. Dissolve sugar slowly.
4. Boil rapidly until setting point is reached.
5. Cool, stir, pot, cover and label.
Yield: 10 lbs

Apple Jelly

    unripe apples and windfalls
    water to cover
    800 g sugar to each litre of juice
      (see method)
    1 or 2 lemons

1. Wash apples. Remove damaged and bruised parts.
2. Cut in two or four if large. Do not remove skin and cores.
3. Put in greased preserving pan and cover with water.
4. Add thinly pared lemon rind and juice. Bring to the boil.
5. Simmer until soft — 45 minutes to 1 hour.
6. Strain juice through a scalded cloth or jelly bag for a few hours.
7. Add 800 g sugar to each litre of juice. Stir until dissolved.
8. Boil rapidly until setting point — 20 minutes approximately.
9. Pour quickly into hot jars. Cover and label.

Marmalade (normal method)

    1 kg Seville oranges
    1 lemon
    2.5 litres water
    2 kg sugar

1. Scald fruit. Peel, remove some pith if wished, shred peel.
2. Chop flesh roughly. Tie pips and pith in muslin.

3. Put fruit, water, peel and pips in bowl and steep for 12 hours. (This speeds up the softening of the fruit but is not essential.)
4. Simmer in preserving pan for 1½ to 2 hours.
5. Add warmed sugar and stir until dissolved.
6. Boil rapidly until settling point is reached.
7. Stand for 10–15 minutes. Stir, pot, cover and label.

Marmalade (pressure cooker)
>1 kg Seville oranges
>1 lemon
>1.25 litres water
>2 kg sugar

1. Prepare fruit as for normal marmalade.
2. Put pulp, rind, pips tied in muslin and half water into pressure cooker.
3. Pressure cook for 10 minutes at 10 lb pressure.
4. Reduce pressure at room temperature.
5. Remove pips, add remaining water.
6. Add warmed sugar, changing to larger saucepan if necessary.
7. Finish as for normal marmalade.

Use of Pressure Cooker in Jam Making

A pressure cooker is useful in jam making when it is necessary to soften the fruit before adding the sugar — as when making blackcurrant or gooseberry jam or marmalade. Once the sugar is added, the lid is not used; in fact it is often necessary to transfer the jam to a larger saucepan at this stage.

Note. Follow the instructions and recipes supplied with the pressure cooker. Other recipes may be used but the amount of water should be halved as little is lost by evaporation.

Rules for making jam in a pressure cooker
1. The trivet is not used.
2. The base of the cooker should not be more than half full when the fruit and water are added. If necessary, reduce the amount of water and add it later with the sugar.
3. Jam should be cooked at 10 lb pressure.
4. Allow pressure to reduce at room temperature.
5. Do not attempt to open the lid until the pressure is reduced.

Advantages of using a pressure cooker for preserving
1. Pressure cookers soften fruit and vegetables quickly.
2. Steam and odours are reduced.
3. Fruit bottling is quicker.

BOTTLING

Principle of Bottling

This method of preserving involves sterilising *the food and the container at the same time*. The heat should be high enough, and maintained for long enough, to destroy micro-organisms and enzymes. (Bacteria are not usually found in fruit because of its high acidity.) If the temperature is too high, the food will go mushy, but if it is too low, the contents of the jar will not keep and in certain cases may be toxic. To avoid recontamination of the food, the jars must be sealed while hot in order to create a vacuum.

Special Equipment
1. A large saucepan or pressure cooker.
2. A false base such as latticed wood or cardboard to prevent the jars from touching the base of the saucepan. The trivet is used in the pressure cooker.
3. A sugar thermometer.
4. Vacuum bottles.

Bottles. There are two types of glass vacuum jar available for bottling — clip-top bottles and screwband bottles. They can be bought in many sizes. *Clip bottles* have a glass or lacquered metal lid, a rubber ring which helps to create an airtight seal, and a strong metal clip. This spring clip allows the steam to escape during processing. *Screwband bottles* also have a glass or lacquered metal lid with either a separate rubber ring or a thin rubber rim fitted to the inside of the metal lid. The lid with the fitted rim may be only used once. Both types have a metal screw band which is screwed on loosely during processing and tightened immediately afterwards. It should be further tightened during cooling if necessary.

Examination

It is essential to check bottles and lids before processing.
1. Rubber rings must be sound and should be soaked in warm water for 10 minutes before use.
2. Metal lids should show no signs of rusting or loss of shape. Rings and lids are scalded just before use.

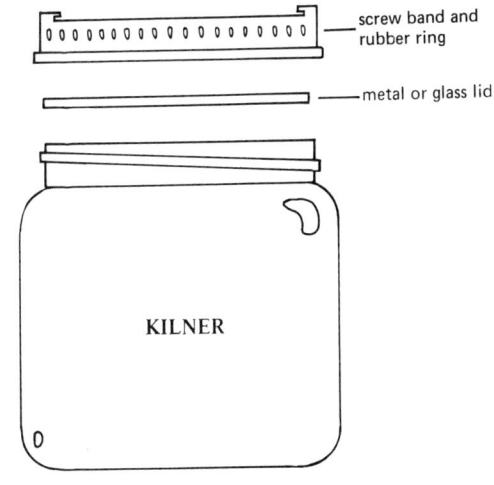

Fig. 7.9 Preserving jar

3. Jars should be examined for cracks or chips. Even a small chip on the rim of the jar will prevent an air tight seal.
4. Bottles should be washed and rinsed in hot water before use. It is not necessary to dry them as it is easier to pack fruit into wet jars.

Jam jars. These may be used for bottling fruit, chutneys or pickles but a special synthetic skin cover or waxed cloth must be used. The cellophane covers which are used in jam making are not suitable, as they allow evaporation of the contents.

Fruit Bottling

1. Fruit should be just ripe, sound and dry, and be of even size. It should be as fresh as possible and have a good colour.

2. Preparation (as for normal cooking). *Hard fruit* (apples, pears): wash, peel, core, leave in halves or quarters. To prevent browning put in salt water and rinse before use. *Soft fruit:* pick over, wash gently if necessary, drain. *Stoned fruit* (plums, damsons, cherries): remove stalks, wash. *Rhubarb:* trim, wash, cut in 3–6 cm lengths. *Gooseberries:* Top and tail, wash, prick skins to prevent bursting. *Tomatoes:* Skin and pack dry in bottles with 2 tsp salt and 1 tsp sugar sprinkled between layers of each kg fruit.

3. Syrup or water. Both liquids may be used in fruit bottling. Syrup gives a better appearance and flavour but causes the fruit to rise in the jars. Strength of syrup may vary according to taste. An average proportion would be *200 g sugar to 500 ml water.* Method: put water in a saucepan, add sugar and stir until dissolved. Boil for 1 minute. 500 ml syrup will be sufficient to pack 2 x 1 kg jars.

4. Packing
(i) Pack fruit closely to the top of the jar, shaking and using the handle of a wooden spoon to ease fruit gently into position. Avoid damaging or crushing the fruit. Fill with hot syrup to 3 mm from top.
Soft fruit: add fruit and liquid alternately to allow penetration of fruit by the liquid and prevent air pockets. Turn jar and use a wooden spoon to ease out air bubbles as these may contain micro-organisms.
(ii) Cover carefully with rubber ring, lid and clip or screw top. Unscrew screwtop jar by a quarter turn to allow for expansion and escape of steam.

Dry pack: Fruits such as strawberries, raspberries and blackberries can be rolled in caster sugar and packed dry. There will be some shrinkage by this method.

Pack jars closely and fill with hot syrup almost to the brim (Bernard Alfieri)

Bottling: prepare fruit as for normal cooking (Bernard Alfieri)

Saucepan method: place jars in pan of warm water and heat slowly to simmering point (Bernard Alfieri)

5. Sterilisation

	Oven	Saucepan	Pressure Cooker
	Preheat oven to 150°C (300°F) Gas 1. Stand bottles at least 50 mm apart on a large tin containing 10 mm water. Lids are laid loosely on top of jars after filling. *Process 2 kg fruit or under for shorter time shown below. Process up to 6 kg for the longer time.	Stand covered jars in a saucepan of water heated to 38°C (100°F). Jars should not be touching and a false bottom should be used on the saucepan. Cover jars with more warm water and heat slowly to simmering point – 85°C (185°F). Maintain as below.	Place trivet in cooker. Add 3 cm warm water. Warm jars before filling. Put filled jars into cooker. Put on lid, heat until steady flow of steam appears. Put on weight and bring to 5 lb pressure. Maintain as below. Reduce pressure for 10 minutes at room temperature.
Fruit	Boiling Liquid*	Hot Liquid	Boiling Liquid
a. Soft fruits (e.g. blackberries, raspberries, strawberries)	35–60 mins	5 mins	1 min
b. Apple slices	30–60 mins	2 mins	1 min
c. Gooseberries Rhubarb	40–60 mins	10 mins	1 min
d. Stone fruits (e.g. peaches, plums)	55–75 mins	20 mins	3–4 mins
e. Pears	65–85 mins	40 mins	5 mins
f. Tomatoes	80–100 mins	50 mins	15 mins
Finishing	Wipe outside of jar. Put on clips/screwbands. Test in 12–24 hours.	Using a jug empty out some water. Lift out bottles on to wooden surface. Tighten screwbands. Test in 12–24 hours.	Lift pressure cooker carefully. Open when pressure has reduced. Tighten screwbands. Test in 12–24 hours.

6. *Testing.* When bottles are quite *cold* remove clips or screw-bands and lift carefully by the lid. If the lid comes off, check the cause of the fault e.g. perished rubber ring, chip on rim of jar or a piece of fruit lodged in the rim of the jar. The food must be used up quickly or reprocessed. Do not test synthetic skin (Porosan) in this way; when tapped the lids should give a taut ringing sound.

7. *Labelling and Storing.* Label all jars with the name of the fruit, date and method used. Store in a cool dry dark place. Wash, dry and rub oil into screwbands and clips to prevent rusting. Open bottles by inserting a knife carefully under the rubber ring and levering up.

Fruit Purées

Fruits such as apples and tomatoes may be preserved more compactly in pulp form. Imperfect fruit may be used and it is quicker than the normal method of bottling fruit.

Method. Prepare as for cooking. Simmer in very little water until soft and thick. Put through a nylon sieve if a smooth purée is required. Sugar may be added at this stage according to taste. Return to saucepan, bring to the boil and *while boiling* pour into heated jars. Rubber rings and lids, which have been scalded just before use, should be *quickly* applied and clip or screw-bands put on. Immerse in warm water (saucepan method), bring to the boil and process for 5 minutes. Cool and test in 12 hours.

Vegetable Bottling

As many fatalities have occurred in countries where home bottling and canning of vegetables is practised it is now considered unwise to preserve vegetables by these methods. More reliable methods such as freezing and pickling should be used. Most vegetables are contaminated with thermophillic (heat resisting) soil bacteria, and a temperature of 240°F (117°C) is required in order to destroy their spores efficiently. Spores can survive several hours of boiling. Where acid is present, however the heat resistance of these bacteria is reduced. This is why fruit (which is strongly acid) can be satisfactorily preserved at the temperature of boiling water.

Other Preserving Recipes

Spiced Vinegar (for pickles)

    1 litre vinegar
    ½ tsp. cinnamon bark
    1 tsp. cloves
    1 tsp. whole allspice
    1 blade mace
    1 tsp. peppercorns

Method 1. Put vinegar and spices in bottle. Leave for 1–2 months, shaking now and then. Strain through double muslin.

Method 2. Put ingredients in a bowl which is placed in a saucepan of water. Bring to boiling point. Remove from heat. Stand for at least one hour. Strain. Use hot or cold.

Mixed Pickle

1 small cauliflower
500 g–1 kg pickling onions
1 small cucumber
500 g French beans or green tomatoes

Method. Peel onions, wash and cut other vegetables into large dice. Sprinkle with salt and leave for two days. Wash and break cauliflower into small florets. Steep in brine (50 g salt – 500 ml water). Cover and leave for 24 hours. Wash all vegetables thoroughly. Pack into sterilised jars, cover with cold spiced vinegar and seal with waxed or synthetic skin tops. Do not use for 2–3 months to allow flavour to develop.

Piccalilli (Mustard Pickle)

1 cauliflower
1.5 kg vegetable marrow
300 g onions
1 small cucumber
1 litre vinegar
150 g white sugar
1 tsp. turmeric
2 tsp. dry mustard
2 tsp. ground ginger
15 g cornflour
350 g salt

Method. Prepare and cut vegetables into small pieces. Place in a bowl, sprinkling salt between the layers. Leave for 12–48 hours. Blend flour and spices in 3–4 tablespoons vinegar. Rinse vegetables and drain. Simmer for 20 minutes with the sugar and vinegar. Stir in blended spices and boil for 3–5 minutes. Fill into sterilised jars. Cover with waxed cloth or synthetic skins. Do not use for 2–3 months to allow flavour to develop.

Apple Chutney

2 kg cooking apples
300 g sultanas
2 large onions
2 tsp. salt
200 g brown sugar
½ tsp. mixed spice
½ tsp. ground ginger
¼ tsp. cayenne
500 ml brown vinegar

Chutney can be made from other fruit such as gooseberries, blackberries, rhubarb and red or green tomatoes. Substitute any of these fruits for the apples in the above recipe.

Method. Wash, peel, core and chop apples into large dice. Peel and chop onion. Stew all ingredients gently for 1–2 hours, until thick and jam-like in consistency. Stir regularly as chutney is likely to stick to the saucepan and burn.

Pour into heated clean jars. Cover with waxed cloth or synthetic skin covers. If screwtops are used, avoid letting the chutney come in contact with the metal. Do not use for 3 months to allow chutney to mellow and mature.

Salted Beans

175 g salt to 500 g French beans

Method. Wash, string and slice beans as for cooking. Pack beans and salt into deep bowls or an earthenware jar in alternate layers. Cover top with a thick layer of salt. As contents soften and shrink, top up with more beans and salt. Pack down, tighten and cover with an airtight plastic cover.

To use: wash thoroughly. Soak for 1–2 hours in warm water. Cook until soft, approximately 30 minutes.

FREEZING

Freezing is the simplest and most successful method of home preservation. When handled correctly, frozen food retains its appearance, flavour, colour and, in most cases, texture. It involves minimal reduction of food value and is quick and simple to do. A far greater variety of foods can be frozen than can be preserved by other methods. The only disadvantages are the initial cost of buying a freezer and the subsequent running costs. These are gradually offset by the savings made by bulk buying, freezing garden produce and eliminating waste.

Foods to Freeze

Almost all foods are suitable for deep freezing:
Fresh fruit and vegetables in whole and puréed form.
Raw and cooked meat and fish.
Reheated dishes (e.g. fish pies, croquettes, meat cakes).
Complete meals on a foil tray.
Baking (e.g. bread, cakes).
Partly prepared dishes (e.g. dough, pastry).
Sauces, puddings, cold sweets.
Advance cooking for parties and Christmas entertaining.
Prepared baby foods.
Breadcrumbs, stuffing etc.
Sandwiches and packed lunches.
Bulk buys of butter, ice cream, cheese.

Do Not Freeze:

1. Bananas – they blacken.
2. Lettuce and other salad greens – they go limp.
3. Whole tomatoes. Freeze them puréed instead.
4. Whole melon. It is best to cut it up and freeze it in syrup.
5. Milk, cream and plain yoghurt – they separate. Fruit yoghurt and whipped cream freeze reasonably well.
6. Whole eggs (but it is possible to freeze separated egg white and yolk).
7. Mayonnaise and baked custard – they separate.

8. Whole potatoes do not freeze well. Mashed potatoes or partly cooked chips can be frozen.
9. Jelly and dishes with a high gelatine content.

Preparation for Freezing
1. Only freeze best quality foods; fruit and vegetables should be in peak condition. Freezing inferior produce is a waste of time and space.
2. Make sure meat, poultry and fish are absolutely fresh and not already frozen and thawed out.
3. Freeze in reasonably small quantities. Never freeze more than one-tenth of the capacity of the freezer in 24 hours. More than this will raise the temperature of the freezer, slow down the freezing process and result in badly frozen food.
4. Freeze in meal-sized portions. It is better to freeze in small amounts as these are quicker to freeze, thaw and reheat than larger portions.

Packaging
All packaging for deep freezing must be air-tight, water and vapour proof. Failure to pack food correctly will result in loss of colour, flavour, texture and in some cases 'freezer burn' will occur. This causes toughening and discolouration of foods, particularly protein foods such as meat and fish; the 'burn' appears as a greyish patch. Suitable packaging materials can be bought in supermarkets and freezer centres, and most can be washed and re-used. Glass containers must not be used, as they crack.

Rules for Packaging
1. Use moisture and vapour-proof containers (see above).
2. Pack foods tightly and expel all air. (A drinking straw is handy for drawing out air.)
3. Seal tightly with wire ties or a warm iron.
4. Leave 10–20 mm headspace over all liquid foods to allow for expansion.
5. Liquid foods may be 'preformed' if polythene bags are placed in straight-sided boxes before the liquid is poured into them. Square and rectangular shapes save space.
6. Overwrap sharp bones with foil.

Suitable containers

Polythene boxes with well fitting lids. These are expensive but long-lasting.

Polythene bags, heavy guage, with gussets. These are sealed with plastic and wire 'ties' or a warm iron.

Foil containers, which come in various shapes and sizes. Lids are available, or they can be covered with aluminium foil. They can be put directly from the freezer into the oven.

Aluminium foil, heavy-duty, is useful for overwrapping small items. It is also used to line casserole dishes when preparing stews etc. for freezing. When frozen, the contents can be removed in the foil and overwrapped, leaving the casserole free for other uses.

Waxed cartons, tubs. Used cream, yoghurt, margarine and other plastic food containers are useful for freezing small quantities. Use tinfoil to cover if there is no lid.

Waxed paper (used to hold cereals) or polythene squares may be used to separate sets of chops, fish cutlets, rissoles and such like. Foil is not suitable for this.

Freezer tape is necessary to seal containers.

Labels. Label all frozen food clearly with weight, contents, date etc. Use a crayon or special freezer marker as most other writing materials are indecipherable after a time.

Keep a record of the contents of the freezer: the food, date frozen, amount and any extra information necessary about thawing or reheating. A notebook or chart can be used and kept near the freezer. As each item is removed, it should be ticked off. Use foods in rotation – first in first out.

Shopping for the Freezer
1. Buy frozen food items last thing on shopping expeditions.
2. Keep frozen foods together and, if possible, wrap them in several layers of paper.
3. Pack in the centre of other goods to insulate them or use special insulated bags.
4. On returning home place them immediately in the freezer.

Packaging for home freezing (National Magazine Co.)

Foods well positioned in the freezing compartment (Syndication International)

Blanching Times for Vegetables

Asparagus	2–4 mins
Aubergines, sliced	7–10 mins
Beans, French, broad	3 mins
Beetroot, small, whole	5–10 mins
Brussels sprouts	3–4 mins
Cabbage, young	1½ mins
Carrots	3–5 mins
Cauliflower in florets	3 mins
Celery	3 mins
Corn on the cob, medium	6 mins
Courgettes, sliced	1 min
Mushrooms, sautéed in butter	1 min
Onions, small	4 mins
Parsnips, turnips	2–3 mins
Peas	1 min
Peppers	3 mins
Potatoes	cook fully
Spinach	2 mins

Freezing Chart

FOOD	STORAGE	PREPARATION
Beef	8–12 months	Bone, trim and joint.
Lamb, veal, pork	6 months	Bone, trim and joint.
Minced meat	3 months	Pack in 200 or 400 g packs.
Offal	3 months	Wash thoroughly, pack in small quantities. Freeze quickly.
Sausages	3 months	Open-freeze, then place in polythene bags.
Bacon	1–2 months	Will keep better if vacuum sealed.
Stews etc.	2 months	Freeze, thaw and reheat in foil dishes.
Pâté	1 month	Freeze in small quantities.
Chicken	12 months	Must be very fresh. Remove giblets. Do not stuff.
Other poultry	6 months	As chicken.
Fish, oily	4 months	Must be freshly caught. Separate fillets with waxed paper.
Fish, white	6 months	As oily fish.
Fish, shell	2 months	Advisable to buy ready-frozen.
Fish, pies etc.	2 months	Use foil dishes.
Stock, soup, sauces	3 months	Freeze in 250, 500 ml drums, cartons. Allow head space.
Pastry, raw	3 months	Roll out to required size if possible.
Pastry cooked	6 months	Fragile; wrap and store carefully.
Reheats (e.g. fish cakes)	2 months	Use foil dishes.
Pies, savouries	3 months	Use foil dishes.
Cakes	3–6 months	Cut down on spices and essences. Pack when cold.
Bread	1 month	Wrap in polythene.
Fruit, vegetables	up to 12 months	Pack fruit in polythene bags. Blanch vegetables before packing.
Eggs, separated	6 months	Note number of whites and yolks and label accordingly.

Rules for Freezing

1. Chill all food before freezing to avoid excessive rise in the freezer temperature.
2. Turn freezer to coldest setting 2–3 hours before freezing. Most freezers have a 'fast freeze' switch which reduces the temperature from $-18°C$ to $-25°C$.
3. Place food in the special freezing compartment or shelf with the packages touching the base or sides if possible. Store similar foods together. Allow circulation space between the packages.
4. Open-freeze foods which are likely to stick together (e.g. peas, soft fruits, sprouts, sausages) spread out on flat trays and place in polythene bags as soon as they are frozen.
5. Frozen foods should be used within 12 months if not sooner (see chart). After that time they deteriorate in colour, flavour and nutritive value.
6. Fatty foods such as pork, bacon, and oily fish tend to become rancid and must be used up more quickly.
7. Bought frozen foods can be stored for up to 3 months. This makes allowances for time spent in the factory and the shop.

Thawing

1. Never refreeze frozen foods which have thawed.
2. Leave food in its container while thawing.
3. Many dishes may be cooked straight from the freezer (e.g. stews, meat pies, chops, steaks, fish fillets), but be sure they are *thoroughly* cooked.
4. Vegetables should always be cooked from frozen to preserve nutrients.
5. Joints of meat should be thawed before cooking unless a meat thermometer is used to ensure the joint is cooked through.
6. Poultry must always be fully thawed before cooking.
7. Fruit should be thawed slowly in the refrigerator.
8. All food, once thawed, should be used up quickly, as bacterial action proceeds more quickly in soft food.
9. Most food should be reheated quickly, either in the oven or in a double saucepan.
10. Frozen food can be thawed, *cooked* and refrozen – for example, frozen beef can be thawed, made into a meat pie, baked and refrozen.

Special Points

Meat
1. Ensure that it is well hung prior to freezing (beef 7–10 days).
2. Joint into suitably sized pieces.
3. Trim fat to avoid rancidity, and either bone completely or wrap any remaining bones in foil.
4. Freeze at lowest possible temperature to avoid excessive drip-loss.

Poultry
1. Chickens should be frozen *unstuffed*.
2. Giblets are frozen separately.

Fruit
1. Prepare as for cooking. Pack in polythene bags and sprinkle sugar between layers if wished.
2. Soft fruits can be open-frozen until they are hard and then quickly packed into polythene bags. This method ensures that the fruit does not stick together, and small amounts can easily be removed when required.
3. Fruit may also be frozen in a syrup (300 g sugar/500 ml water) or as a purée.

Vegetables
1. Prepare vegetables as for cooking. Carrots and parsnips are better diced after blanching.
2. It is essential to blanch vegetables before freezing in order to destroy the destructive enzymes. Use boiling salted water (1 tsp. salt/500 ml water).
3. Open freeze if required and pack in polythene bags or waxed cartons.

Fish
1. Should be frozen within 12 hours of catching.
2. Shellfish should be blast frozen $(-35°C)$
3. Prepare fish as for cooking, i.e. remove head, gut and bone. Divide into fillets or cutlets.

Stews and casseroles
1. Prepare as usual. Avoid using garlic as it tends to develop a musty taste.
2. Avoid overseasoning as freezing intensifies flavours.
3. Consistency of sauce should be thin as it thickens on freezing. Make enough sauce to cover meat and vegetables.
4. Shorten cooking time by 20–30 minutes, as cooking will be completed in the reheating time.
5. Make sure the stew is cold before transferring into tinfoil containers or foil-lined casserole.
6. Reheat for 1 hour at low temperature.

Bread and cakes
1. Dough can be frozen but fully baked bread gives better results.
2. Pastry freezes well, both cooked and uncooked.
3. Pack bread and cakes in polythene bags, but remember that decorated cakes or puddings should be open frozen and then packed in a rigid container.
4. Most sandwiches freeze well, but avoid egg, banana, tomato, lettuce and mayonnaise in fillings.

QUESTIONS
1. What are the underlying principles of food preservation?

How may these principles be applied in practice?
Describe briefly how food is (a) frozen and (b) canned on a commercial basis.
2. Explain the principle of accelerated freeze drying. Describe briefly the methods used for freeze drying vegetables and include references to the packaging of such products.
What are the advantages of freeze-dried foods?
3. Write a note on food freezing under the following headings:
   a. principle
   b. commercial freezing
   c. nutritive value
   d. ill effects of careless freezing
   e. modern developments
   f. temperatures used for freezing and storing frozen food
4. How do the scientific principles of (a) osmosis and (b) sublimation relate to the preservation of food?
List some permitted preservatives and discuss their effects on foods.
Describe briefly how dehydration of food may be carried out at home.
5. List the chemical factors which affect the setting of homemade jams and jellies.
Describe how to test jam for (a) setting point; (b) gelling qualities and (c) vitamin C content.
What is the effect of an incorrect proportion of sugar on a homemade jam?

8 Human Physiology

All living things are composed of small units called cells. Although many of the simpler forms of life such as bacteria and yeasts consist of a single cell, the human body is a complicated multi-cellular organism consisting of thousands of millions of cells. Each cell within the body is interdependent and must co-operate with other cells in order to ensure the health of the whole body.

Most cells are so small that it is only possible to study them in detail by the use of an electron microscope. One of the few cells which can be seen without a microscope is an egg — whether the human egg, which is about the size of a pinhead, or the huge ostrich egg, the largest known cell. The relatively large size of the egg cell is due to the great number of nutrients it contains. Most tissue cells are

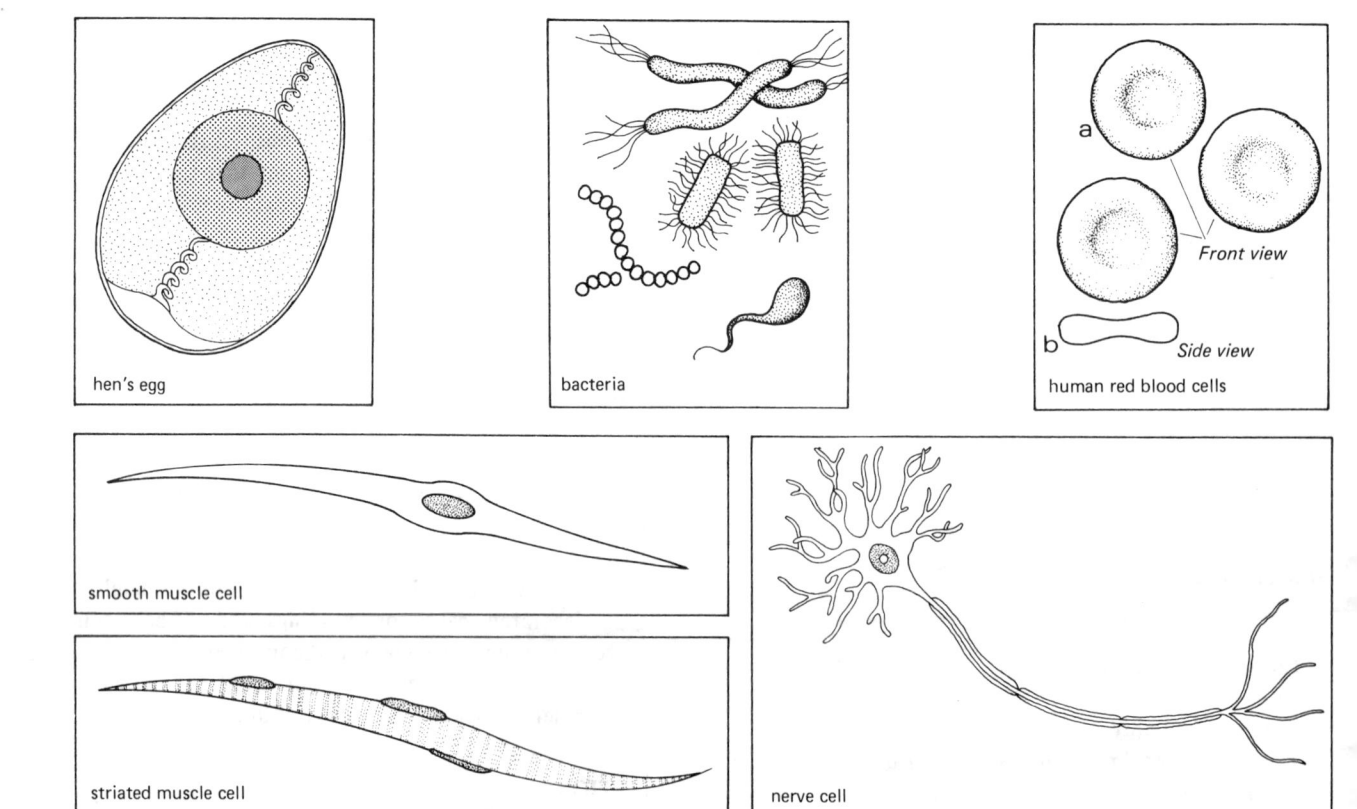

Fig. 8.1   Cell types

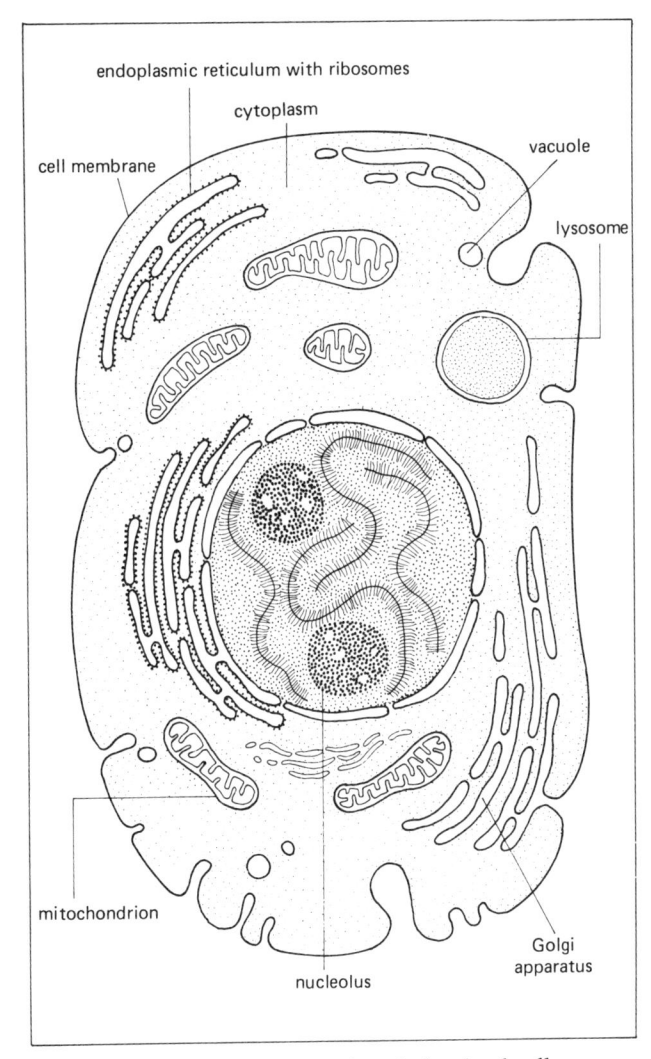

Fig. 8.2 Structure of typical animal cell

b. *Cell wall.* A thick outer wall of cellulose present only in plant cells, which gives plants their rigidity and firmness.

c. *Cytoplasm.* A jelly-like liquid medium in which numerous structures are suspended. The largest of these is the nucleus.

d. *Nucleus.* The life centre of the cell. It is the area of greatest chemical activity and controls growth and cell division. The greater part of the nucleus consists of nucleoplasm, a liquid medium which contains the chromosomes. These rod-shaped structures containing DNA (deoxyribonucleic acid) are thought to constitute the genes, the heredity factors which determine the characteristics of the organism. (In humans this would include size, colour of hair and skin and so on.) Surrounding the nucleoplasm is a double walled membrane similar to the cell membrane.

e. *Other structures* collectively called *organelles* are found in the cytoplasm and include:

Ribosomes. These appear as tiny dot-like bodies which are involved in protein synthesis and are usually associated with the endoplasmic reticulum.

Endoplasmic reticulum. A complicated structure of tube-like membranes found all over the cytoplasm with ribosomes attached to its outer surfaces. This is also concerned with protein synthesis.

Golgi apparatus. Concerned with secreting many vital substances including specific proteins. It is similar in appearance to the endoplasmic reticulum.

Mitochondria. Present in almost all cells except red blood cells and bacteria, these are the energy centres of the cell and are plentiful in muscle, liver and nerve cells and areas of the body where cells are particularly active. They contain enzymes concerned with the oxidation of food, converting potential energy into a form which can be utilised by the cell (ATP).

Plastids. Found in *plant cells only,* many contain pigments which give plants their colour. The most important plastid is chloroplast, containing chlorophyll, the green pigment which can absorb the energy from sunlight in order to manufacture the starch in the plant (photosynthesis — see p. 7).

Lysosomes. These contain destructive enzymes which break down some cell nutrients, mainly proteins, converting them into amino acids. They destroy the cell at the end of its life and for this reason the lysosome of the cell is sometimes called the 'suicide sac'.

microscopic in size and micro-organisms such as bacteria are even smaller than these.

Cells also vary considerably in shape. While many are spherical or oval, others are elongated and tightly packed together, such as those in muscles. Nerve cells have long branched extensions, sometimes several feet in length. Some bacteria have a spiral shape and others have tail-like extensions (flagellae) which propel them about.

Structure of a Cell

Despite the many variations in size and shape, the similarities between cells far outweigh the differences. Because of this, it is possible to make a diagrammatic representation of a typical cell.

a. *Cell membrane.* A thin, double-walled membrane which forms the cell boundary. It is composed of a layer of lipid sandwiched between two layers of protein. Some small molecules can pass through its walls but it restricts the passage of larger molecules. This is known as a selectively permeable or semipermeable membrane.

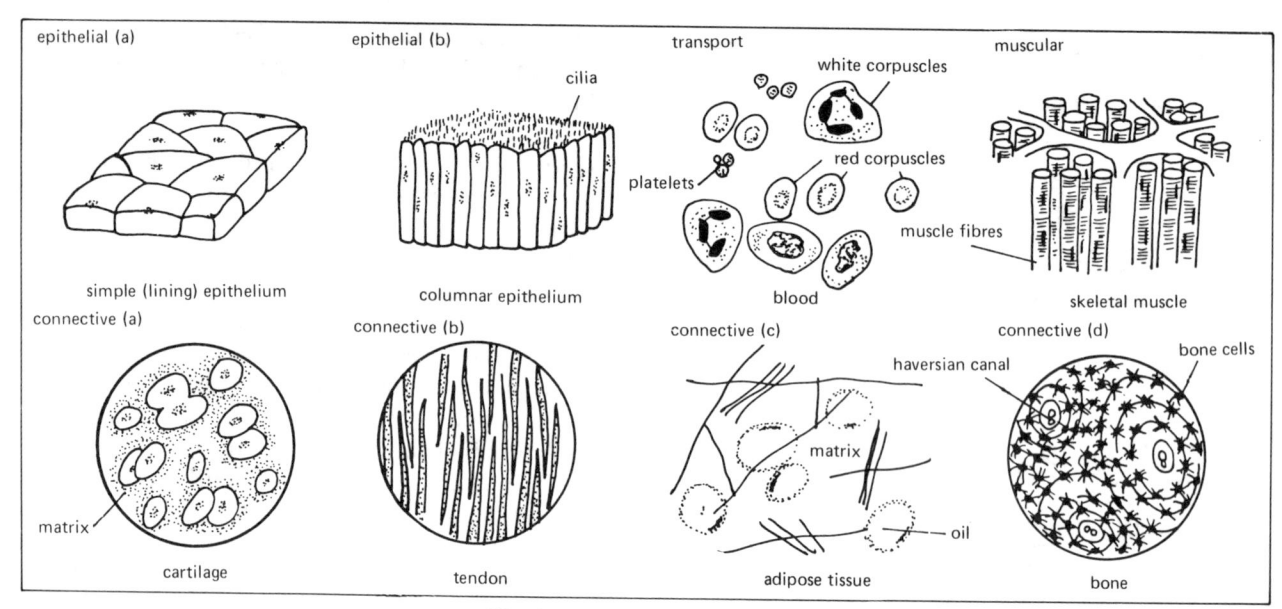

Fig. 8.3 Types of animal tissue

Vacuoles. These are fluid-filled bubbles within the cell which contain nutrients and sometimes waste matter.

Animals and plants grow and reproduce by cell division — a complicated procedure involving the initial division of the nucleus, followed by the growth of a dividing wall and subsequent growth of the new cell until it contains all of the structures found in the parent cell. This type of cell division is called *mitosis*.

TISSUES

Although some single cells are self-sufficient organisms, multicellular organisms such as the human body contain specialised cells, each of which has a particular function. Some, for example, are capable of contraction, others can secrete substances from the blood and so on. Groups of identical cells which together perform a certain function are called tissues.

There are several types of tissues which vary according to the structure of their cells, their position and their function.

1. Connective Tissue

This joins different kinds of tissue together, encloses groups of cells and holds organs in place. It consists of specialised cells surrounded by non-cellular material called a matrix. Much connective tissue contains collagen, a strong fibrous protein.

a. **Cartilage** is a strong connective tissue often found at the ends of bones where it prevents friction. Outgrowths of cartilage form a large part of the nose and ear and it is found in pads or discs between the vertebrae of the back bone.

b. **Bone** is another form of connective tissue, which gives the body shape and firmness. It consists of a hard matrix containing large deposits of calcium phosphate. Interspersed here and there are cells which are connected to one another and are fed by blood vessels which pass into the bone tissue.

c. **Tendons and ligaments** connect muscle to bone and one bone to another. They contain parallel bands of collagen and elastin.

d. **Elastic tissue** is found in the walls of arteries, the stomach, and bladder and forms a large part of the lung tissue.

e. **Adipose tissue** consists of cells which can store many times their own volume of oil or fat.

2. Epithelial Tissue

This consists of flat sheets of closely packed cells. It is found in the skin, lining membranes and outer membranes of many organs. Because of the porosity of this tissue, small molecules can pass through it; this occurs in the alveoli of the lungs and in the small intestine. Some epithelial tissue possesses hair-like projections, called cilia, which propel substances past the cells; this occurs in the trachea when mucus is brushed upwards towards the mouth. Many epithelial cells produce secretions such as mucus which keep surfaces and linings moist.

3. Nervous Tissue

The brain and spinal cord are mainly composed of nervous tissue. This consists of cells with long, branched extensions which carry impulses to other tissues, e.g. to muscles (p. 184).

4. Liquid Tissue

The blood is a fluid tissue consisting of a liquid matrix called plasma in which are suspended many types of blood cells (see p. 160).

5. Muscle Tissue

Muscle is necessary for all movement in the body. The muscle cell is specialised for contraction, which it does when stimulated to do so by nervous impulses. Nearly all bodily processes depend on movement. The blood circulation and transport of nutrients depend on the contraction of the heart. Intake of oxygen depends on contraction of the respiratory muscles. Food is propelled along the digestive tract by peristaltic contraction. Even our entrance into the world is brought about by the powerful contractions of the uterus.

Skeletal (or voluntary) muscle is made up of bundles of fibres running parallel to one another, which are bound together with connective tissue. The fibres are elongated multinucleated cells, which are covered with visible cross-markings or striations, which give the muscle an alternative name, *striated muscle*. These striations consist of thick and thin filaments containing two proteins called actin and myosin respectively.

Contraction occurs in response to stimuli from branches of neurons. Each fibre shortens as the filaments slide over one another; this may continue for a few seconds or relaxation may follow. As the fibres become fatigued quickly, the muscle cannot be held in a state of contraction for long. Energy for contraction comes from the ATP in the cells. When large amounts of energy are required, for example for a quick sprint, the glycogen stored in the muscle is converted into lactic acid, releasing energy. Only one quarter of the energy thus supplied is in the form of mechanical energy and the remaining three-quarters is given off as heat, a normal product of muscular activity. Skeletal muscle contracts much faster than other types of muscle, but as it tires easily, it works best in short bursts.

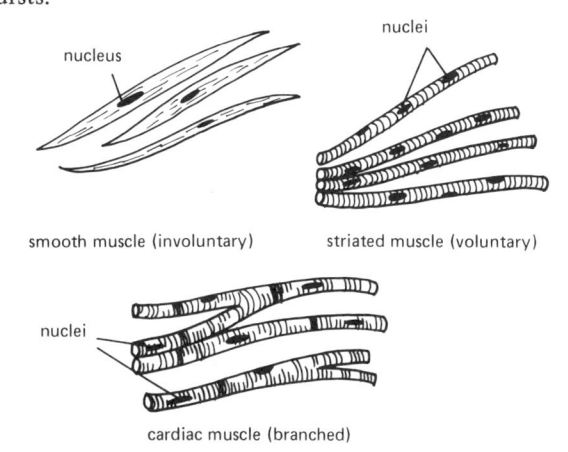

Fig. 8.4 Types of muscle tissue

Skeletal muscle makes up the fleshy part of the body and is found on the limbs and trunk. As it is under the control of the will (the cerebral cortex) it is also called voluntary muscle.

Smooth (or involuntary) muscle. This consists of sheets of spindle-shaped cells (wide at the middle and narrow at each end). Each cell contains a single nucleus. Smooth muscle has no striations and is very elastic. It is not under voluntary control but is controlled by neurons of the sympathetic and parasympathetic nervous system. Smooth muscle is found on the walls of hollow organs and blood vessels. It causes contraction of the arteries, alimentary canal, uterus and bladder; the contractions brought about are long and slow, unlike the stronger quick contractions of the voluntary muscles.

Cardiac muscle consists of short, irregularly striated fibres. Each is a true cell which interlocks with the next, forming parallel lines. Its structure is similar to that of skeletal muscle except for some branching of fibres. Cardiac muscle is capable of initiating its own contractions. The sino-atrial nodes (see p. 157), which are specialised groups of muscle cells, stimulate the contractions of the heart. Although some sympathetic nerves pass into the heart, they serve only to speed up or slow down the heart beat. With each beat of the heart, all of the available energy in the heart muscle is utilised. As there is no energy reserve, loss of oxygen, even for a few seconds could have fatal consequences.

THE HEART

The heart is a strong muscular organ. The cardiac muscle of which it is composed consists of a network of short, interlocking fibres which contract and dilate rhythmically about seventy times a minute. Unlike other muscular organs, the heart works continuously, pumping about 14,000 litres of blood around the body every day.

Position

The heart is situated in the centre of the thoracic cavity, to the front, with the apex pointing downwards and to the left. It lies between the two lungs and above the diaphragm, a strong muscle which separates the contents of the thoracic or chest cavity from the abdominal organs. The heart and lungs are protected in front by the sternum or breast bone, and the rib cage and intercostal muscles afford protection at the sides and the back.

Shape and Structure

The heart could be described as a pear-shaped hollow organ, about the size of the clenched fist of its owner. It consists of four chambers. The upper two, called auricles or atria, are thin-walled chambers which receive incoming blood, whereas the lower chambers, called ventricles,

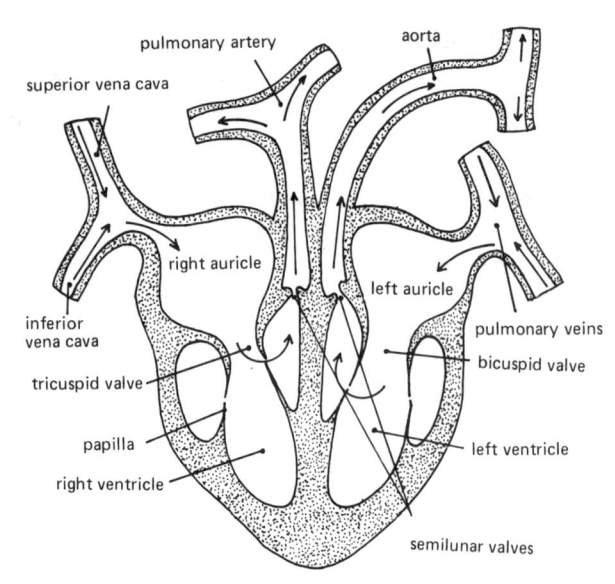

Fig. 8.5 Blood flow through the heart

pump blood out of the heart. In order that the ventricles can perform their pumping action efficiently, their walls are thicker than the auricles; the left ventricle has a particularly thick muscular wall which enables it to pump large quantities of blood right round the body.

A central wall (septum) divides the right side of the heart from the left side. There is no connection at all between the two sides – an important factor because the right side of the heart contains impure or deoxygenated blood which has a high level of carbon dioxide whereas the left side carries only pure oxygenated blood. If the two were to mix (as sometimes happens in the case of a baby born with a hole in the heart) the body would neither get rid of its carbon dioxide efficiently nor obtain sufficient oxygen. Children with this defect lack energy and tire quickly through lack of oxygen. Their skin has a white pallor and their lips appear bluish because of the abnormal amount of carbon dioxide in their blood.

A one-way valve is situated between the right auricle and the right ventricle. It is called the *tricuspid valve* because it has three 'flaps', each of which is attached to the walls of the ventricle below by tendons. These *papillary tendons* are attached to tiny muscles which contract at the same time as the ventricle, preventing the valves from opening upwards. This ensures that the blood flows only in a downward direction, i.e. from auricle to ventricle. A similar valve called the *mitral* or *bicuspid valve* separates the left auricle from the left ventricle.

The outside of the heart is surrounded by a protective membrane called the *pericardium*. A certain amount of fat also accumulates around the heart, especially in older people.

Blood Vessels of the Heart
1. Superior and inferior venae cavae. These are large veins which return deoxygenated blood to the heart from the body tissues. The superior vena cava brings blood from the head, neck, upper chest and arms, while the inferior vena cava returns blood from the trunk and lower limbs. Both open into the right auricle.

2. Pulmonary artery. This is the large blood vessel which takes impure blood from the right ventricle to the lungs to be purified. A one-way valve called the *pulmonary* or *semi-lunar valve* guards the opening to this blood vessel. Immediately upon leaving the heart, the pulmonary artery divides in two and one branch goes to each lung. In the lungs the blood is purified (see p. 164).

3. Pulmonary veins. These return oxygenated blood from the lungs to the heart. There are four of them, two springing from each lung, and all four enter the heart through the left auricle.

4. Aorta. A strong, elastic artery which originates in the left ventricle. It is the largest artery in the body and carries oxygenated blood from the heart to all body tissues.

5. Coronary arteries. Just before the aorta leaves the heart, two small arteries, the coronary arteries, branch out from the aorta and open out on to the surface of the heart. The right coronary artery breaks into branches feeding the underside and back of the heart, while the left coronary artery supplies the left and frontal areas of the heart. They divide into smaller branches (arterioles), eventually merging into capillaries which bring oxygen and nourishment to each part of the muscle of the heart itself. The capillaries are drained by small veins which unite to form the two coronary veins which empty directly into the right auricle. This is known as the *coronary system*.

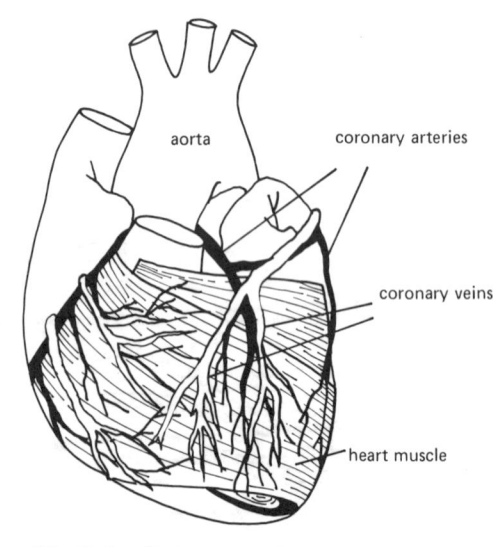

Fig. 8.6 Coronary system

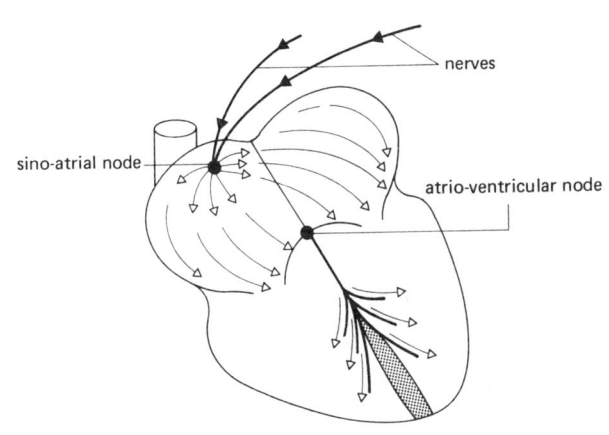

Fig. 8.7 Origin and conduction of the heartbeat

It is when one of the coronary arteries becomes silted up with fatty deposits such as cholesterol that the danger of coronaries (heart attacks) arises. When the artery is either blocked by a clot or closed over completely, one part of the heart muscle is deprived of oxygen so that it ceases to function. If a large vessel becomes blocked, it may damage such a large portion of the heart muscle that the heart cannot refunction and the patient dies.

The condition which causes blockage or narrowing of these arteries is known as *arteriosclerosis*. *Angina* is a painful heart condition which occurs when the oxygen supply to the heart is inadequate. Heart disease is aggravated by an incorrect diet, excess weight, stress, smoking and excess alcohol.

Heartbeat

Most muscles in the body only contract when an electrical impulse from a nerve stimulates them to do so. The heart contains a special group of muscle fibres which make up the *sino-atrial node* or *pacemaker* which is situated near the entry point of the superior vena cava. It sends its own impulses to the walls of the atria (auricles), causing them to contract. During this contraction, a second node, the *atrio-ventricular node,* conveys impulses to the muscles of the ventricles causing them, in turn, to contract. These contractions occur about seventy times a minute in the average healthy person. Electrocardiograms (ECGs) record the impulses from each part of the heart on a special machine. Diseases or faults within the heart are detected when a change in the normal wave pattern occurs.

It is now possible for those with serious heart defects to have artificial pacemakers fitted beside the heart. These regulate the heartbeat by transmitting electrical impulses to the auricles and ventricles.

Factors affecting heartbeat. Although normal heatbeat is maintained by the pacemaker within the heart, changes in the speed of the heartbeat are controlled by nerve impulses transmitted from the medulla in the brain. The *accelerator nerves* increase the strength and rate of heart beat, thus increasing blood flow and oxygen transport to all parts of the body. These nerves come into play when a person is frightened or when the body undergoes physical exertion. The *vagus nerves* have an inhibitory effect, slowing down the heartbeat to its normal rate. As already explained, the normal heart contracts about seventy times a minute when the body is calm and resting. Several factors can cause a change in the rate of contraction:

1. *Activity.* Obviously a person who is running is using up more energy than somebody who is sitting or standing still, so a greater supply of blood is required to bring extra oxygen to the tissues.

2. *Age and body size.* A baby's heart beats faster than an adult's and in old age the heart rate decreases further.

3. *Illness, excitement, drugs, stimulants* can all cause an increase in the heart rate, while other factors such as *cold* and *depressant drugs* cause a decrease.

Circulation

Although the heart is a single organ, it pumps blood around two circuits simultaneously. The *systemic circulation* carries blood to all parts of the body except the lungs and brings it back to the heart. The *pulmonary circulation* is concerned with collecting oxygen and removing carbon dioxide from the blood. This brings the blood to the lungs to be purified and then returns it to the heart.

The heart is, in effect, a double pump. Here is how it works:

1. The superior and inferior venae cavae empty deoxygenated blood from the whole body (except the lungs) into the right auricle. This contracts, forcing blood through the tricuspid valve into the right ventricle.

2. This ventricle contracts, closing the tricuspid valve and opening the semi-lunar valve which guards the entrance

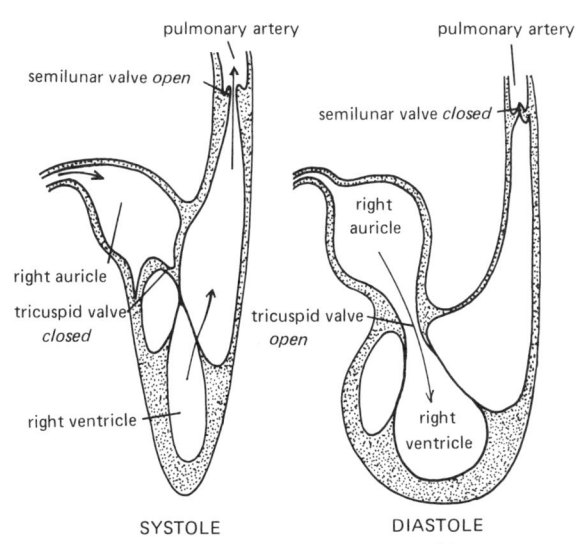

Fig. 8.8 Heartbeat – right side of heart

to the pulmonary artery, allowing the blood to pass outward to the lungs.

3. On relaxation of the ventricle the semi-lunar valve closes and the auricle begins to fill again.
4. The purified, oxygenated blood returns from the lungs to the left auricle through the four pulmonary veins. The left auricle contracts, forcing the blood through the mitral (bicuspid) valve into the left ventricle.
5. This in turn contracts, closing the mitral valve and forcing open the semi-lunar valve at the entrance to the aorta, which brings the blood out of the heart and all around the body.
6. On relaxation of the left ventricle, the semi-lunar (aortic) valve closes to prevent a backward flow of blood and the left auricle begins to fill again.

Although six points are listed in this step-by-step description, there are three main stages in a heartbeat:
1. Both auricles contract (1 and 4 above).
2. Both ventricles contract (2 and 5 above). This strong contraction is called *systole*.
3. There follows a period of relaxation during which the cardiac muscle recharges itself and the auricles fill up again with blood. (3 and 6 above). This resting period is called *diastole*.

Heart sounds. Normal heart sounds are caused by:
1. The closure of the tricuspid and bicuspid valves.
2. The closure of the semi-lunar valves.

This makes a double beat, the second being sharper and quicker than the first. These sounds are easily picked up by a stethoscope.

Systemic circulation. At the point where the aorta leaves the heart, it divides into three branches, one leading to the head and one to each arm. The main branch of the aorta does a 'U' turn and passes behind the heart and through the diaphragm, bringing oxygenated blood to the tissues of the trunk, digestive and reproductive organs and the legs. As it passes through the body it branches again and again, an arterial branch entering each organ or muscle. These *arterioles,* as they are called, continue to branch off becoming smaller and smaller until they eventually penetrate every portion of the body. Within the tissues they form a network of tiny vessels called *capillaries,* whose walls are so thin that they allow oxygen and food molecules to pass through them into the cells. The capillaries take up the waste products of cell metabolism (e.g. urea, carbon dioxide) and subsequently join up to form tiny veins (venules). Eventually these join with other veins, all of which flow into the superior or inferior venae cavae and from there into the right auricle to begin the circuit once again.

Pulmonary circulation. The pulmonary arteries enter the lungs where they branch and rebranch in the normal way, eventually forming capillaries. At this stage an exchange of gases takes place (see p. 165) so that oxygen is exchanged for the carbon dioxide in the blood. Small veins stem from the capillaries, eventually giving rise to the two main pulmonary veins which leave each lung and pass into the left auricle of the heart.

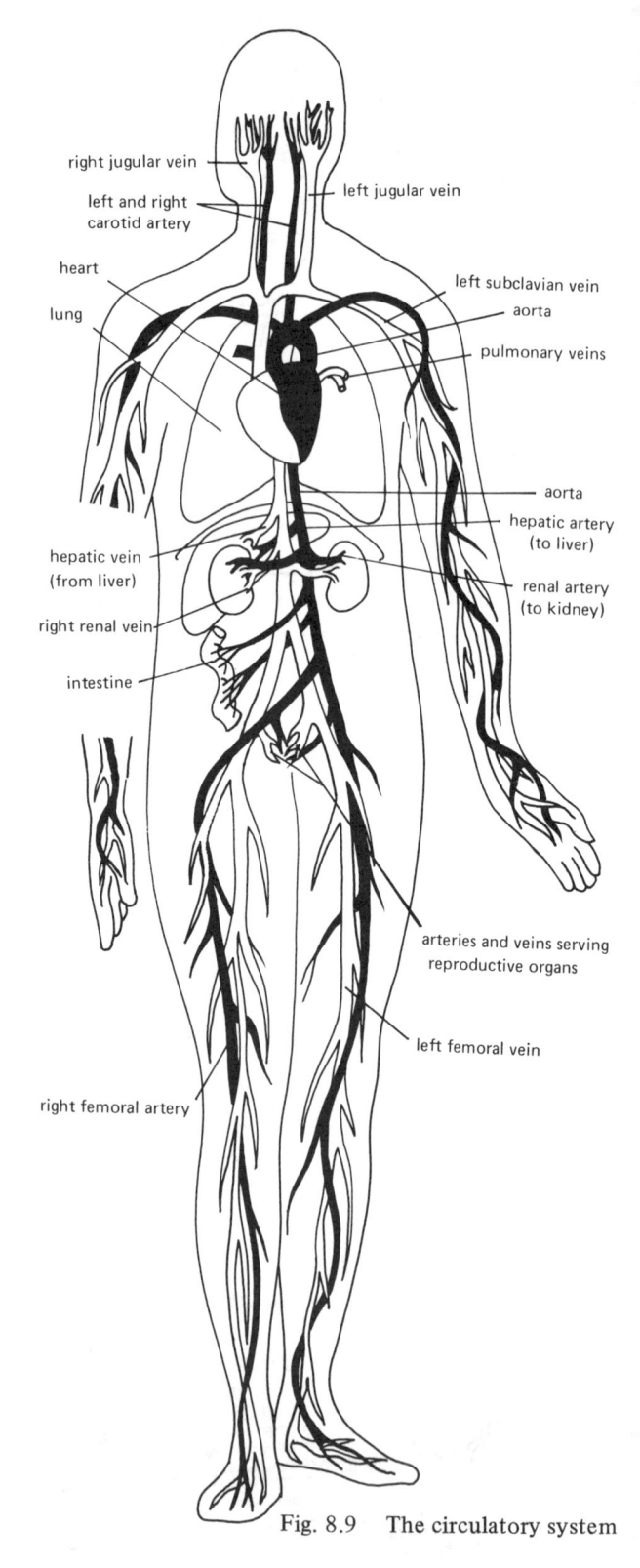

Fig. 8.9 The circulatory system

Pulse

The powerful contraction of the left ventricle sends the blood surging through the main arteries which stretch to accommodate this rhythmic increase in flow. By applying pressure on an artery, for example the wrist or temple, a definite throb or pulse can be felt. When an artery is severed blood gushes out in spurts and pressure must be applied on a part of that artery nearer the heart in order to reduce blood loss. Luckily our arteries are usually so deep in the flesh that only a severe cut causes arterial bleeding.

Blood Pressure

The force of pressure which the blood exerts against the arterial walls can be measured on a mercury scale. Normal blood pressure is expressed as a fraction 120/80, 120 being the strong pressure during systole, which raises the mercury 120 mm, and 80 being the lower pressure during diastole, which raises the mercury 80 mm on the pressure gauge.

Factors influencing blood pressure

a. Cardiac output or the volume of blood pumped by the heart.
b. Resistance to the blood flow provided by the capillaries.

The brain monitors the pressure on arterial walls continuously and if the pressure rises either (i) there is a subsequent decrease in the rate of heartbeat or (ii) the capillaries become dilated, reducing the resistance to the blood flow. When blood pressure falls the heart beats more quickly and the capillaries contract, increasing the resistance to blood flow and thereby raising the pressure.

High blood pressure, often called hypertension, occurs when there is increased resistance to the blood flow. This may be caused temporarily by excitement, stress or annoyance. When the blood vessels become coated with deposits which narrow the bore of the vessels, high blood pressure becomes a permanent complaint, but drugs which help to dilate the blood vessels and ease the situation may be prescribed. Those who suffer from hypertension should avoid stressful situations which are likely to raise the blood pressure.

Shock occurs when the flow of blood is diverted from less important areas such as the skin and limbs and extra blood is circulated to critical organs such as the brain, heart and kidneys. This might occur as the result of an accident, for example. Symptoms of shock include low blood pressure, rapid pulse, sweating, cold hands and feet and eventually loss of consciousness.

Fainting occurs when the blood flow to the brain is inadequate, usually because of stagnation of the blood in the legs. This causes a fall in the flow of blood to the heart and subsequently to the brain. When a person faints, the body becomes horizontal, blood flows more easily from the legs to the heart and brain, and the patient quickly regains consciousness.

Blood Vessels

The vessels through which the blood circulates are of three types: arteries, capillaries and veins.

Arteries are large vessels which carry blood away from the heart to every part of the body. They have thick walls consisting of an outer fibrous coat, an inner lining of endothelium and between them a thick layer of muscular and elastic tissue. The elasticity of arterial walls allows them to expand with the pressure of the blood as it pulsates from the heart. Arteries divide into smaller branches called arterioles. These, in turn, form smaller branches as they move further away from the heart until they penetrate every portion of body tissue. The smallest branches are capillaries.

Capillaries are microscopic blood vessels which connect arteries with veins. They are so narrow in places that the red blood cells have to flatten to pass through. Capil-

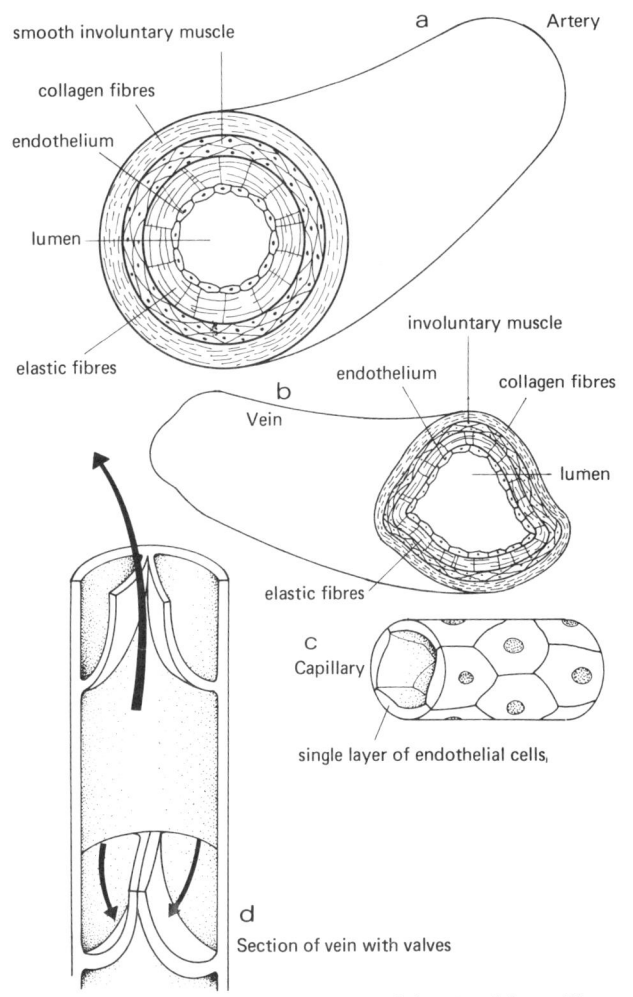

Fig. 8.10 Structure of (a) artery, (b) vein, (c) capillary and (d) section of vein and valves

lary walls, which are only one cell thick, are semi-permeable, thus allowing plasma and small molecules such as oxygen, carbon dioxide and glucose to pass in and out. Having released oxygen and nutrients to the cells, and collected the waste materials of cell metabolism, the blood in the capillaries flows into tiny veins called venules. These unite to form larger veins which join the two main veins, the superior and inferior venae cavae, emptying into the heart.

Veins are blood vessels which return blood to the heart from the tissues. They have thinner walls and a wider bore than the arteries, a factor which assists the return flow of blood to the heart. Because the blood in the veins is flowing against gravity and because the pressure from the pumping action of the heart is lessened by the time the blood reaches the veins, there could be a tendency for the blood to slow down or flow backwards. To prevent this many veins, especially those in the limbs, have valves which prevent a backward flow. Muscular contraction also assists the upward flow by squeezing the blood along the veins. Because most venous blood has less oxygen than arterial blood, it has a darker red colour. The exception to this is the blood in the pulmonary veins, which carries oxygen and is bright red.

THE BLOOD

The blood is the principle medium of transport in the human body. It is a liquid tissue, red in colour, and is 55 per cent fluid (plasma) and 45 per cent solid substances, mainly cells. The average adult has five to six litres of blood in his body.

Plasma

This is a pale yellow fluid in which are dissolved many substances including sodium chloride (salt), mineral salts, sodium bicarbonate which keeps the blood alkaline, glucose and lipids for energy production and amino acids necessary for the building and repair of cells. It also contains complex proteins such as hormones, enzymes, antibodies and fibrinogen which have specialised functions. Suspended in the plasma are platelets and blood cells, both red and white. The plasma carries all these substances to the tissues where many of them are concerned with metabolism. Waste matter such as carbon dioxide and urea are collected from the cells for excretion.

Composition of plasma
 Water 90%
 Protein 7%
 Salts 1%
 Other substances (e.g. nutrients, hormones and waste) 2%

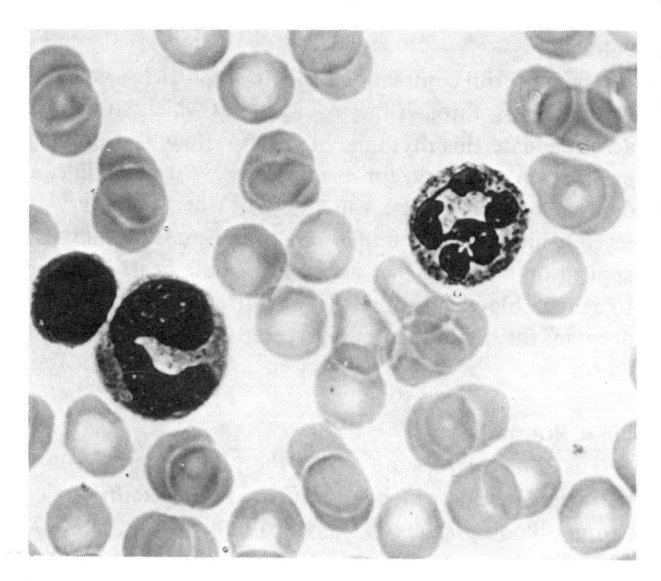

Blood seen under the microscope. The large white corpuscles with their distinctive nucleii stand out from the smaller and more numerous red corpuscles. (Gene Cox)

Red Cells (Erythrocytes)

These are minute discs composed of spongy cytoplasm filled with haemoglobin – an oxygen-carrying pigment. They are biconcave in shape (thinner in the centre), which increases their surface area and thus their capacity for absorbing oxygen. During the formation of red blood cells in the bone marrow, the nucleus, then present in the cell, is ejected, leaving them without the ability to reproduce themselves. About 200,000 million red blood cells (1 per cent of the total) are thought to be replaced daily by the bone marrow. They have a short life span of three to four months as they are easily damaged by blood pressure or friction when passing through capillaries. After that time they are removed from circulation and broken down by special cells in the spleen and bone marrow into their components, the iron-rich 'haem' and the protein 'globin' (see below). The protein is utilised in the manufacture of new cells, while the iron passes to the liver, where it is recycled. Some of it is converted into strongly coloured bile pigments but most of it returns to the bone marrow where it is used to make more haemoglobin. Some B group vitamins including B_{12} and folic acid are required in the manufacture of haemoglobin.

Haemoglobin

This is a large, complex protein which contains iron at the centre of its molecule. It is important in the blood because it has the ability to combine readily with oxygen. In the lung capillaries it combines loosely with the oxygen breathed in to form oxyhaemoglobin, each atom of iron taking up two atoms of oxygen, which gives the blood a bright red colour. Just as it readily takes up oxygen, it equally quickly releases it in parts of the body where the oxygen concentration is low – i.e. in the tissues.

Anaemia occurs when the blood suffers a deficiency of haemoglobin. *Iron deficiency anaemia* occurs when dietary intake of iron is less than the amount lost from the body. This causes a reduction in the amount of haemoglobin and the red cells are reduced in size and number. Usual causes are prolonged bleeding, frequent heavy periods or an inadequate diet. Dietary factors which affect haemoglobin formation are iron, folic acid, vitamin B_{12} and vitamin C. *Pernicious anaemia* occurs when abnormalities in the stomach prevent normal absorption of iron. Over-large, immature cells which are deficient in haemoglobin are released from the bone marrow. This causes tiredness and lack of energy.

White Cells (Leucocytes)

These are nucleated and are considerably larger than red cells, but are less plentiful than them: there are over five hundred red cells to a single white one. The life of the white cell is extremely short, varying from one to fourteen days. White cells are made in the bone marrow, lymph nodes (see p. 163) and spleen. They are of several types, which vary according to the shape of the nucleus present; this variation shows up in laboratory stains. Some have granules in their cytoplasm (granulocytes): these include neutrophils, basophils and eosinophils. Others are without granules (agranulocytes) and include lymphocytes and monocytes. All are capable of moving about and changing their shape and can squeeze out between capillary walls in order to fight infection.

The principle function of white cells is to destroy invading organisms such as viruses and bacteria. Most do this by surrounding the organism and digesting it, and are therefore known as phagocytic (organism-digesting) cells. White cells are always found on the site of an infection or injury, where they destroy infecting organisms and dead tissue. Pus consists of millions of dead white cells. Lymphocytes help to protect the body by forming *antibodies* (see p. 133) which neutralise or destroy foreign substances.

Inflammation of an injury is caused by an increased blood supply to the infected area.

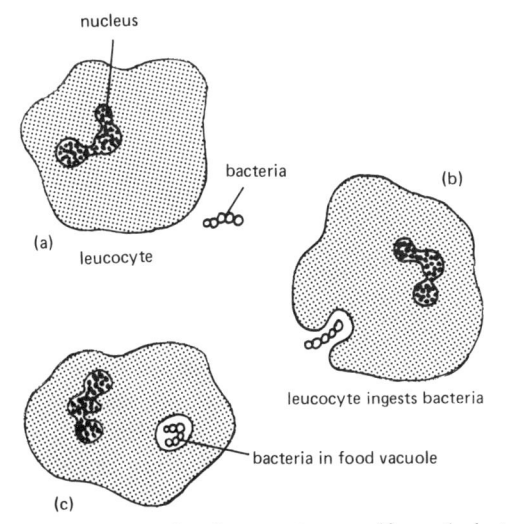

Fig. 8.11 Phagocytosis – leucocyte engulfs and destroys bacteria

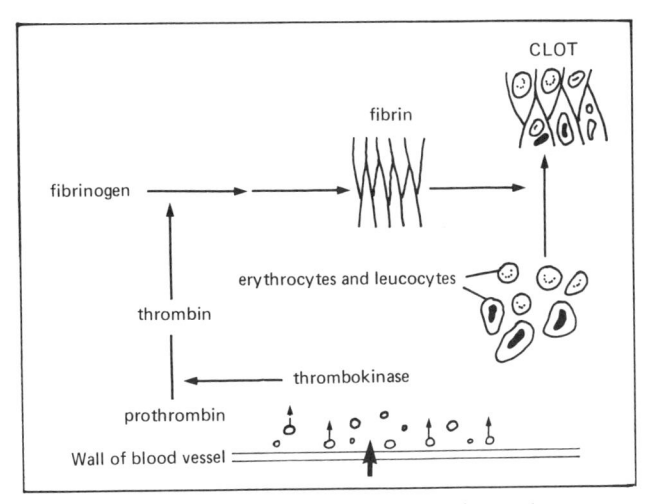

Fig. 8.12 Summary of blood clot formation

Platelets (Thrombocytes)

These are the smallest blood cells; they are, in fact, fragments of larger cells of the bone marrow. They are concerned with blood clotting.

Clotting

An essential function of the blood is its clotting mechanism, for without this even a simple cut could make a person bleed to death. Plasma contains small amounts of a protein called *fibrinogen*. When a blood vessel is cut or damaged, a protein (thrombokinase) is released from the damaged walls and platelets. These activate an enzyme called prothrombin, present in the blood, changing it to *thrombin*. Thrombin converts fibrinogen into fibrin, an insoluble stringy substance which forms a network around the wound in which red cells and platelets become entangled. This *clot* seals the opening and stops the bleeding. Some liquid, called serum, oozes out of the clot which eventually dries out to form a scab. That protects the wound while new skin grows underneath.

Serum. Plasma minus its fibrinogen.

Haemophilia. A disease usually found in males which causes uncontrolled bleeding. It is due to a deficiency of some of the plasma proteins required for clotting.

Thrombosis. A condition in which clots are formed within the blood vessels. These can cause blockages.

Coronary thrombosis. A blockage of the coronary blood vessels due to clotting.

Functions of the Blood

1. *Oxygen transport* is the chief function of the blood. Haemoglobin in the red cells takes up oxygen in the lungs and carries it to the tissues where it readily diffuses out of the capillaries and into the body cells.

2. *Carbon dioxide transport.* Carbon dioxide, formed during cell respiration, diffuses into the capillaries and is carried by plasma and red cells to the lungs where it is removed.

3. *Transport of waste.* Toxic substances such as urea pass into the plasma and are carried to the kidneys where they are removed. Excess salts and glucose are also carried in the blood to the kidneys for removal.

4. *Temperature control.* Heat generated by metabolism is distributed by the circulating blood. Secretion and evaporation of sweat also play a part in maintaining an even temperature and prevent overheating. Blood can be diverted to the skin when the body is warm so that excess heat is lost by convection. When the body is too cold, surface blood vessels contract and less blood circulates to the skin, so we get pale.

5. *Food transport.* Digested food such as amino acids and glucose pass into the capillaries in the small intestine, are carried by the portal vein to the liver and from there into the general circulation.

6. *Distribution of hormones and enzymes.* Hormones produced in the endocrine glands are carried in the blood to the areas where they are required. Some enzymes are also transported by the blood.

7. *Prevention of infection.* The function of white cells in the destruction of invading bacteria has already been explained. Lymphocytes manufacture antibodies — proteins which destroy the more virulent forms of bacteria by causing them to clump together so that they can neither multiply nor invade the tissues. They also produce antitoxins which neutralise toxins formed by bacteria.

8. *Homeostasis.* The blood is the medium whereby the correct balance of oxygen, pH, temperature, nutrients etc. is achieved in order to maintain an environment in which the body cells can function. The creating of these optimum conditions is known as homeostasis.

Blood Groups

Since the blood of all human beings is not alike, a person requiring a blood transfusion cannot be given blood from any donor. Antigens present on the surface of the red blood cells of many people determine the blood group;

Summary of Blood Functions
1. Transport of oxygen
2. Transport of waste materials — carbon dioxide and urea
3. Temperature control
4. Transport of nutrients
5. Distribution of hormones and enzymes
6. Prevention of infection
7. Homeostasis — maintenance of ideal environment

---

these are proteins which stimulate the production of the antibodies that help to destroy bacteria. In blood grouping, the antigens involved are called A and B: some people have antigen A in the blood (blood group A) while others have antigen B (blood group B). A small percentage of people have both antigens in their blood (blood group AB). The blood of the largest percentage, however, (46 per cent) contains neither antigen and is called blood group O.

When red blood cells, which carry one or both antigens, come in contact with the appropriate antibody, the cells agglutinate or clump together and cannot function normally. It is obvious, therefore, that matching antigens and antibodies are never found in the same blood. For this reason, the bloods of a donor and a recipient are carefully tested before transfusion to ensure that they are compatible. If type B blood was given to someone with anti-B antibodies, the blood cells would mass together, blocking blood vessels, causing kidney damage and eventually death.

Blood Group	Antigens Present	Antibodies	Can donate to	Can receive from
A	A	anti-B	A, AB	O, A
B	B	anti-A	B, AB	O, B
AB	A & B	neither	AB	O, A, B, AB
O	none	anti-A & anti-B	A, B, AB, O	O

Blood groups are inherited genetically and do not change.

Rhesus factor. In addition to the main blood groups, another group of antigens is found in 85 per cent of humans. It is called the Rhesus factor (after the Rhesus monkey in whose cells these antigens were first discovered). People whose blood contains such antigens are grouped Rhesus positive (Rh+) while those without are called Rhesus negative (Rh−). This factor may cause problems during pregnancy. If, for example, a Rh+ father and a Rh− mother conceive an Rh+ baby, the Rh+ cells in the foetus would begin to form antibodies in the mother's blood which would eventually re-enter the blood of the foetus and begin to destroy its red cells. A Rhesus baby is usually born prematurely and may have congenital diseases, but it is usually possible to replace the abnormal blood by transfusions and save the baby's life.

Blood count. This is a measurement of the amount of blood cells present in one cubic millimetre of blood. Healthy blood contains approximately five-and-a-half million red cells per mm^3, and about 10,000–12,000 white cells.

Blood banks. These are usually non-profit-making organisations whose function is to collect blood from voluntary donors and to store it until it is required by hospitals. Both blood and plasma can be stored.

THE LYMPHATIC SYSTEM

Because the circulatory system is a completely enclosed system, blood has no direct contact with the tissues. The lymph system forms a bridge between the blood and the cells, which enables it to collect substances from the tissues and return them to the general blood circulation. In some ways it resembles the circulatory system, in that lymph is a fluid similar to plasma which flows through small, tube-like vessels resembling veins. Unlike blood, lymph does not form a complete circuit but flows in one direction only, outward from the tissues towards the neck, where it empties into the blood stream. Because the lymph system has no pumping mechanism similar to the heart, it depends, like veins, on muscular contractions to help the lymph along the vessels. Like veins, lymph vessels also contain many valves which prevent a backward flow and give the lymph vessels a beaded appearance.

How Lymph is Formed

Lymph vessels begin in the *intercellular* spaces between the tissues. Deep in the tissues the capillaries release plasma through the thin endothelial wall. This contains glucose, salts and amino acids, but few large protein molecules, as these are too large to pass through the capillary wall. The fluid, at this stage, is called the tissue fluid or extra-cellular fluid (ECF) and bathes the cells which extract from it the oxygen and nutrients they require. This diffusion occurs due to the high blood pressure in the arteriole end of the capillary which still exists due to ventricular contraction. Waste materials from the cells, such as carbon dioxide and urea, diffuse from the cells into the tissue fluid. Because the pressure in the tissue fluid is greater than that of the blood at the venule end of the capillary, most of the tissue fluid returns to the capillaries.

The small amount which does not drains into blind-ended lymph vessels in between the cells and they carry the fluid (now called lymph) to larger vessels, eventually forming a system, most of which collects into a major vessel, the *thoracic duct.* This passes up the centre of the trunk, through the diaphragm and empties its contents into the left subclavian vein.

Lymph from the upper right-hand portion of the body flows into the right lymphatic duct and returns to the blood stream at the right subclavian vein.

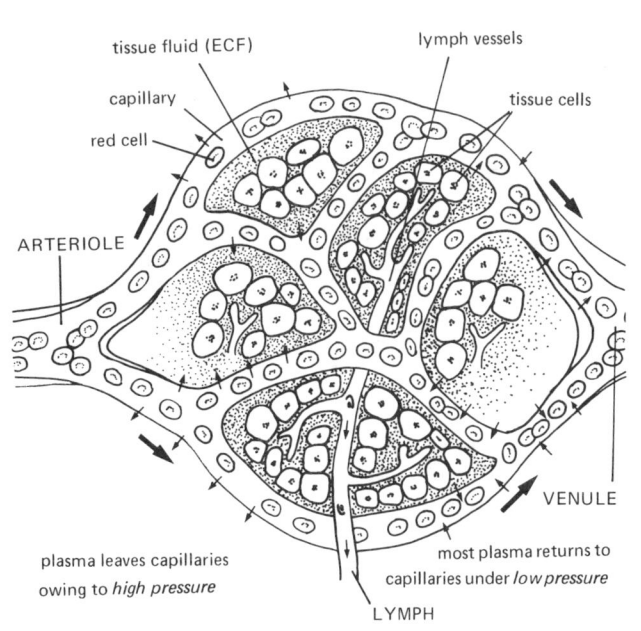

Fig. 8.13 Formation of lymph

Lacteals in the villi of the small intestine (see p. 174) are part of the lymph system. They absorb digested fats which return to the general circulation with the lymph.

Lymph nodes (glands). Here and there over the lymph system hundreds of lymph nodes or glands are distributed. They are plentiful in the groin, abdomen, thorax, armpits and neck. Lymph nodes are composed mainly of connective tissue and special cells within the gland tissue destroy foreign particles such as bacteria present in the lymph by engulfing them as the lymph filters through the gland. When there is a major infection, the lymph nodes swell considerably causing what is known as 'swollen glands'. A septic finger can cause tenderness at the axilla (underarm) where the glands actively try to intercept the invading bacteria and prevent them from infecting the rest of the body. The tonsils, adenoids and spleen are also composed mainly of lymphatic tissue.

Lymph nodes manufacture lymphocytes (white blood cells) which circulate in the lymph and eventually pass into the bloodstream. These produce antibodies which help to fight infection.

Functions of the Lymph

1. It is an exchange medium between the blood and the cells.

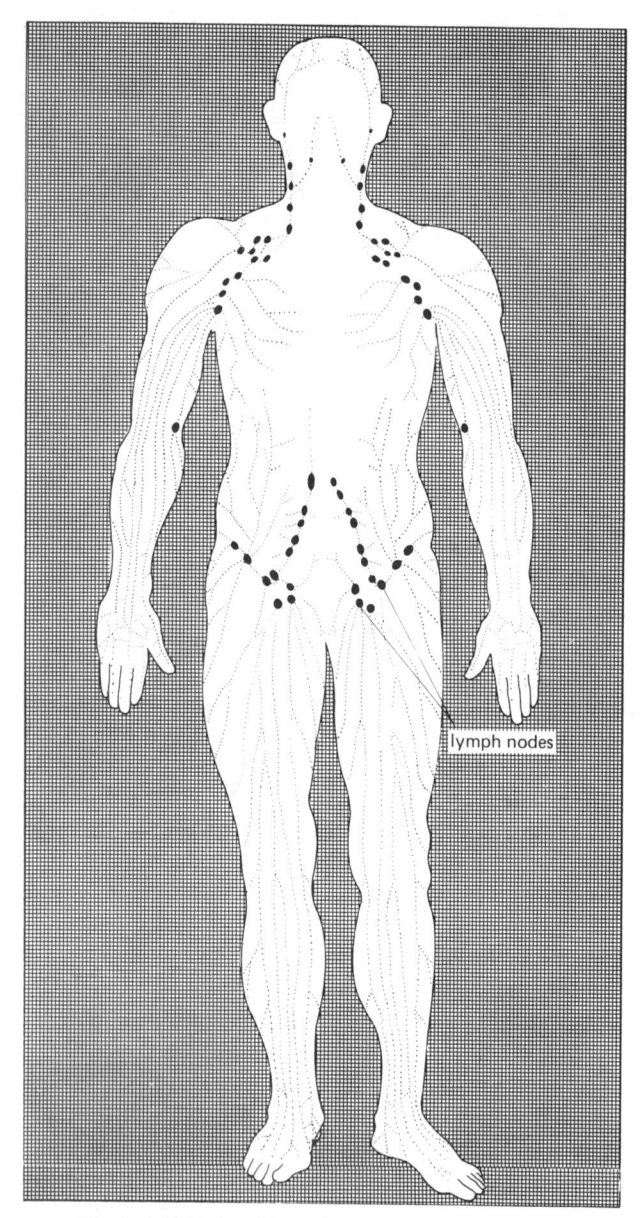

Fig. 8.14 Sites of lymph nodes in the human body

tissues. The colouring matter – haemoglobin – passes to the liver where it forms part of the bile.

Functions of the Spleen
1. It plays a part in the manufacture of blood cells in the foetus.
2. It removes worn-out red blood cells.
3. It manufactures lymphocytes.
4. It destroys bacteria as they pass through in the bloodstream.

BREATHING

Oxygen is required continuously by the body in order to release energy for all vital body processes. The system whereby oxygen is taken into the body through the air passages and lungs is often called the *respiratory system*, although this may be confused with *respiration* which is the series of chemical changes involved in the release of energy from food. For this reason in this chapter we will simply talk about *breathing*.

The Mechanism of Breathing
Ventilation of the lungs – that is the introduction of oxygen and removal of carbon dioxide – is brought about by muscular movements of the chest which rhythmically increase and decrease its volume. These movements are:
1. Inspiration or inhaling – breathing in.
2. Expiration or exhaling – breathing out.

The thoracic cavity is an airtight box consisting of the ribcage and intercostal muscles and the arched muscular diaphragm beneath. The inner surface of the thoracic cavity and the outer surface of the lungs are covered with thin membranes, the pleural membranes, which adhere to each other closely with the help of the thin film of moisture between them. It follows that any movement of the chest walls will also affect the lungs, which adhere to them.

2. It drains tissue spaces and helps maintain homeostasis.
3. White blood cells (lymphocytes) are manufactured in the lymph nodes.
4. Lymph nodes reduce the spread of infection by filtering dangerous organisms from the lymph and destroying them.
5. Digested fats are absorbed and transported from the lacteals to the bloodstream.

The Spleen
This is a small spongy organ situated in the top left-hand corner of the abdomen, beside the stomach. It is composed of fibrous elastic tissue and is supplied with blood by the splenic artery. Its main function is to remove damaged or worn-out red cells by entangling them in its

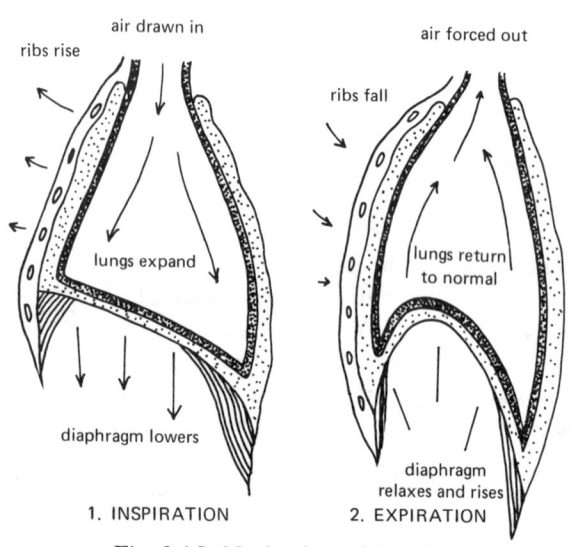

Fig. 8.15 Mechanism of breathing

1. Inspiration. During inhalation, the intercostal muscles contract, lifting the rib-cage upwards and outwards. At the same time the diaphragm, which is normally dome-shaped, contracts and flattens. Both of these movements have the effect of increasing the volume of the thoracic cavity. This stretches the lungs, reducing the pressure within. As atmospheric pressure outside is now greater, air rushes in through the air passages to equalise the pressure.

2. Expiration. When we breathe out or exhale the opposite occurs. The intercostal muscles and diaphragm relax, returning to their normal position and thus reducing the size of the thoracic cavity. The lungs follow this movement. The smaller capacity of the thorax increases the pressure of air within, so the air is forced out until pressure inside and outside is equal.

Inhaling and exhaling occur normally about fifteen to twenty times a minute in an average adult and more often in babies and small children. The rate of breathing can be increased when necessary: for example, exercise demands extra supplies of oxygen, and the rate of breathing quickens during muscular activity in order to meet this need.

Not all the air in the lungs is removed with each breath: if this happened, the lungs would collapse. The total capacity of the lungs of an average adult is about five and-a-half litres, and only about half a litre of air is breathed in and out during normal breathing. The air breathed in and out is called the *tidal volume* of air, while the air remaining in the lungs at all times (at least 1,200 ml) is called *residual air*. During very deep breathing extra air is forced in and out of the lungs, and this new total volume (up to four litres) is known as the *vital capacity* of the lungs. Athletes can increase their vital capacity considerably through exacting training programmes. Respiratory disease and smoking can reduce the vital capacity.

The Pathway of Air

Air flowing in and out of the lungs passes through the respiratory tract, consisting of the mouth or nose, the pharynx, the larynx, the trachea, bronchii and their branches within the lungs, the bronchioles.

The nose. As air passes through the nose it is 'conditioned', i.e. warmed and filtered by the cilia (hairs) of the nasal cavities. Mucus, a sticky substance secreted by cells in the nose, entraps dust, destroys bacteria and moistens the air passing through.

The mouth. Breathing through the mouth is not as satisfactory because the air is neither filtered nor conditioned and lung infection may ensue through bacterial action or inflammation caused by dryness. The air passes into the *pharynx,* a wide cavity at the back of the nose and mouth. This divides into two tubes, the *oesophagus* (gullet) and the *trachea* (windpipe). A piece of tissue called the *epiglottis* acts as a sort of lid which closes over the trachea during swallowing so that food cannot pass into the trachea. Sometimes (often when a person is talking and eating at the same time) a particle of food literally 'goes down the wrong way'. This brings about a violent fit of coughing which usually expels the offending particle.

The larynx or voice box is a wide area at the beginning of the trachea. It is protected at the front by a piece of cartilage. During male puberty the voice box enlarges, deepening the voice and pushing the cartilage outwards to form the 'Adam's apple'.

The vocal cords are two strips of tissue which stretch across the larynx. As air passes through them they vibrate. Muscles adjust the tension of these cords, causing changes in pitch which are utilised in speech and singing. Inflammation of the vocal cords is known as *laryngitis.*

The trachea is a tube of fibrous tissue about 10 cm long. It is kept open by 'C' shaped bands of cartilage, so that pressure from surrounding organs does not close it. The trachea is lined with epithelial cells which secrete mucus to keep the passage moist and entrap any dust or bacteria which may have eluded the mucus of the nose. It is lined with cilia — hair-like projections whose function is to sweep mucus, dust and bacteria upwards toward the mouth. This mucus mixture, called sputum, is generally infected with bacteria.

The trachea divides into two branches as it approaches the lungs, the right bronchus entering the right lung and the left bronchus, the left lung. The structure of the bronchii is similar to that of the trachea: the 'C' shaped bands continue but become thinner as they pass further into the lung tissue. In each lung the bronchus branches into several bronchioles, penetrating each portion of the lungs until eventually microscopic bronchioles end in bundles of air cells or alveoli. These act as reservoirs for air, each resembling a tiny bubble. The walls of the air cells are extremely thin; they are, in fact, formed from a single layer of endothelial cells. They are completely surrounded by a dense network of capillaries, the walls of which are also very thin. The millions of alveoli in the lungs provide a huge surface area (about 100 sq metres — the size of a tennis court) through which exchange of gases can take place.

Gaseous Exchange

The alveoli are filled with air which contains a relatively high level of oxygen. The blood in the capillaries has a low concentration of oxygen. Because the walls of both alveoli and capillaries are extremely thin, the oxygen diffuses firstly into the moisture which bathes the alveoli and then through the capillary walls into the bloodstream

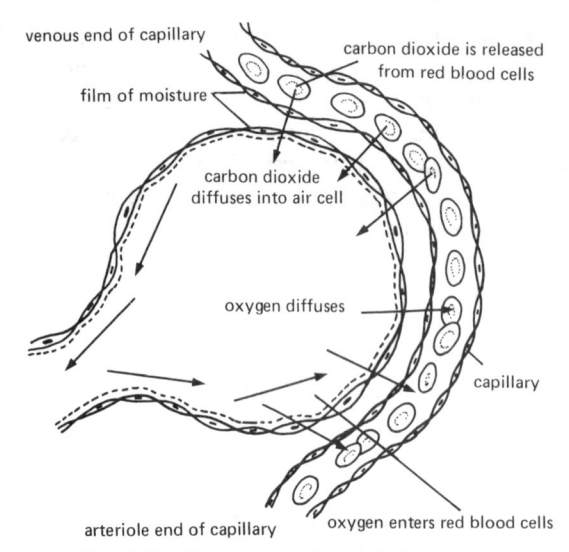

Fig. 8.16 Exchange of gases in lung tissue

where it combines with haemoglobin.

Carbon dioxide is carried from the tissues as bicarbonate salts. As a result of stimulation from enzymes present in the alveoli, the bicarbonate breaks down into free carbon dioxide and water. The concentration of carbon dioxide in the capillaries is now higher than in the air cells. Diffusion of carbon dioxide takes place into the alveoli and it is breathed out through the same passages through which 'fresh' air reached the air cells.

The pulmonary veins carry the oxygenated blood back to the heart from where it is pumped to every tissue in the body. In the tissues it is released in areas where oxygen is deficient. Carbon dioxide is collected and carried to the lungs where the process is repeated.

Structure of the Lungs

The lungs practically fill the thoracic cavity, allowing just sufficient space for the heart which lies between them. They are cone or pyramidal in shape with the pointed end uppermost, are light and spongy in texture and are covered with a smooth membrane called *pleura*. The lungs are protected at the back by the back bone, and the ribs and intercostal muscles provide protection at the sides and in front. The diaphragm separates them from the abdominal organs. The colour of the human lung varies from mottled pink to a grey-brown colour, the latter probably in a city dweller or cigarette smoker. The right lung is folded into three *lobes,* the left one into two lobes. The only opening into the atmosphere is through the trachea to the nose and mouth. The main blood vessels and bronchii enter each lung near the heart, at the *root* of the lung. As well as bronchii, bronchioles and air cells, the lungs contain numerous arteries, veins and capillaries, lymphatics and nerves.

Control of Breathing

Although we do have some conscious control over the

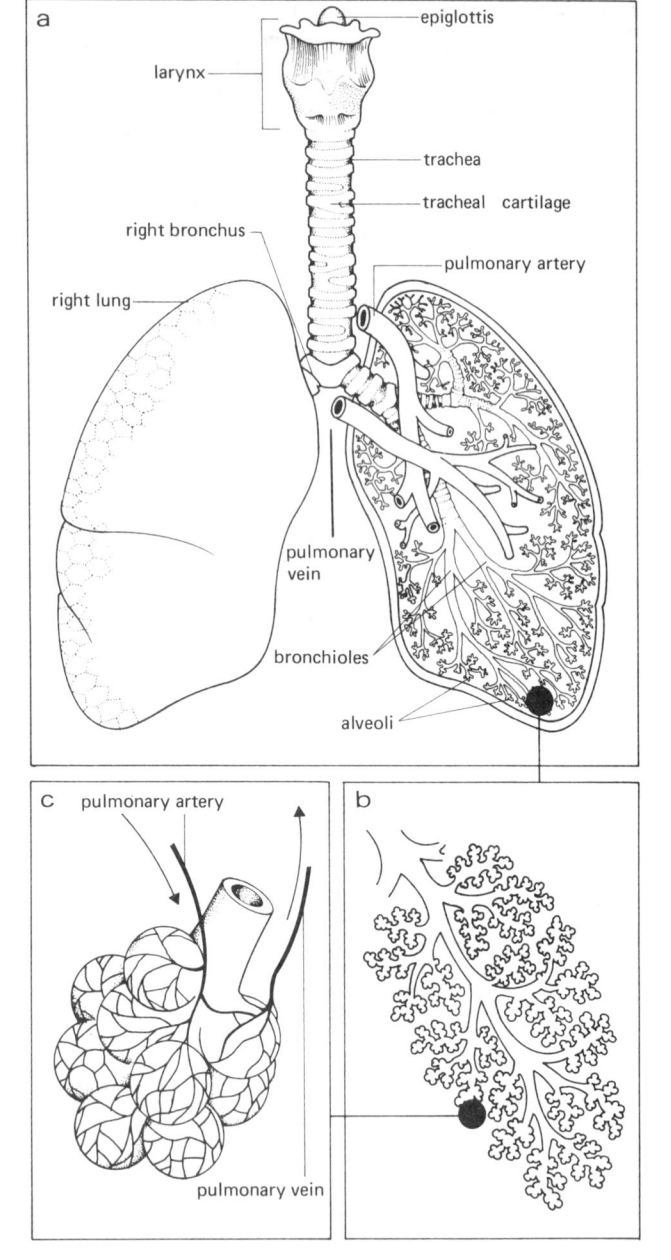

Fig. 8.17 (a) Breathing organs in man; (b) bronchioles; (c) alveoli

rate and depth of our breathing, we find if we try to hold a breath that before long we have to give in to the involuntary system which controls our breathing. As already explained, vigorous exercise increases the demand for oxygen. How do the lungs know when to increase the rate of breathing? This is controlled by chemical and nervous factors.

1. The medulla oblongata is the area of the brain controlling the breathing mechanism. Cells in this part of the brain are particularly sensitive to the concentration of carbon dioxide in the blood: when this rises above normal level (e.g. when a person begins to run), the medulla oblongata increases the nervous impulses sent down the spinal cord to the intercostal muscles and diaphragm, causing them to increase the rate of breathing so that more carbon dioxide is expelled. This automatically means that more oxygen is taken in to help increase the energy output. When the carbon dioxide concentration in the lungs is low (e.g. during sleep), the medulla decreases the rate and depth of breathing.

2. *Certain blood vessels* operate a similar breathing control mechanism. The aorta and *carotid artery*, below the neck, contain sensitive receptors which can detect a reduction in the oxygen level of the blood. These send impulses to the medulla oblongata stimulating it to increase the level of breathing.

3. *Alveoli* are also concerned with breath control. When they are stretched owing to expansion of the lungs, they stimulate the nerves in the respiratory centre

Air Pollution

Pollutant	Origin	Results	Methods of Prevention
Tar, smoke, nicotine	Cigarette smoking	Heart disease, lung cancer bronchitis and other respiratory diseases	No smoking
Pathogenic micro-organisms, (e.g. bacteria) and viruses	Unhygienic habits, careless coughing, sneezing – droplet infection	Respiratory infections, tuberculosis, various infectious diseases	Improved ventilation, vaccination, more fresh air and exercise
Carbon dioxide	All forms of combustion including breathing	Increased susceptibility to respiratory infections, pallor; poisonous in large amounts	Improved ventilation, use of solar and wind energy instead of fossil fuels
Minute particles such as dust, soot, ash	Domestic and industrial combustion, mining, certain industrial occupations	Respiratory infections Damage to lungs – emphysema etc.	Smokeless fuels, use of solar/wind energy
Sulphur dioxide	Industrial combustion oil refineries etc.	Lung damage, respiratory infections. Damage to plant life and buildings	Remove sulphur from fuels. Decentralise industry
Carbon monoxide	Incomplete combustion in industrial and domestic furnaces. Car Exhausts	Combines with haemoglobin in blood, reducing oxygen level – eventually causing unconsciousness and death	Decentralisation. Banning or restricting use of private cars in built up areas. Modify engines and exhausts
Hydrocarbons	Car exhausts and industrial combustion.	Lung and eye irritation Smog. Damage to plant life	Modify car engines
Lead	Car exhausts	Affects nervous system	Restrict use of cars Modify engines
Radioactive fallout	Nuclear power plants. Leaking radiation from radioactive scanners and other medical/industrial equipment	Leukaemia, affects growth. Various illnesses and death	Utilise fossil fuels economically Use of solar/wind energy instead of nuclear energy

which stop contraction, allowing the respiratory muscles to relax and the lungs to reduce in size.

4. *The hormone adrenaline* (see p. 187) also influences the rate of breathing by narrowing or widening the bronchioles, thus reducing or increasing air supply according to circumstances.

Composition of Air (approximate)

	Inspired Air	Expired Air
Oxygen	21%	16%
Carbon dioxide	0.04%	4.04%
Nitrogen	79%	79%
Water vapour	varies	saturated

Ventilation

It goes without saying that if the lungs are to function efficiently the air which is breathed in must have a good supply of oxygen (21 per cent). This brings us to the subject of ventilation, which is covered in more detail in Part 2. Badly ventilated homes and buildings result in an inadequate supply of oxygen to the lungs. As carbon dioxide continues to accumulate, a feeling of drowsiness will result, with increased susceptibility to respiratory complaints such as colds and influenza.

Diseases of the Respiratory Organs

These include:

Tuberculosis (TB) – caused by a bacillus.

Pleurisy – an infection of the pleurae.

Colds and sore throats – caused by viruses and bacteria.

Pneumonia – a viral or bacterial infection which causes an accumulation of mucus and lymph in the bronchioles and alveoli, reducing the surface available for gas exchange. Severe cases can be fatal but the disease can be successfully treated with antibiotics.

Emphysema – a disease generally associated with polluted air. The walls between the alveoli become inflamed and break down, again reducing the surface area exposed to the air.

Bronchitis – caused by inflammation of the bronchii.

THE DIGESTIVE SYSTEM

Food, which is essential to life, is of no use to the body until it is broken down into a form which can be absorbed by the body cells. Digestion is the process whereby large molecules of food are broken down into small, soluble molecules. This occurs physically (or mechanically) by grinding with the teeth and the squeezing movements of the muscular walls of the digestive tract.

Chemical breakdown of food involves the breaking of links in food molecules by the addition of water. This is called *hydrolysis* (from the Greek hydro – water and lysis – loosening). Chemical breakdown is speeded up by the action of enzymes. There is also a psychological factor in the digestion of foods. The flow of saliva and gastric juices can be triggered off by the sight, smell or even the thought of food. The *hypothalamus,* a small area at the base of the brain, controls our appetite, initiating the feeling of hunger when the stomach is empty. It is thought that the hypothalamus is controlled by blood sugar level and by nervous impulses from the stomach. When we have eaten sufficient food, the hypothalamus switches off our appetite so that we do not eat too much and become overweight. When we are ill or upset, the hypothalamus reacts in such a way that we 'lose our appetite'. Long before biochemistry research uncovered the function of the hypothalamus, Shakespeare wrote

Now good digestion wait upon appetite
and health on both.
MACBETH

As we have seen from our study of nutrients, many food molecules – proteins, carbohydrates and fats – are made up of large and complicated molecules and are insoluble in water. Not all foods must be digested. Water, salts, glucose, alcohol and many drugs are made up of simpler soluble molecules which can be absorbed without digestion. On the other hand it is not possible to digest certain foodstuffs such as cellulose. These pass through the body without being digested and are eliminated in the faeces.

The Alimentary Canal

This is a continuous muscular tube through which the food passes from the time it is ingested through the mouth until the waste is eliminated at the anus. Most of the canal is lined with mucous membrane, the glands of which produce many digestive fluids containing one or more enzymes. The muscular structure of the walls of the alimentary canal enables it to contract and relax alternately, pushing the food through in a wave like movement called peristalsis.

Enzymes

What are enzymes? They can be defined as organic catalysts which control or accelerate chemical changes within living organisms, yet remaining unchanged themselves.

Thousands of chemical reactions take place in the body. These changes are collectively known as metabolism and include the anabolism or building up of complex molecules and the catabolic reactions, breaking large

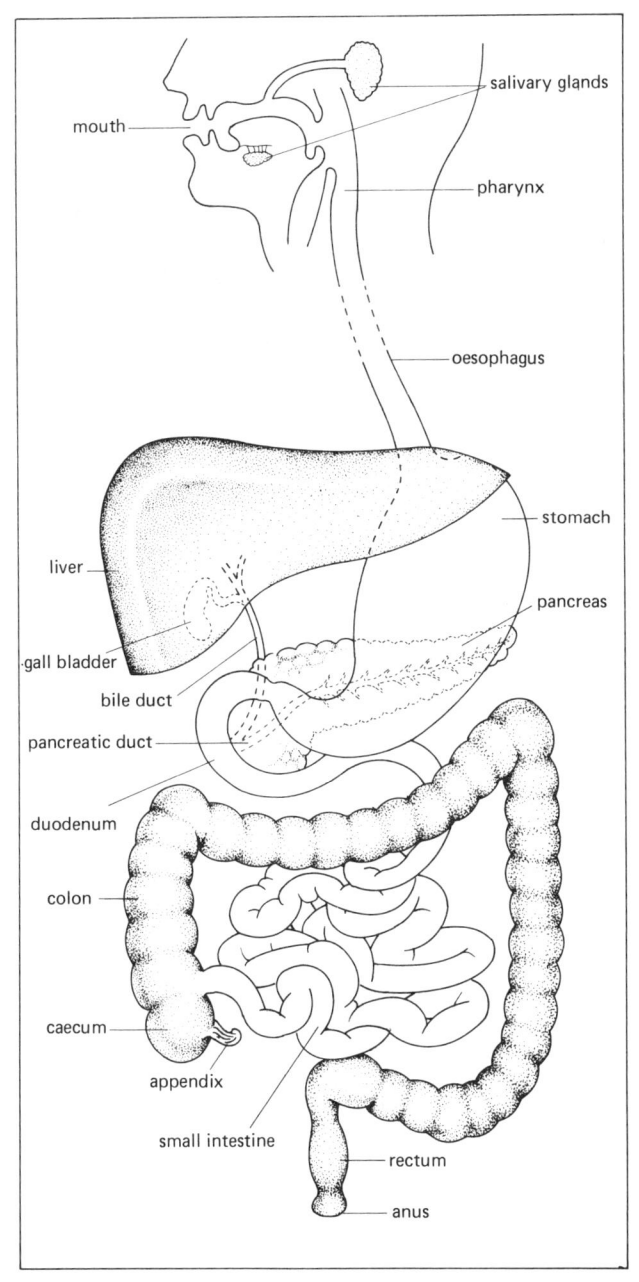

Fig. 8.18 Alimentary canal

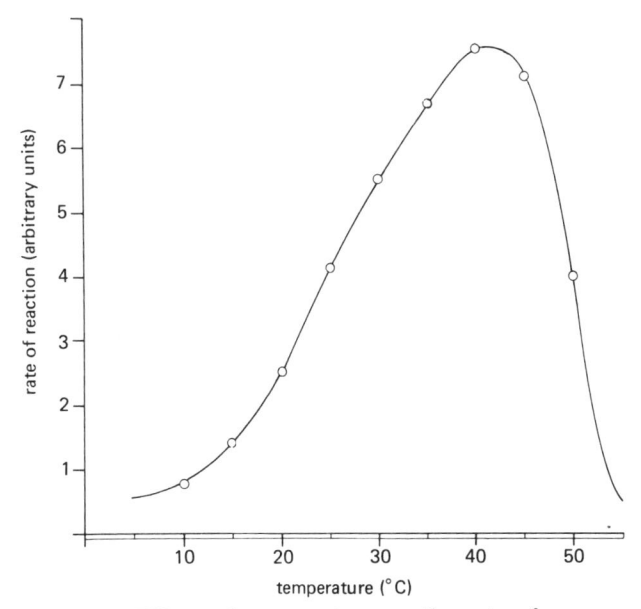

Fig. 8.19 Effect of temperature on the rate of enzyme activity

molecules into simple substances with the release of energy. Each change involves many steps, and each step is controlled by a particular enzyme. The enzymes also speed up processes which would otherwise take place more slowly – for example, one enzyme can catalyse a reaction millions of times each minute. Without enzymes, body chemistry would run amok.

The substance on which an enzyme acts is called the *substrate* and each enzyme is usually named after its substrate – for example, the enzyme which acts on maltose is called maltase. (Most enzymes end with 'ase' but there are exceptions, e.g. pepsin and trypsin.)

Enzymes are proteins and as such are easily affected by heat. Most will work best at a particular temperature and pH. Enzymes present in animals have an optimum temperature of 37°C (normal body temperature), but plant enzymes work at a lower temperature of about 25°C. All enzymes are inactivated by cold temperatures and destroyed (denatured) by extreme heat. Most will not react at temperatures above 60°C and boiling point destroys them. (We have already mentioned that vegetables are blanched before freezing to prevent enzyme action.)

pH. Although many enzymes react in a neutral environment and will not react in extremely acid or alkaline conditions, some prefer a specifically acid or alkaline medium. Digestive enzymes vary. The enzymes pepsin and rennin work best in an acid medium, which is made possible by the plentiful secretion of hydrochloric acid in the stomach, but intestinal enzymes favour a slightly alkaline medium.

Catalytic action of enzymes. Most chemical reactions demand a great deal of energy, and if the required amount of energy were supplied at one time, the cells would burn up. Enzymes reduce the amount of energy required by changing the reaction into several steps or stages, each of which requires very little energy.

Selectivity of enzymes. The action of each enzyme is specific: it will accelerate one particular action only and have no effect on any other. This selectivity can be compared to a lock and key. The enzyme is the lock and the substrate molecules together form the key which fits it. Both molecules fit exactly into the active site of the enzyme, 'unlocking' them and enabling them to react with each another. After this brief union, the molecules leave the enzyme, which remains unchanged and ready to act as a 'lock' to a similar 'key'.

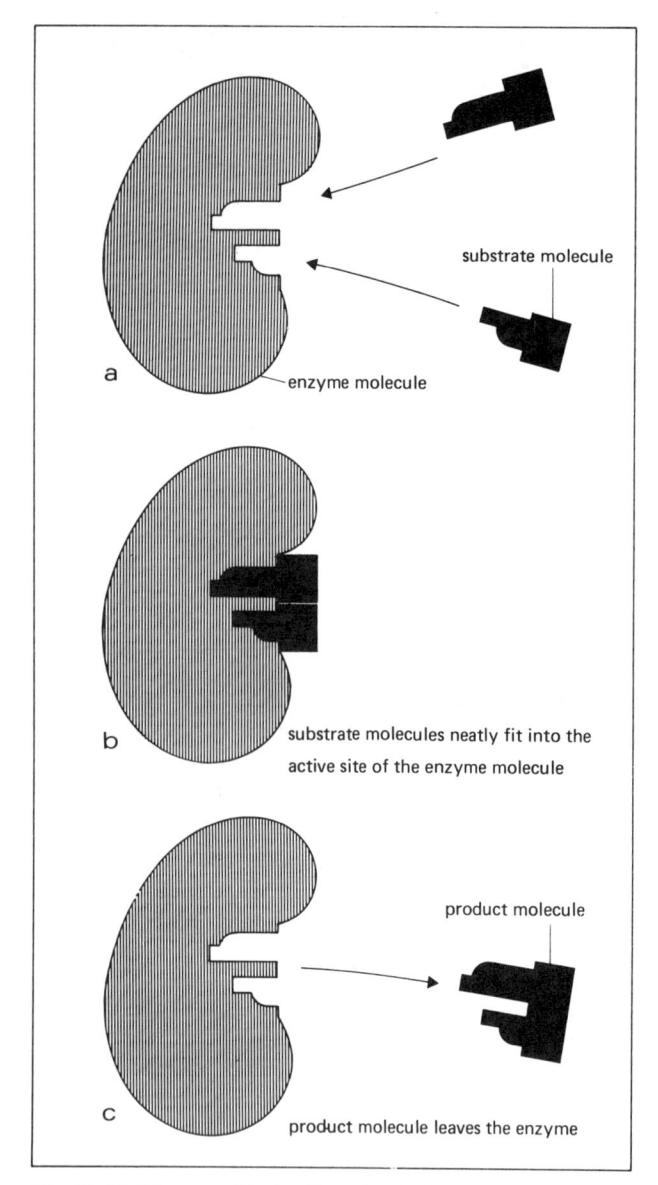

Fig. 8.20 Diagram illustrating the 'lock and key' action of enzymes and substrate molecules

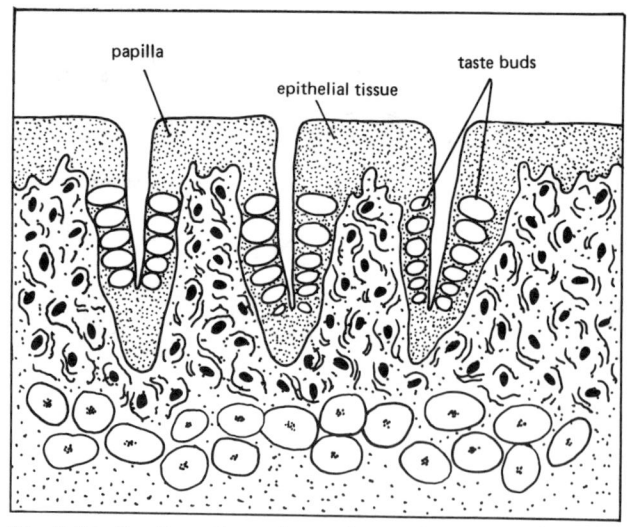

Fig. 8.21 Section through surface of tongue, showing taste buds

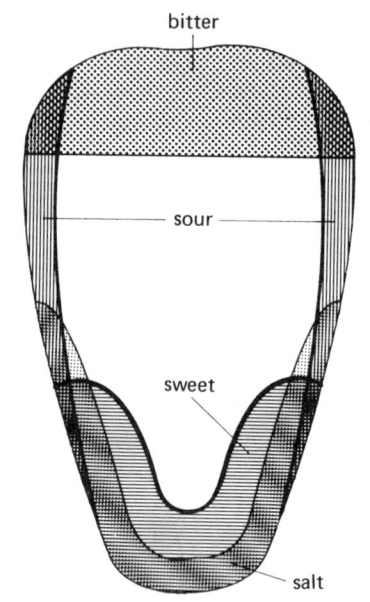

Fig. 8.22 Tongue and distribution of taste buds.

Co-enzymes. Many enzymes require the presence of a co-enzyme in order to function. Co-enzymes are not proteins but their presence is essential if the catalytic action is to proceed. Many co-enzymes are related to the vitamin B group. Two well-known ones are ATP and NAD (see p. 176).

The Mouth

Digestion starts in the mouth. The food is masticated or ground down by the teeth in order to facilitate swallowing and to increase the surface area which will be exposed to the action of the digestive juices. Saliva softens and begins to dissolve the food, assisting in the sensation of taste.

Taste. The sensations of taste and smell play a major part in digestion by stimulating the flow of digestive juices. The sight or smell of food increases the amount of saliva secreted, making the mouth 'water'. Special sensory cells called taste buds are found on the surface of the tongue. These are stimulated by chemicals in food and send messages to the brain which interprets the 'taste'. Our sense of taste can only distinguish between chemicals which taste bitter, salty, sweet or sour. Taste and smell are closely related. Similar sensory cells are found in the epithelial lining of the nose and these can more readily distinguish the subtle differences between flavours when the odours from food in the mouth pass into the back of the nose.

Salivary glands. Saliva is secreted by three pairs of glands:
1. The *parotid glands* at the back of the cheeks under each ear.

Fig. 8.23 Salivary glands

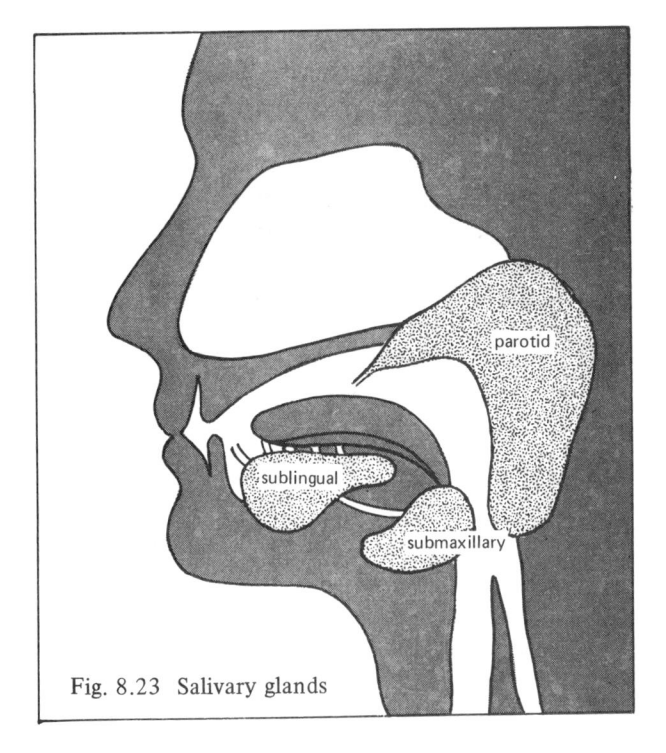

2. The *submaxillary glands* under the jaw.
3. The *sublingual glands* under the tongue.

They consist of secretory cells grouped around tiny ducts leading into a main duct which leaves the gland and passes into the mouth. Saliva is secreted continuously in order to keep the mouth moist but secretion is increased as a result of reflex action caused by the odour, taste or even the thought of food.

Saliva is an alkaline fluid which contains a sticky secretion called mucus, common to all organs in the alimentary tract. Saliva contains an enzyme ptyalin or salivary amylase, which acts on cooked starch, breaking it down into dextrin, a shorter chain molecule, and maltose, a disaccharide. Both require further digestion before they can be absorbed into the bloodstream.

Test to show the action of saliva. Wash out the mouth with warm water and then obtain a copious secretion of saliva by a chewing action. Rinse again and collect 10 cm^3 of saliva solution in a test tube. Divide solution into two halves, boiling one half for one minute. Label both test tubes.

Make up a solution of starch with hot water. Leave to cool. Put 5 cm^3 of starch solution into each test tube of saliva. Shake well. Place both test tubes in a beaker of water at 37°C for 30 minutes.

Test both for starch (see p. 11).
Test both for sugar (Fehlings test — see p. 11).

The tube containing boiled saliva will give a positive result for starch and a negative result for sugar because boiling inactivates or kills the enzyme, therefore no change occurs.

The tube containing fresh saliva will give a negative result for starch but a positive one for sugar, showing that the enzyme in saliva changes starch to sugar.

Swallowing. When the food has been sufficiently masticated and moistened by saliva, the tongue forms it into a ball (bolus) and passes it into the throat.

The Oesophagus

After passing through the pharynx, the food enters the oesophagus or gullet, which lies behind the trachea, running parallel with it. During the act of swallowing the epiglottis closes over the top of the larynx to the front, in order to prevent food from passing into the trachea and causing a fit of choking. The oesophagus is a straight, muscular tube, about 25 cm long, which extends from the pharynx to the stomach. As soon as the bolus of food enters, the oesophagus contracts rhythmically, first in front of the food and then behind it, so that the food is gradually pushed towards the stomach. This involuntary wave-like movement is brought about by circular muscles in the walls of this and all organs of the digestive tract and, as already mentioned, is called peristalsis. Peristaltic contraction continues throughout the length of the alimentary canal. There is no digestion in the oesophagus except a continuation of the salivary digestion of starch.

The Stomach

The stomach is situated on the left of the abdomen immediately below the diaphragm. It lies beside the liver and in front of the pancreas. It is a large, hollow organ which acts as a reservoir for food so that our food intake can be limited to three or four times a day instead of every 20—30 minutes.

In common with other digestive organs, the stomach has three layers: an outer covering of peritoneum, a thick layer of smooth (involuntary) muscle and an inner lining. The muscles in the middle layer are arranged alternately in longitudinal, circular and oblique fashion. Such an arrangement gives the muscles elasticity, enabling the stomach to expand considerably after meals in order to hold all the food eaten. It also produces strong contractions which churn the food about so that it mixes with the gastric juice.

The inner layer is divided into a sub-mucous and mucous coat; this mucous coat is deeply folded when the stomach is empty. Between the folds are millions of gastric glands which secrete gastric juice.

At the point where the oesophagus enters the stomach a tight ring of muscle, the cardiac sphincter, controls the entry of food and prevents the contents of the stomach from passing back up into the oesophagus. When harmful food is eaten, or when the stomach is inflamed, the powerful muscular contractions of the stomach, together with reverse peristalsis in the gullet, cause vomiting.

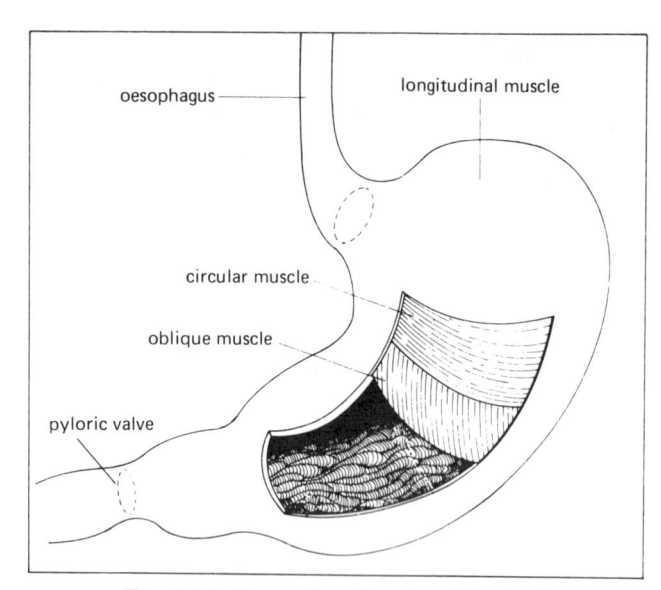

Fig. 8.24 Stomach wall and gastric glands

At the lower end of the stomach, a similar sphincter muscle, the pylorus, gradually releases the contents of the stomach into the duodenum.

When food reaches the stomach, the strong muscular walls contract and relax alternately, churning the food about and mixing it thoroughly with the gastric juice until it reaches a thick creamy consistency, when it is known as *chyme*. The length of time food remains in the stomach varies between two and five hours, depending on the type of food eaten. Water and some liquids pass through the stomach quickly, but potatoes and other carbohydrate foods remain much longer and proteins and fats stay in the stomach longest of all. The presence of food causes the blood vessels in the stomach lining to dilate and supply extra blood for digestion.

Digestion in the Stomach

Gastric glands secrete many substances into the stomach including hydrochloric acid, mucus and three enzymes — pepsinogen, rennin and gastric lipase.

1. Pepsinogen. When the walls of the stomach become distended with food, the hormone gastrin is secreted by the stomach lining and passes into the blood stream. When it reaches the stomach again it stimulates the gastric glands to produce pepsinogen. This becomes active when mixed with hydrochloric acid and changes to pepsin — a proteolytic (protein-splitting) enzyme which breaks down the large protein molecules into shorter chains called peptones.

2. Rennin causes the milk protein caseinogen to clot — i.e. it changes to casein and this combines with calcium to form calcium caseinate which is readily digested by pepsin. This enzyme is most active in young infants.

3. Gastric lipase. Very small amounts are secreted and they have little effect on the fats present in food.

Hydrochloric acid
1. This is secreted by oxyntic cells in the mucous lining of the stomach. It has a very low pH (1–2) producing the degree of acidity necessary for the action of gastric enzymes.
2. Its strong acid content destroys bacteria unless they are present in extremely large numbers.
3. Hydrochloric acid neutralises the action of ptyalin soon after it enters the stomach.

Mucus is a viscous secretion which adheres to the wall of the stomach, preventing the acid from destroying and digesting it. Over-acidity causes ulceration of the stomach (gastric ulcers) and the duodenum (duodenal ulcers) and less severe, but unpleasant symptoms such as heartburn and indigestion. Nervous tension increases the secretion of hydrolchloric acid.

Other functions of the stomach
1. The heat of the stomach melts fats and releases them from any proteins to which they may be attached.
2. The stomach breaks down the cellulose covering on many plant foods, releasing the contents to the action of enzymes.
3. A little absorption takes place in the stomach. Soluble foods of simple molecular structure such as salt, water, glucose, water-soluble vitamins and alcohol are absorbed into the blood flowing through the stomach's capillaries.

When its liquid contents leave the stomach, they pass into the small intestine. The first 25 cm of this are known as the duodenum and into this section pass the secretions of two important glands — the pancreas and the liver.

The Pancreas

The pancreas is a long, pinkish-white gland which lies under the stomach, stretching from the bend in the duodenum to the spleen. It is similar to the salivary glands in structure, being made up of groups of secretory cells called alveoli, which produce pancreatic juice. These open into minute ducts which join up to form the main pancreatic duct, and it is this which brings the pancreatic juice to the duodenum, entering at the same point as the bile duct from the liver. Both bile duct and pancreatic duct pour their secretions into the duodenum. The pancreas also contains groups of specialised cells, called *Islets of Langerhans,* which are involved in the production of insulin. They have no connection with digestion and form part of the endocrine system (see p. 186).

Pancreatic juice. This is a clear, colourless alkaline fluid containing water, salts and enzymes. The most important salt is sodium bicarbonate ($NaHCO_3$), which neutralises

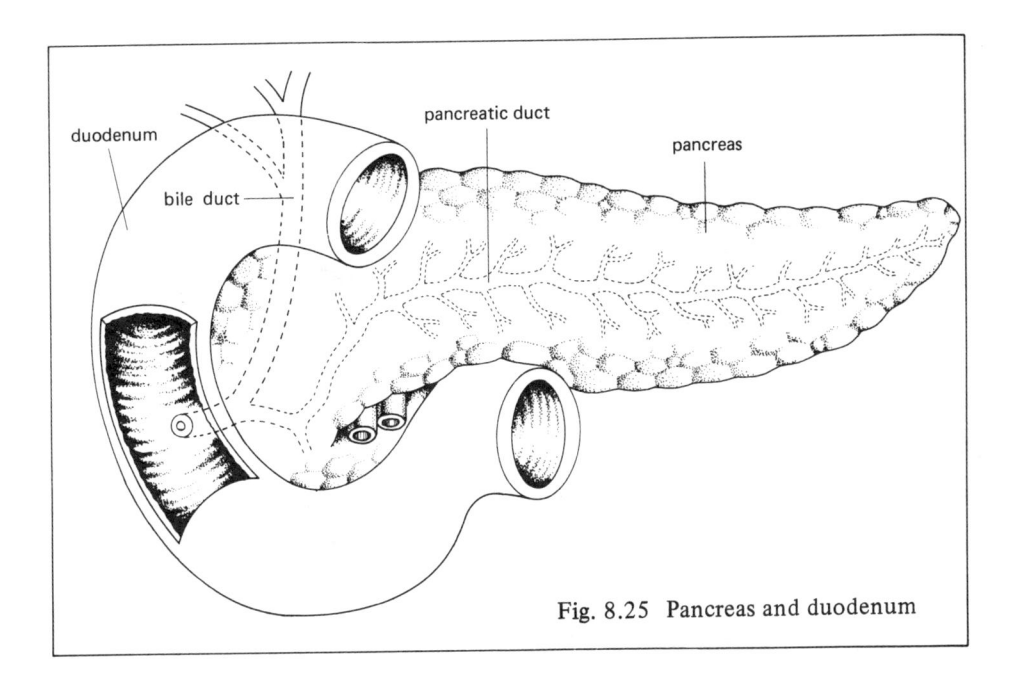

Fig. 8.25 Pancreas and duodenum

hydrochloric acid bringing the contents of the duodenum to pH8.

Pancreatic enzymes. Three important enzymes are secreted by the pancreas:

Trypsin is secreted in an inactive form and is activated when it mixes with the alkaline contents of the duodenum. It is a powerful proteolytic (protein-splitting) enzyme, which attacks some of the peptide links within large protein molecules, breaking them into smaller chains. Proteins which were partially digested by the stomach are further broken down in the duodenum, some of them to absorbable amino acids.

Amylase breaks down carbohydrate, and in particular starch, by hydrolysis, converting it to maltose.

Lipase splits fats into fatty acids and glycerol. The alkalinity of the pancreatic juice assists the breakdown but the work of lipase depends to a large extent on the action of bile.

Bile (see p. 177). Although this secretion produces no enzymes, it has an important part to play in the digestion of fats. Bile salts, like all emulsifiers, are both hydrophilic (water-loving) and hydrophobic (water-hating). The hydrophilic end of the molecule attaches itself to the water molecules in the food while the hydrophobic end is attracted to the fats. This emulsification breaks down the large globules of fat into tiny droplets, creating a greater surface for the fat-splitting enzymes to work on. Hormones secreted by the duodenum stimulate the release of bile during digestion.

Structure of the Small Intestine

The small intestine is a narrow tube about six metres long, which coils around the abdomen. It is held in place by a fine membrane called the mesentery. Its structure is similar to other areas of the digestive tract: an outer coat of peritoneum, a middle muscular coat and an inner lining. This epithelial layer is deeply folded and on its surface are millions of tiny projections called villi, which give it a velvety texture. Each contains a lacteal, which is a blind-

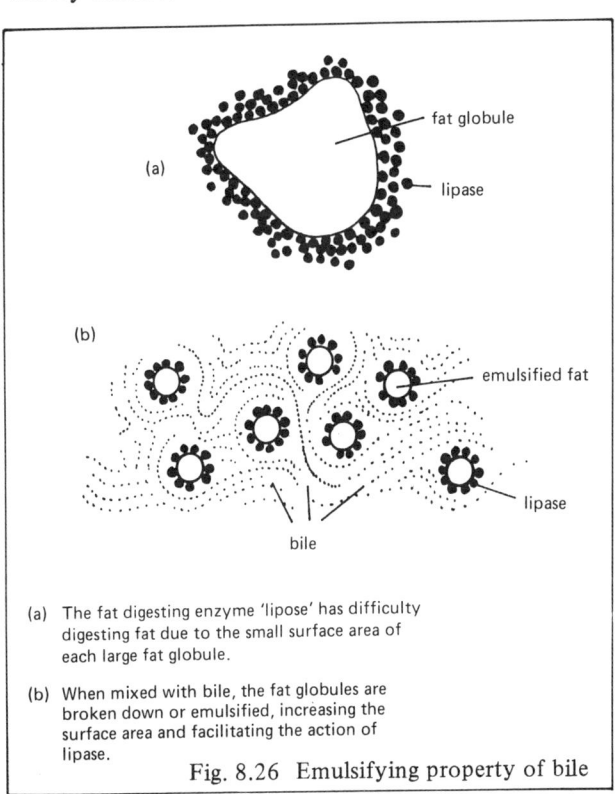

(a) The fat digesting enzyme 'lipose' has difficulty digesting fat due to the small surface area of each large fat globule.

(b) When mixed with bile, the fat globules are broken down or emulsified, increasing the surface area and facilitating the action of lipase.

Fig. 8.26 Emulsifying property of bile

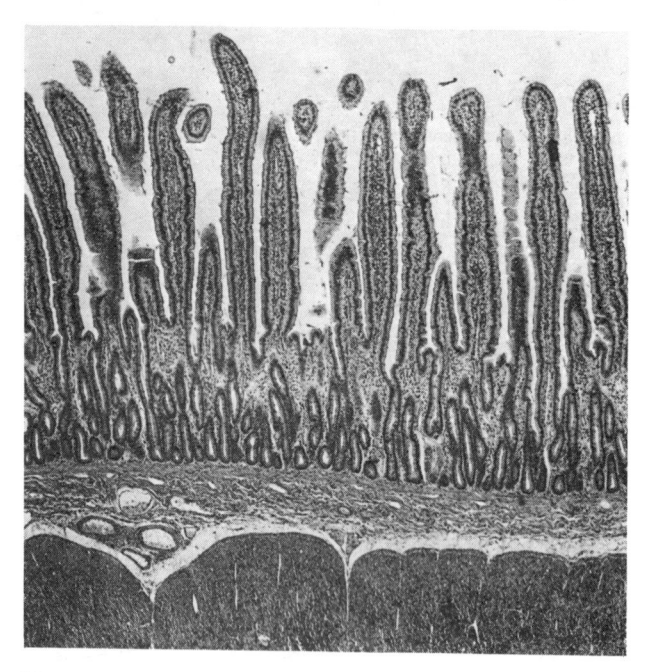

Transverse section of the small intestine showing villi (Gene Cox)

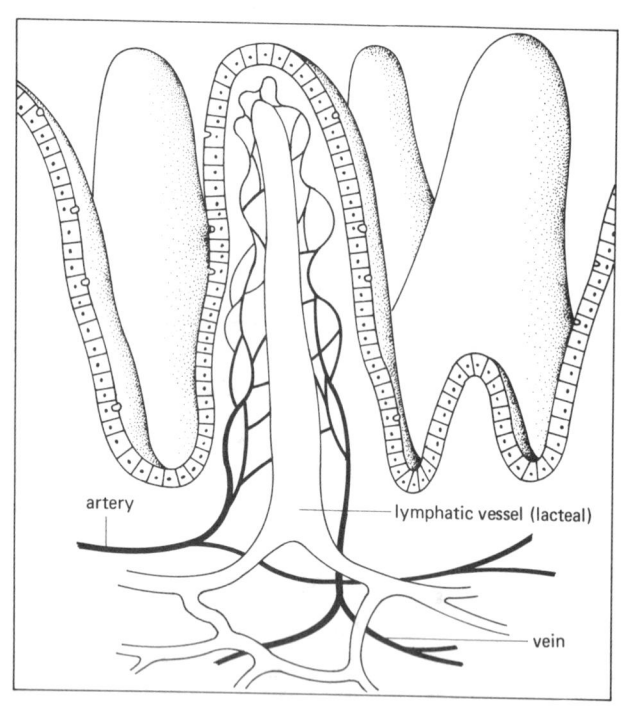

Fig. 8.27 Structure of villi

ended duct of the lymphatic system, and a network of tiny capillaries. These increase enormously the surface area of the intestine, an important factor in the absorption of foods. Between the villi are tiny openings which secrete intestinal juice *(succus entericus)*. The digestion of proteins and carbohydrates is completed in the *ileum*, the second part of the small intestine, by enzymes present in intestinal fluid.

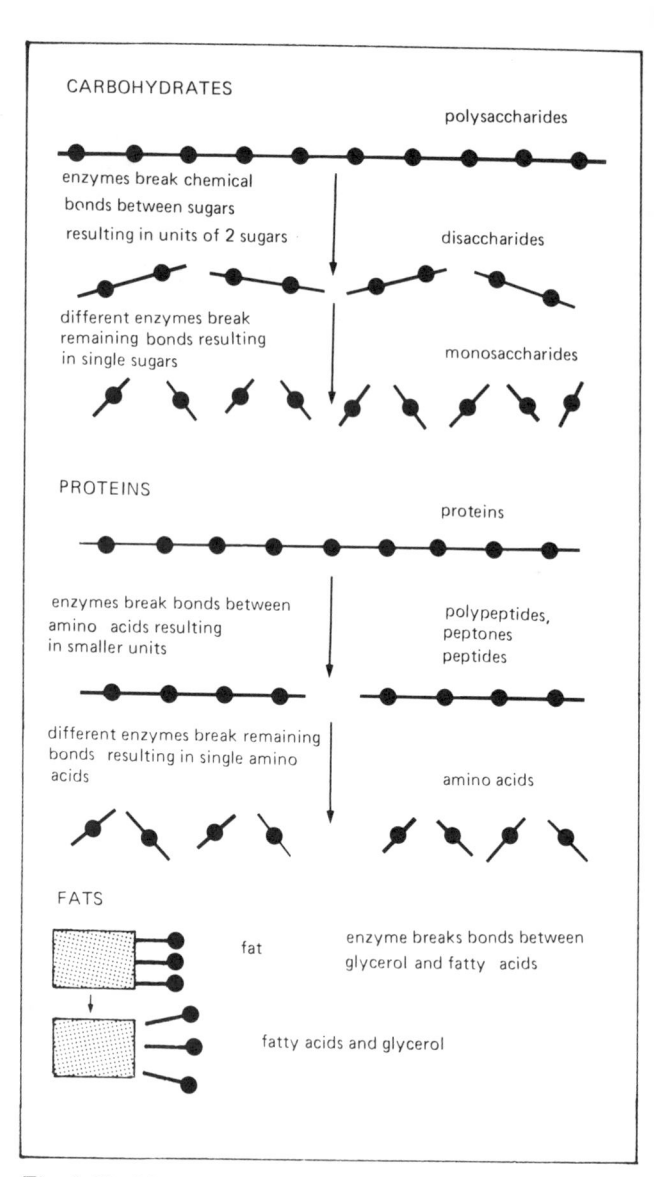

Fig. 8.28 Diagram showing chemical changes that take place during digestion

Digestion in the ileum

1. Proteolytic enzymes – peptidases (also known as eripsin) – break the links of any remaining polypeptides, degrading them to diffusible amino acids.
2. Carbohydrate-splitting enzymes convert disaccharides to monosaccharides:
   Maltase converts maltose to glucose.
   Invertase or *sucrase* converts sucrose to glucose and fructose. Lactase converts lactose (milk sugar) to glucose and galactose.
3. Intestinal lipase breaks down the few remaining fats into fatty acids and glycerol.

Digestion is now completed. All nutrients have been broken down into soluble, diffusible products, amino acids, monosaccharides, fatty acids and glycerol.

Absorption. The structure of the ileum suits the process of absorption perfectly. It is very long and the villi on its inner surface increase the absorption area to many times what it would be if it were a smooth tube. It is thought to have a surface area of about 100 sq. ft. As the epithelial walls are extremely thin, fluids and small molecules can readily pass through. The ileum is richly supplied with blood vessels and lacteals which carry away the products of digestion.

Monosaccharides and amino acids diffuse through the epithelial walls into the capillaries and are carried to the liver by the portal vein. Emulsified glycerides and fatty acids pass into the lacteals, where many are rearranged to form different triglycerides. The lacteals join lymphatic vessels in the intestine which pass up through the thoracic duct and empty into the bloodstream at the left jugular vein.

Absorption is both a process of diffusion and of active transport. Diffusion occurs at first when the concentration of nutrients in the intestine is greater than that in the lining cells, so the nutrients pass into the cells. The concentration in the cells then becomes greater than that in the blood vessels and lacteals, so nutrients pass from the cells into these vessels. Further on in the intestine, where the concentration is reversed and most of the nutrients are absorbed, the remainder pass through the lining walls by active transport. This process requires energy, which is taken from digested food.

Food remains about four to five hours in the small intestine, by which time all digestible nutrients have been absorbed. Any substances which have not been absorbed pass through the *ileocaecal valve* into the large intestine.

The Large Intestine (Colon)

This is a wide tube which begins at the *caecum,* passes up the right side of the abdomen, crosses the body and passes down the left side and into the rectum, a strong muscular tube. Material entering the large intestine contains large amounts of water, which have been poured into the alimentary tract during digestion. The colon also receives worn-out cells from the digestive tract, undigested food, particularly cellulose, and huge numbers of relatively harmless bacteria. The main function of the large intestine is to reclaim water and, by absorbing it, to concentrate the faeces. The bacteria present break down some undigested substances such as cellulose, and in the process they manufacture some vitamin K and B group vitamins, particularly B_{12}. These are absorbed through the inner lining.

The contents of the large intestine take about 12–24 hours to pass through to the rectum. The faeces are stored in the intestine until, as a result of strong peristaltic contractions, they are eliminated from the body.

When the large intestine becomes infected, it tends to pass the contents through more quickly so that insufficient water is absorbed. This is known as diarrhoea and is a common symptom of food poisoning. Diarrhoea is usually accompanied by dehydration. The reverse occurs when the contents remain too long in the large intestine: too much water is absorbed and the faeces become hard and difficult to pass. This is known as constipation.

Utilisation and Storage of Food

All digested nutrients are transported by the blood to wherever they are required. In the tissues, cells absorb and metabolise glucose, amino acids and fats.

Glucose. Much of the glucose which passes through the portal vein to the liver is absorbed immediately by the cells, where it is oxidised to provide heat and energy:

$$\underset{\text{glucose}}{C_6H_{12}O_6} + \underset{\text{oxygen}}{6O_2} \rightarrow \underset{\text{carbon dioxide}}{6CO_2} + \underset{\text{water vapour}}{6H_2O} + \underset{\text{energy}}{\text{energy}}$$

This does not take place all at once but in a gradual series of steps involving about thirty different reactions (see p. 176). The carbon dioxide and vapour produced are excreted by the lungs during breathing.

Excess glucose is converted into glycogen, an insoluble starch-like chain of glucose molecules, which is stored in the liver and muscles until required. When the blood sugar level falls, liver glycogen is converted back into glucose and passes into the bloodstream. Muscle glycogen is used as a source of energy for muscular contraction. If more glucose is available than can be stored by the body, it is converted into fat and stored as adipose tissue (see below).

Fat. Some digested fat is oxidised to produce energy, a little is used for structural purposes (e.g. to form part of the cell walls) and the remainder is stored as an energy reserve. Certain cells have the ability to absorb large amounts of oil within the cytoplasm. Fat is stored in these cells which group together to form adipose tissue, under the skin and around certain organs such as the kidneys. Unlimited amounts of fat can be stored by the body but large amounts make the body obese and cause many medical problems including heart disease and various respiratory complaints.

Amino acids. After digestion, amino acids are carried to the liver cells. Many are immediately released into the bloodstream and subsequently pass into the body tissues where they are built up again (synthesised) in order to form new cells, to repair tissues and to manufacture protein substances such as hormones and enzymes. Amino acids cannot be stored by the body except in a limited way as part of the structure of muscles and glands. Excess amino acids are deaminated (the nitrogen is removed) in the liver (see p. 178). The remaining part of the molecule is either oxidised to produce energy or stored as glycogen.

Metabolism

This term describes all the chemical changes which occur in the body. It involves *anabolism,* the building up of molecules into more complicated substances, and *catabolism,* the breaking down of complex molecules into simpler substances. Anabolism requires energy; catabolism releases energy.

Energy, which is required in order to maintain normal body activities, is released from every cell during the process of oxidation. The method by which this occurs is called *cellular or tissue respiration*. It involves a series of chemical reactions, the first stage of which takes place in the cell cytoplasm. It starts with the formation of glucose phosphate, which is formed when glucose, a six carbon compound, reacts with a phosphorus compound. Eventually this is broken down to a three carbon compound, *pyruvic acid*. The whole process is called *glycolysis* and utilises ten enzymes in its various reactions.

In the second stage, which occurs in the mitochondria of the cell, pyruvic acid is oxidised by a cycle of reactions, the *citric acid* or *Krebs cycle,* to produce carbon dioxide and water. Again ten enzymes are used to bring about the series of changes. A far greater amount of energy is released during this second stage. During the cycle the energy present in the food is transferred to a substance called ATP (adenosine triphosphate) which becomes the energy 'bank' of the cell, releasing energy on demand. When energy is required one of the phosphate groups of ATP breaks off, producing ADP (adenosine diphosphate), and releases energy.

The Liver

Although not an organ of digestion, the liver plays a part in the digestion of fats. It is also involved in the sorting of the products of digestion before they are released into the general circulation.

The liver is the largest and most versatile gland in the body and it carries out many important functions. It is situated at the top of the abdominal cavity, to the right, with its convex upper surface fitting into the underside of the diaphragm. It lies beside the stomach and is partly covered by the ribs.

The liver is dark red in colour and has an outer covering of peritoneum. It is divided into two lobes, and on the under side, in the cleft between the lobes, is the entry/exit point of most of the liver vessels: the hepatic artery, the portal vein, the bile duct and the hepatic vein. The portal vein is unusual in that it carries blood from one organ to another – i.e. from intestine to liver.

Detailed structure. The liver is composed of identical cells

Digestion

Digestive gland	Secretion	Enzymes and other substances	Substances acted upon (Substrate)	Product
Salivary glands (Mouth)	Saliva (alkaline)	Ptyalin or salivary amylase	Cooked starch	Dextrin Maltose
Gastric glands (Stomach)	Gastric juice (acid)	Pepsin Rennin Gastric lipase Hydrochloric acid (not an enzyme)	Protein Caseinogen (milk protein) Fats	Peptones Casein Fatty acids, glycerol
Liver	Bile (alkaline)	Bile salts (not an enzyme)	Fats	Emulsified fat
Pancreas	Pancreatic juice (alkaline)	Trypsin Amylase Lipase	Protein Starch Fats	Polypeptides Maltose Fatty acids, glycerol
Ileum	Succus entericus (alkaline)	Peptidase (eripsin) Lipase Maltase Sucrase (invertase) Lactase	Peptides Fats Maltose Sucrose Lactose (milk sugar)	Amino acids Fatty acids, glycerol Glucose Glucose, fructose Glucose, galactose

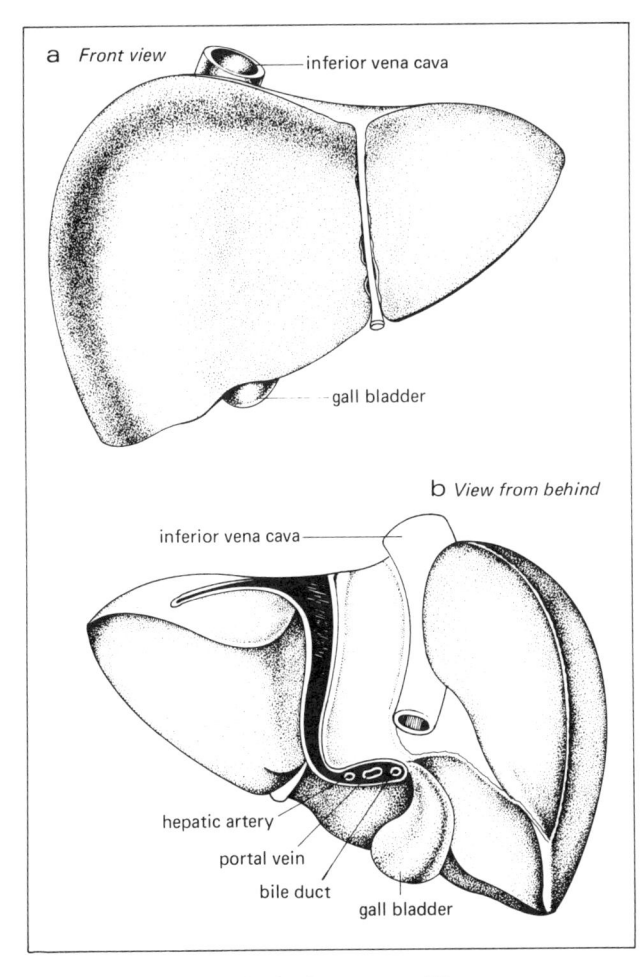

Fig. 8.29 Structure of liver

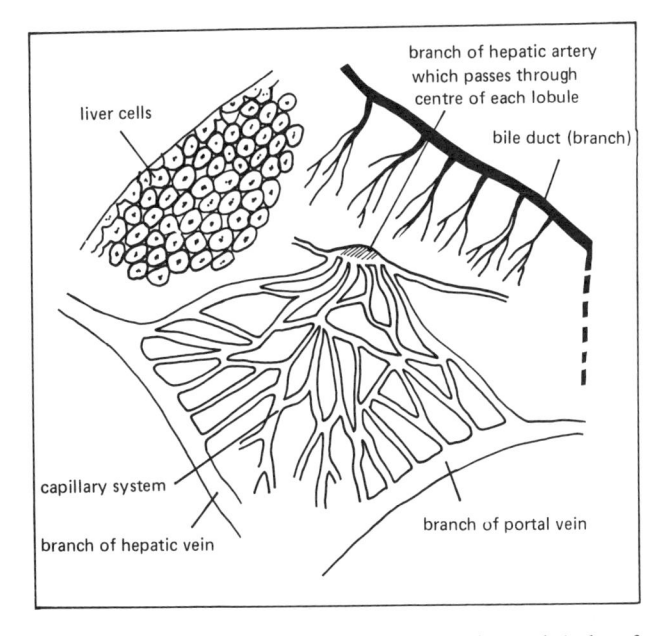

Fig. 8.30 Diagrammatic representation of one lobule of the liver

which are arranged in columnar groups called hepatic lobules. An elaborate system of channels and blood capillaries passes through and around these lobules. Small channels run through the liver cells collecting bile. This passes into larger ducts which flow between the lobules and collect to form a main duct bringing bile to the gall bladder, a pear-shaped bag lying in a groove under the liver. Here the bile is concentrated and stored. When food enters the duodenum from the stomach, bile is released from the gall bladder into the duodenum where it assists the pancreatic enzyme lipase in the emulsification of fats.

Bile is a bright greenish-yellow fluid which contains water, salts, pigments and mucus. The pigment *bilirubin* is formed from haemoglobin, which is released from dead blood cells. Bile salts, which do not contain enzymes, assist the fat-splitting action of lipase by emulsifying fats. Bile is strongly alkaline because of its sodium bicarbonate content. This helps to neutralise the acidic contents of the stomach and provides the correct pH for the action of intestinal enzymes.

Hepatic circulation. The liver has a double blood supply. First, a branch of the aorta, the *hepatic artery*, brings blood rich in oxygen to the liver just like any other organ. This divides into small capillaries which run between the lobules, providing oxygen for cellular respiration. Secondly the *portal vein* carries blood rich in nutrients from the gut to the liver, where they are acted upon or stored by the liver cells. Capillaries from the portal vein flow between the liver lobules, releasing their nutrients into the liver cells.

Running down through each lobule is a small vein which collects deoxygenated blood from the liver. Each vein joins with others until they reach the main hepatic vein emptying into the inferior vena cava. The blood from the portal vein releases most of its nutrients so that by the time it reaches the hepatic vein the blood carries the correct concentration of nutrients required by the cells of the body. These are released in a steady flow, twenty-four hours a day (see below).

Functions of the Liver

It is said that the liver has more than fifty functions, many of which are concerned with the screening of digested food. Before the blood which leaves the villi in the intestine reaches the general circulation, it passes through the liver. Some nutrients are utilised immediately, others are altered before release into the bloodstream and many are stored in the liver itself.

1. *Regulation of nutrients.* Digested nutrients are sorted out by the liver cells. This screening prevents the system from becoming overloaded and allows the correct balance of nutrients to be released into the blood-

stream in a continuous steady flow. For example, after a carbohydrate-rich meal, the blood in the portal vein has a high proportion of glucose – but after leaving the liver, the amount of glucose in the blood is the normal 80–150 mg per 100 cm^3. The liver thus controls the amount of sugar in the blood.

2. *Formation of glycogen.* As glucose is soluble in water it is not easily stored in the body. In order to overcome this, the liver cells convert glucose to an insoluble polysaccharide, a form of animal starch called glycogen, which is stored by the liver cells and muscle and converted by enzyme action back into glucose when the blood sugar level drops. When blood sugar level rises above 160 mg per 100 cm^3, glucose is excreted by the kidneys.

3. *Deamination.* Excess amino acids are not stored in the body; those which are not required for protein synthesis are *deaminated*. This means that the amino group (NH_2), containing nitrogen, is removed from the molecule and converted into ammonia (NH_3). This, in turn, is converted into the less toxic *urea* –$CO(NH_2)_2$– which is released into the bloodstream and eventually excreted from the body by the kidneys. The rest of the molecule is converted into glucose which, if not required immediately, is stored as glycogen.

4. *Fats.* When glycogen stores are full, the liver cells convert surplus glucose into fat which is subsequently deposited around the body as adipose tissue – beneath the skin, around the kidneys and so on. This is a primitive defence mechanism which stores food for times when it may be scarce. As food scarcities are rare in western countries, over-indulgence in either fats or carbohydrates results in obesity. However, this mechanism is essential for the greater proportion of the world's population, who live perilously close to starvation.

By reversing the process, stored fats can be converted back to glucose by the liver, when the level of glucose in the bloodstream is low. This would occur during a slimming diet, for example.

5. *Secretion of bile.* See above.

6. *Maintenance of body temperature.* The many and varied chemical reactions which take place in the liver result in the release of large amounts of energy in the form of heat. This is distributed about the body by the circulating blood.

7. *Storage.* The liver stores some fat, vitamins A and D, vitamin B group and certain minerals such as iron, potassium and copper. These act as reserves and are released when required by the body.

8. *Synthesis of plasma proteins.* The liver manufactures components of the blood such as fibrogen, albumin and prothrombin.

9. *Destruction of red cells.* The liver completes the destruction of red cells and recycles their components. Haemoglobin is broken down and the pigment in it is used in the formation of bile, most of which is eventually excreted in the faeces. Iron is stored and used again to form new red cells.

10. *Formation of antibodies.* The liver is involved in the manufacture of antibodies which destroy invading organisms.

11. *Detoxication.* Another defensive function of the liver is to destroy dangerous toxins which may develop in the blood. It also neutralises many drugs and chemicals, such as hormones, which are no longer required by the body.

12. *Homeostasis.* Many of the functions of the liver play a part in maintaining the concentration and composition of the blood and hence all body fluids, ensuring that they are kept at the optimum conditions for the life of the cells and ultimately for the body's survival.

Summary of Liver Functions
1. Controls release of nutrients
2. Formation of glycogen
3. Deamination
4. Converts excess glucose to fat for storage
5. Secretion of bile
6. Maintenance of body temperature
7. Stores some fat, vitamins A, D, B group and certain minerals
8. Manufactures fibrinogen and albumin
9. Recycles components of red cells
10. Manufactures antibodies
11. Destroys toxins

EXCRETION

The waste substances produced during the metabolic processes must be removed regularly to enable the body to function properly. Such wastes are removed by the excretory organs: the skin, the kidneys and the liver (see p. 176). (This acts as an excretory organ by altering and storing some waste substances.) The large intestine (see p. 175) is an organ of elimination rather than excretion.

The Skin

This is the largest organ of the body. As well as being an excretory organ, it is also involved in sensation and temperature regulation. The skin is composed of two layers: the epidermis on the outside and the dermis underneath. The thickness of each depends on its position in the body.

Epidermis. This has an area of active growth called the *malpighian layer* which contains the pigment *melanin*, determining the colour of the skin. Above this is a layer of living cells; at the surface these flatten into hard, dried cells forming the *corneous layer,* which protects the dermis from infection and damage. These surface cells are constantly being worn away and replaced by an upward growth from beneath.

Dermis. A thick layer containing connective tissue in which are distributed sweat glands, hair follicles, capillaries, nerves and lymph vessels.

Sweat glands. Each gland consists of a coiled tube lined with secretory cells which take fluid from the capillaries nearby and release it through *ducts* which open onto the skin surface as *pores*. This fluid, known as sweat, consists mainly of water, with traces of salt and organic waste (e.g. urea) dissolved in it.

Hair follicles are indentations of endothelial cells into the dermis. Active cells at the base of the follicle multiply rapidly, forming an outward growth of hair. Opening into each hair follicle are *sebaceous glands* which lubricate each hair with an oily secretion, making it waterproof and supple. The activity of these glands determines whether the hair and skin are greasy or dry. Attached to each hair follicle is a tiny muscle which contracts when we are cold or frightened, pulling the hair erect and creating what we call 'goose pimples'. This occurs in order to provide extra heat insulation, as air becomes trapped between the upright hairs. The erect hair or plumage on animals and birds makes them appear more menacing when confronted with a dangerous situation. In humans bristling hair is a vestigial reminder of our early evolutionary history.

Capillaries. The dermis is supplied with a large number of capillaries which bring oxygen and nutrients to the skin tissue and remove waste. Each sweat gland is surrounded by a network of capillaries from which fluid is secreted continuously. This plays a part in maintaining normal body temperature.

Nerves. The skin is supplied with numerous nerve endings, particularly on sensitive areas such as the finger tips. These act as receptors, making us aware of variations in temperature, pressure and pain. They form part of the nervous system, and enable us to respond to various stimuli in order to survive.

Subcutaneous fat. Situated beneath the dermis are layers of fat cells — adipose tissue — which act as insulators.

Functions of the Skin
1. To *protect* the inner tissues.
2. To *excrete* waste substances.
3. As a *temperature regulator.*
4. As a *sensory* organ.
5. To manufacture *vitamin D.*
6. To act as an *insulator* by storing fat.
   The first four functions are relevant here.

1. Protection. The corneous layer of the skin provides a waterproof protective covering for the whole body, preventing loss of body fluids and infection by micro-organisms. The skin itself acts as a cushion, preventing injury to the delicate tissues underneath. The malpighian layer screens the body from the harmful ultra-violet rays of the sun.

2. Excretion. Water and traces of waste are removed from the capillaries and pass on to the surface of the skin through the sweat ducts and pores. In this way the skin helps to maintain homeostasis. When sweat is lost rapidly, as in very hot climates, the body may be deficient in salt. Extra salt should then be included in the diet.

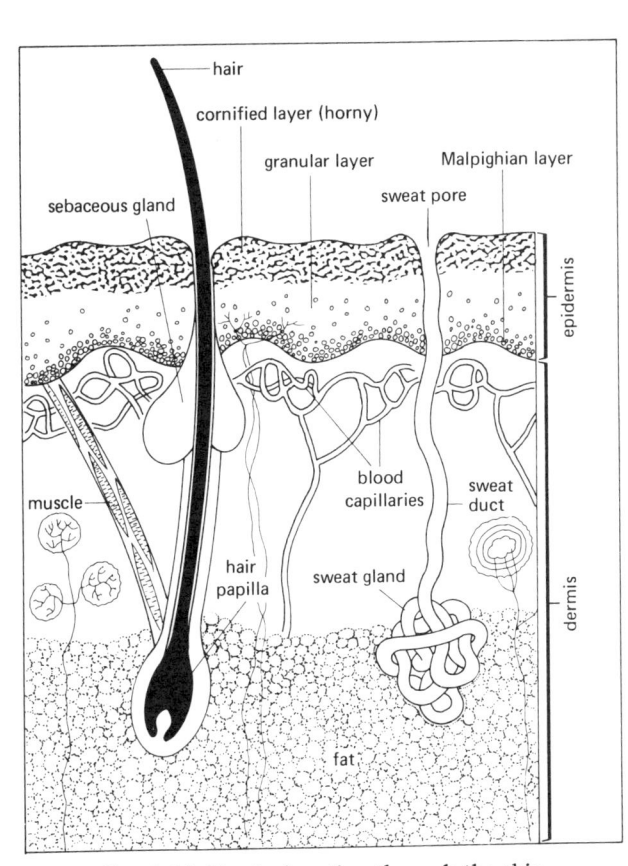

Fig. 8.31 Vertical section through the skin

3. *Temperature regulation.* The skin regulates body temperature in two ways:

a. By *evaporation of sweat.* Heat is taken from the body surface in order to evaporate sweat, which has the effect of cooling the body.

b. *Vasodilation* and *vasoconstriction.* When the body becomes too warm, the brain, which is sensitive to even a slight increase in temperature, sends nervous impulses to the skin causing the tiny arterioles to dilate or widen, allowing extra blood to the surface of the skin. Blushing or reddening of the face is evidence of this taking place. The heat is lost from the blood in these surface capillaries by convection and radiation and thus the body is cooled. The brain can stimulate an increase in sweat production also, as happens when the body overheats during fever. Likewise, when the body becomes too cool, the brain brings about a reduction in sweat production and the arterioles are constricted, reducing the volume of blood to the skin surface and thereby lessening heat loss. Shivering is an involuntary reaction to the cold — a feeble attempt to warm up the body by muscular movement.

4. *Sensory organ.* The skin responds readily to sensations such as heat, cold, pain and texture. The nerves in the skin transmit this information to the brain which either stores it or acts upon it (for example it stimulates the arm and fingers to move quickly from a hot object).

5. *Manufacture of Vitamin D:* see p. 19.

6. *Insulation:* see p. 14.

The Kidneys

These are situated on the back wall of the abdomen

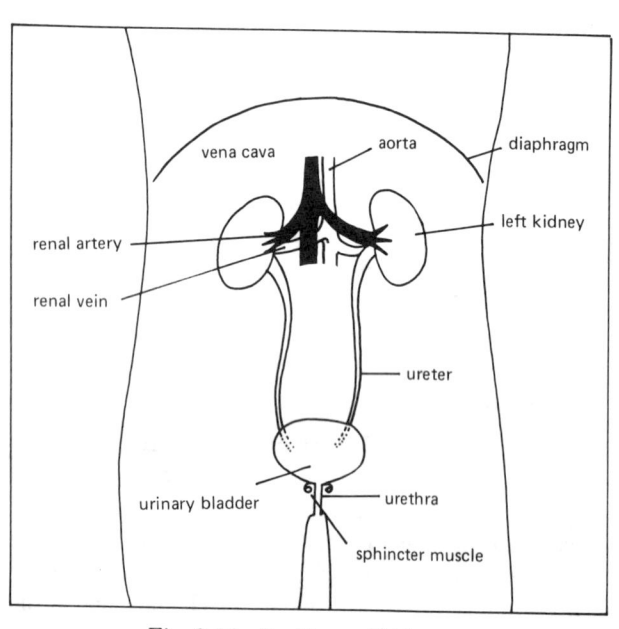

Fig. 8.32 Position of kidneys

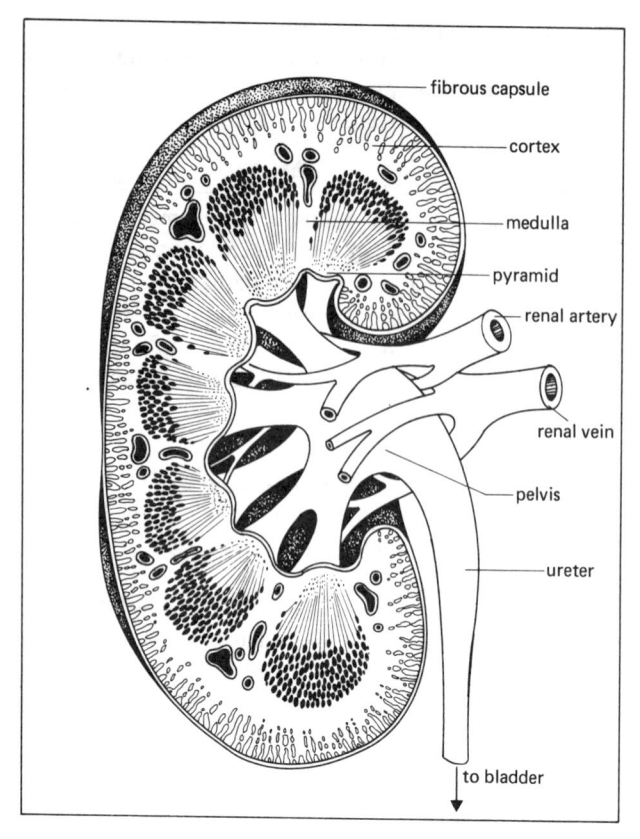

Fig. 8.33 Structure of kidney

just below the diaphragm, one on either side of the lumbar vertebrae. Each is deeply embedded in fat, which protects it and is held in position by connective tissue. Surmounting each kidney is an adrenal gland (see p. 187).

The kidneys are two dark red, bean-shaped organs. Each is about 12 cm long and weighs over 100 grams. The inner border or *hilum* faces the back bone and at this point the renal artery, a branch of the aorta, brings a supply of oxygenated blood from the heart. Each kidney is drained by a renal vein which leaves at the hilum and empties into the inferior vena cava.

Basic structure. When a kidney is cut lengthwise in two (a longitudinal section) it is seen to consist of an outer close-textured tissue called the *cortex* which is dark in colour. This leads into a light area, the *medulla,* which contains shaded areas called *pyramids.* They pass into the *pelvis* of the kidney, a funnel-shaped collecting tube which narrows to form the *ureter,* passing downwards to the base of the bladder.

Detailed structure. Each kidney consists of over a million functional units called *nephrons,* whose purpose is to remove the constituents of the urine from the blood. Each begins as a cup-shaped funnel called *Bowmans capsule,* and filling the inside of this capsule is a tightly packed capillary bed, called the *glomerulus,* which is made up of

180

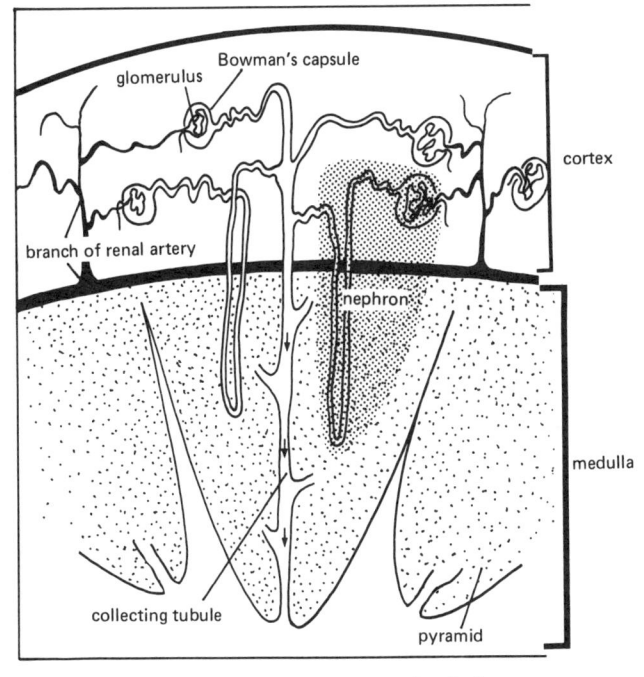

Fig. 8.34 Structure of tubule

about fifty tiny capillaries originating from a branch of the renal artery. From Bowmans capsule a highly convoluted tube called the *proximal* (near) *tubule,* twists its way through the cortex. It continues down into the medulla in a long, narrow U-shaped turn called the *loop of Henle,* returns to the cortex and coils several times again. It is now called the *distal* (distant) *tubule,* as it is further from the capsule. Eventually it joins several other tubules which together form a collecting duct opening into the pelvis of the kidney. Altogether the tubules provide about thirty-seven miles of surface from which urine is filtered.

Renal circulation. The renal artery, carrying a quarter of the total blood volume at each beat of the heart, enters the kidney at the hilum. It passes into the cortex, breaking into capillaries, and eventually forms the knot of tightly packed capillaries which make up the glomerulus. This is almost completely surrounded by Bowmans capsule. The blood entering the glomerulus is under great pressure as it is coming almost directly from the aorta. The walls of both the capillaries and capsule are extremely thin and water, salts, urea and other substances are filtered through the pores of the capsule into the tubule. This mixture, at present called *nephric filtrate,* is very similar to lymph. The blood cells and protein molecules remain in the capillaries.

An efferent arteriole leaves the glomerulus and immediately breaks into a network of capillaries which wind around the proximal tubule and loop. Here 80 per cent of the filtrate is reabsorbed by the capillaries as well as any useful substances such as glucose, amino acids, vitamins and hormones, but harmful substances such as urea are not reabsorbed. The structure of the tubules assists this selective reabsorption as they are lined with cells containing microvilli, tiny projections which increase the surface area.

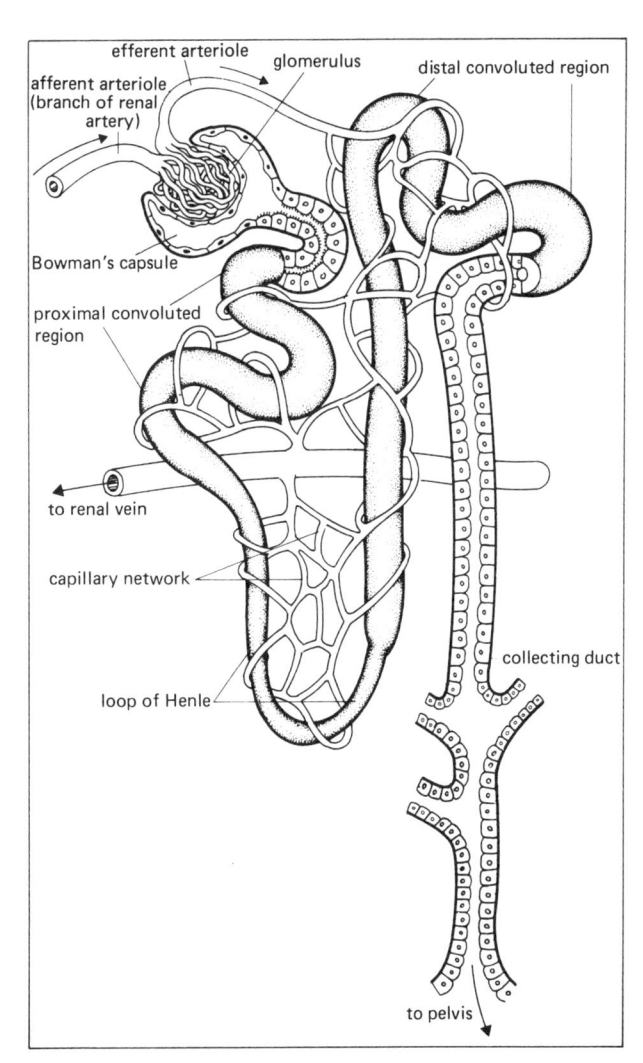

Fig. 8.35 Detailed structure of kidney: cross-section

In the distal tubule a very precisely regulated reabsorption, directly related to the concentration of water, salt and acid/alkali in the blood, takes place. The tubule now containing urine of the correct dilution joins with other tubules and passes into the medulla and out into the pelvis of the kidney. The urine trickles down the ureter and into the *bladder,* a hollow organ with walls of elastic tissue which stores urine until it reaches a certain pressure. At this point the sphincter muscle at the base of the bladder relaxes and the bladder contracts, expelling the urine from the body through the urethra. This is known as *micturition.*

The venous capillaries leaving the nephrons contain blood which has lost oxygen and gained carbon dioxide by normal metabolism. Through the filtering action of the tubules, the blood has lost water, salts, urea and traces of other substances. These capillaries unite to form a major vein which empties into the renal vein. This leaves the kidneys and joins the inferior vena cava which returns the blood to the heart.

Functions of the Kidneys
1. Their major function is the *removal of waste* products from the blood. These are dissolved in water and eliminated from the body.
2. They regulate the amount of *water* in the body.
3. They regulate the amount of *glucose,* salt and other substances in the blood.
4. They are one of the major *homeostatic* devices in the body. They adjust the concentration of body fluids and regulate the pH balance of the blood.

Composition of Urine (average)
Water 96%
Urea 2%
Salt (NaCl) 1%
Other substances (e.g. uric acid, calcium, potassium, ammonia) 1%

Urea is an organic compound containing nitrogen which is produced during the breakdown of protein.

Urine is a clear, pale yellow fluid. Between one and one-and-a-half litres are excreted in 24 hours, although the amount varies according to atmospheric temperature. In cold weather more water is lost through the kidneys and less by perspiration, while in hot weather or in a warm room the order is reversed.

Osmoregulation
Water lost by urine, faeces, sweat and lungs must be balanced by the intake of water in food and drink. This delicate balancing of the water level of the body is controlled by the brain. Sensory receptors in the hypothalamus stimulate the pituitary gland just below to produce the antidiuretic hormone ADH (sometimes called vasopressin). This, on reaching the kidneys, stimulates the tubules to return more water to the blood, thereby increasing the concentration of urine. This would occur after considerable water loss occured, e.g. after sweating or diarrhoea. Likewise when the blood becomes too dilute (when large quantities of liquid have been drunk), the amount of ADH is suppressed and the tubules return less water, increasing the quantity of urine excreted.

Inability to secrete this hormone causes *diabetes insipidus,* the symptoms of which are thirst and frequent heavy urine.

THE NERVOUS SYSTEM
The human body consists of millions of cells, all of which are interdependent. These make up a collection of organs, which are taking part in several different processes at any given moment: the heart beats, the lungs expand and relax, the digestive organs propel the food through the alimentary canal and so on. None of these organs works in isolation — for example the heart could not function without oxygen from the lungs and glucose from the digestive organs to supply energy. In order to control the various body processes and enable the body to work as a unit, there must be a system which co-ordinates all the others. In fact there are two systems — the nervous system and the endocrine system.

The nervous system can be divided into two main systems, the central nervous system and the peripheral nervous system.

Central Nervous System
The central nervous system consists of the brain and spinal cord. Together these receive information, interpret and store it and transmit instructions. In other words the system receives stimuli such as feeling sensations, light, sound, heat and taste, decides to react in a certain way (e.g. to blink in response to light) and then does so. The peripheral system acts in conjunction with the central nervous system by gathering the information, feeding it into the central nervous system and then transmitting the instructions to the organs.

The Brain
The brain is protected by the skull and surrounded by a triple membrane — the meninges. It can be divided into several areas but the three main divisions are:

The cerebrum, which is the largest part.

The cerebellum at the base of the skull.

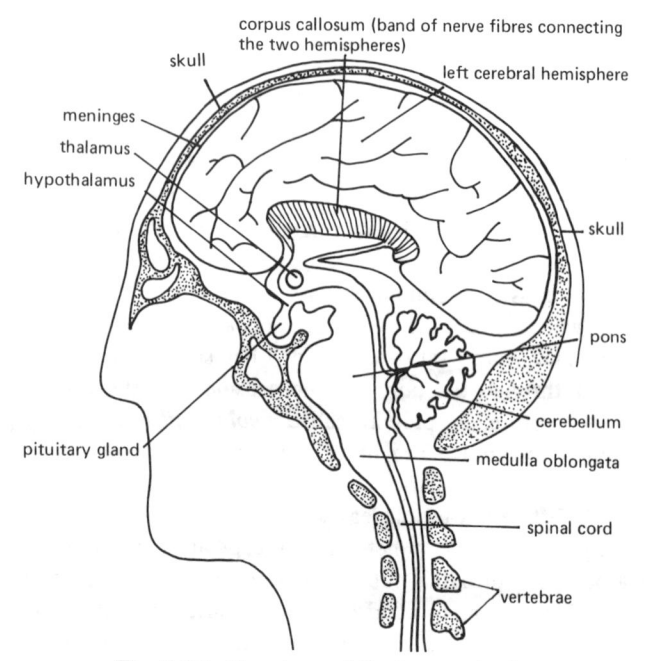

Fig. 8.36 Structure of the human brain

The medulla oblongata which connects the brain with the spinal cord.

The cerebrum is composed of two deeply folded hemispheres. Its exterior, the cerebral cortex, is made up of 'grey matter', composed of billions of tightly packed nerve cell bodies. The white matter beneath this consists of many billions of nerve fibres which are extensions of the cell bodies in the cortex.

Certain areas or 'centres' of the brain are associated with sensations or functions governing different parts of the body.

The *large frontal lobe* of the cerebrum is concerned with personality and behaviour and such indefinable areas as thought, memory, mind, intellect and emotions.

The *side of the frontal lobe* is associated with speech.

The *temporal lobe* at the side controls hearing.

The *occipital lobe* at the back controls vision.

The *motor centres* across the top of the cerebrum are concerned with voluntary movements of the body. The centre on the right side controls movement on the left side of the body and the centre on the left side controls movements on the right side of the body. This occurs because the motor fibres leaving these areas cross to the opposite side.

Beside the motor centres is the *centre which controls conscious sensations*.

The major areas of the brain are connected by nerve fibres so that, for example, if we hear a noise, impulses will pass to the motor centre and the visual centre so that we turn and look for the source of the sound. Thus the brain co-ordinates every act.

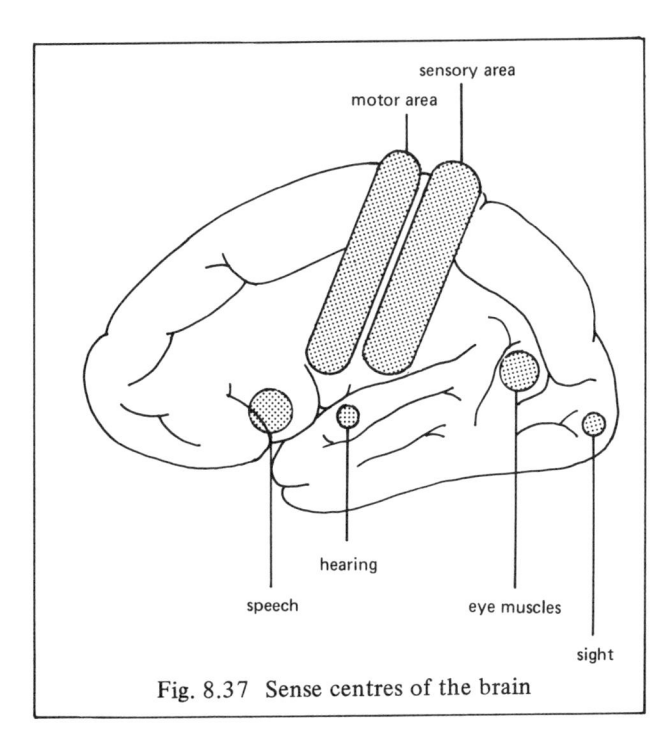

In the centre of the brain are found many more co-ordinating centres:

The *thalamus* processes and interprets messages. It receives sensations of pain and pleasure.

The *hypothalamus* regulates the body's temperature and water content, releases certain hormones and, as already mentioned, controls feelings such as hunger and thirst.

The *reticular formation.* At any one time the brain is on the receiving end of thousands of impulses or messages. This section of the brain 'censors', as it were, these impulses, only allowing the more important ones through to our conscious mind. It also helps us to sleep by inhibiting most of the impulses to our conscious centres.

The cerebellum, which is situated to the back at the base of the brain, is pleated into long parallel folds. It is involved with muscular control and balance and helps to co-ordinate muscular movements. The *pons* (bridge) connects the two halves of the cerebellum.

The medulla oblongata lies under the pons and connects the brain to the spinal cord. This is the centre of *involuntary reactions.* It controls such important processes as breathing, heart beat, and blood pressure. It also plays a part in glandular secretion and digestion. The *vagus nerves* spring from the medulla and assist in regulating the actions of the internal organs.

Leading from the underside of the brain are twelve pairs of cranial nerves. These control movement and sensation in the head and neck and include the optic and auditory nerves. The brain receives impulses from each sensory organ and transforms them into sensations – for example, the ear may send messages to the hearing centre of the brain, but unless this area is functioning normally, we cannot hear; it is not the ear which hears, in effect, but the brain.

Through the centre of the brain and spinal cord flows the *cerebrospinal fluid,* a liquid similar to lymph which bathes the nerve cells and supplies them with nutrients.

The Spinal Cord

This is a long, cylindrical mass of nerve tissue which runs down the back bone to the second lumbar vertebra. The bony structure of the vertabrae protects it from damage. The spinal cord thickens from bottom to top as nerve fibres from different parts of the body join it. On the outside it consists of white matter made up of many nerve fibres which convey impulses to and from the brain. The grey matter, forming an 'H' shape in the centre of the cord, consists of cell bodies and motor neurons. The positioning of grey and white matter is the reverse of that in the brain.

Between the vertebrae, at intervals, the spinal cord

Fig. 8.37 Sense centres of the brain

sends out thirty-one pairs of nerves. These are made up of:

a. A sensory neuron bringing stimuli from sensory organs such as the skin. This enters at the back of the spinal cord forming the *dorsal root* which passes into grey matter.

b. A motor neuron which stimulates an organ or muscle into action. This leaves the front, forming the *ventral root*.

These join together to form the mixed nerves of the peripheral nervous system which pass to all parts of the body. The cell bodies of the sensory neurons form bumps on the dorsal roots called *ganglia*.

Functions of the Spinal Cord
1. It carries impulses to and from the brain.
2. It is the centre for simple reflex actions (e.g. lifting one's fingers from a very hot object).

Reflex. A reflex action is a rapid involuntary response by an organ to a stimulus. It is not necessary for the impulse to pass to the brain before the action takes place, although the impulse does travel upwards along the association neurons to register that the action is occurring. One of the simplest types of reflex action is the knee jerk used by doctors to test reflexes. If one leg is crossed over the other and tapped sharply just below the kneecap, the lower part of the leg should jerk outwards. This is how it works. Sensitive receptors in the leg muscle send impulses through the 'reflex arc' — i.e. impulses pass from the receptor along the sensory fibre. They cross over to the motor fibre (see Synapse p. 185) which stimulates the same muscle to contract.

Reflex actions can be 'conditioned' or learned. Many things we do consciously at first become conditioned reflexes by repetition. For example walking, which is a difficult exercise for a young child, becomes a conditional reflex after some practice, so that the child no longer has to think about it.

Peripheral Nervous System
This consists of the spinal and cranial nerves which carry impulses to and from the central nervous system. These nerves transmit impulses from receptors to the central nervous system and transmit impulses from it to the skeletal muscles.

Autonomic Nervous System
This consists of sensory and motor neurons which run between the central nervous system and various internal organs. Its control is involuntary and deals with the heart, blood vessels, lungs, digestive organs and glands. It can be subdivided into two systems, each of which works in opposition to the other. While one stimulates contraction, the other inhibits contraction.

a. The sympathetic nervous system makes sudden action possible by increasing blood supply to the heart and lungs and reducing the supply to the organs of digestion and excretion. It speeds up conversion of glycogen to glucose.

b. The parasympathetic nervous system reverses the effects of the sympathetic system, restoring the body to normal. It decreases heartbeat and blood supply to lungs and restarts digestive and excretory organs. It also exercises control on the degree of the contractions which occur.

Nerve Cells (Neurons)
In order to understand clearly how the nervous system works, it is necessary to study the basic unit of the system. This is called a neuron or nerve cell.

1. Sensory neurons or afferent neurons convey impulses from sensory organs (e.g. skin, eye) *to* the central nervous system. They consist of a cell body containing a nucleus and surrounded by cytoplasm. Groups of cell bodies together form *ganglions* – swellings here and there on the nerve near the spinal cord. The cell body lies on a long fibre called a *dendron*, covered by a fatty

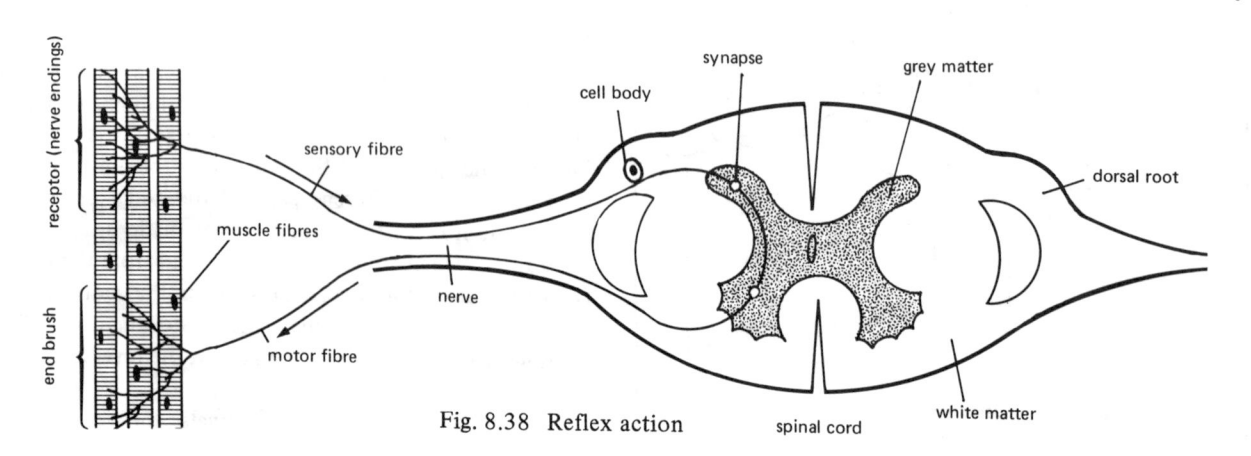

Fig. 8.38 Reflex action

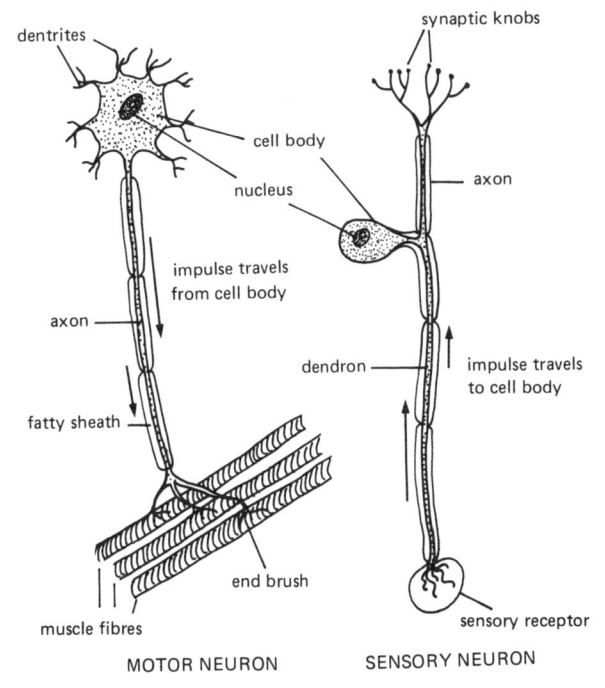

Fig. 8.39 Nerve cells

3. Association neurons link afferent and efferent neurons, carrying impulses between the two. They are found exclusively in the brain and spinal cord.

Nerves

A nerve resembles an electric cable. It consists of a bundle of nerve fibres bound in connective tissue and, like the wires in an electric cable, the nerves are insulated from each other by a fatty sheath. Most nerves contain both motor and sensory fibres.

Nerve impulses are electrochemical reactions in nerve fibres. When a nerve fibre is stimulated, usually at one end, a reaction causes a change in the next part of the fibre. This, in turn, causes another section to react and in this way the impulse travels along the length of the fibre. (It could be compared to a lighting fuse.) The speed of the reaction is roughly 120 metres per second. A supply of oxygen is necessary for the transmission of impulses.

Synapse. A chemical connecting link which occurs only in the central nervous system. Neurons do not connect directly with one another: there is always a small space between the end of one neuron and the beginning of another. This area of near-contact is called a synapse. The space or gap is bridged by a hormone called acetylcholine (ACh). This is secreted by the ends of the axon, diffuses into the synapse and reaches the terminal ends of the adjacent dentrites, transmitting the impulse across the gap.

sheath which both insulates and protects it. It is this which gives the white matter of the brain and spinal cord its colour. The outer end of the sensory neuron contains a sensory receptor which picks up sensations and stimulates the neuron (nerve fibre). This passes an impulse along to the other end where the fibre branches into terminals.

2. Motor neurons or efferent neurons transmit impulses *from* the central nervous system to various organs and glands. They have irregular cell bodies which together form the grey matter of the central nervous system. Branching outwards from the cell body are *dendrites* which carry impulses towards the cell body. A long thin branch called an *axon* passes from the cell body bringing motor impulses from the central nervous system. The axon, like the dendron, is covered in a fatty sheath. The end of each axon branches outwards and terminates in an end brush or end plate which is in close contact with the muscle fibres.

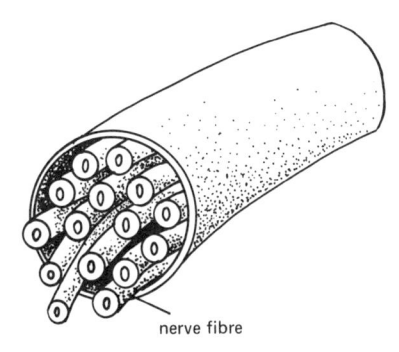

Fig. 8.40 A nerve: containing both sensory and motor fibres

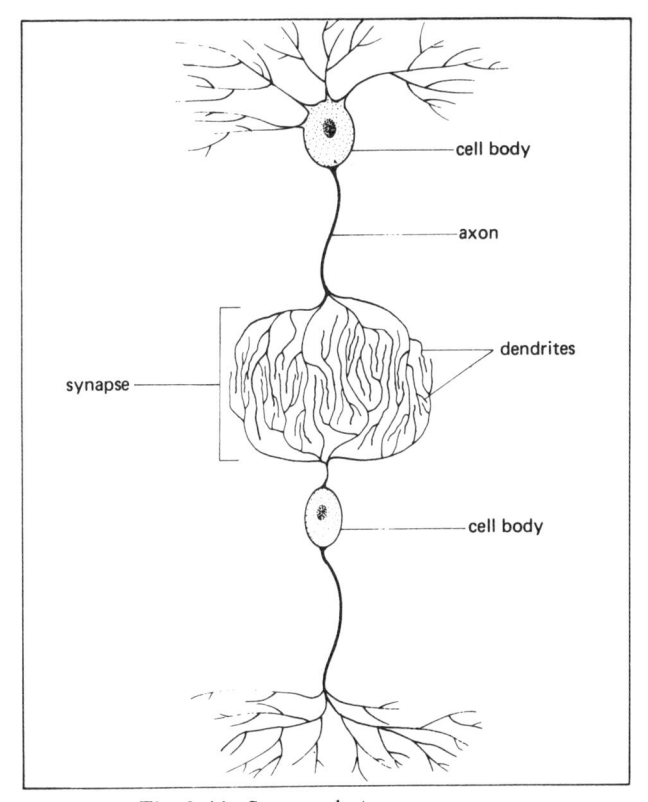

Fig. 8.41 Synapse between neurons

185

An enzyme is subsequently released which destroys ACh, preventing continuing stimulation.

THE ENDOCRINE SYSTEM

Like the nervous system, the endocrine system is a medium of communication and co-ordination within the body. When we studied digestion we saw that glands such as the salivary glands and the liver secreted substances which passed into the alimentary canal through ducts. These are known as *exocrine* or *ducted* glands. The *endocrine* or *ductless* glands release their chemicals directly into the blood flowing through the glands. These chemical messengers are called *hormones*; the word hormone comes from the Greek *hormaein* meaning 'I activate'.

Hormones are carried by the blood to every cell in the body until they reach their target organ where they cause certain changes to take place. Although hormones are slow to cause change, their effects last longer than the effects of nervous impulses. They regulate such long-term processes as growth, sexual maturity and ageing.

The amount of hormone secreted is extremely important as an imbalance leads to various disorders. Normally, the amount of hormone is adjusted to suit the body's current requirements.

Many hormones are neutralised in the liver and eventually excreted by the kidneys.

Pituitary Gland

This 'master' gland influences the activity of many endocrine glands. It is a pea-sized structure at the base of the brain which is divided into two parts, the *anterior pituitary* and the *posterior pituitary*.

The pituitary gland is closely associated with the hypothalamus which lies just above it. The hypothalamus is very sensitive to changes in the body and can respond by stimulating the pituitary into the production of hormones.
Anterior pituitary. The anterior pituitary secretes at least seven hormones:

1. *Human growth hormone* (HGH) influences the growth of cells. Excess causes gigantism, deficiency results in dwarfism.

2. *Lactogenic hormone* (LTH) stimulates milk production in females after birth.

3. *Thyroid stimulating hormone* (TSH) or thyrotropic hormone stimulates the thyroid gland to secrete its hormone thyroxine.

4. *Adrenocorticotropic hormone* (ACTH) stimulates the adrenal cortex to release its hormones.

5. *Follicle stimulating hormone* (FSH) acts on the gonads (sex organs). In the female it stimulates egg ripening and oestrogen production; in males it stimulates the development of the seminiferous tubules and sperm production.

6. *Luteinising hormone* (LH) stimulates production of sex hormones – progesterone in females and androgens in males.

7. *Melanocyte stimulating hormone* (MSH) affects the pigmentation of the skin.

Posterior lobe. Oxytocin stimulates the contraction of the uterus thus beginning labour in childbirth. When birth is artificially induced it is injected into the bloodstream.

Antidiuretic Hormone (ADH)
a. Stimulates reabsorption of water from kidney tubules.
b. Affects blood pressure by constricting the walls of small arterioles.

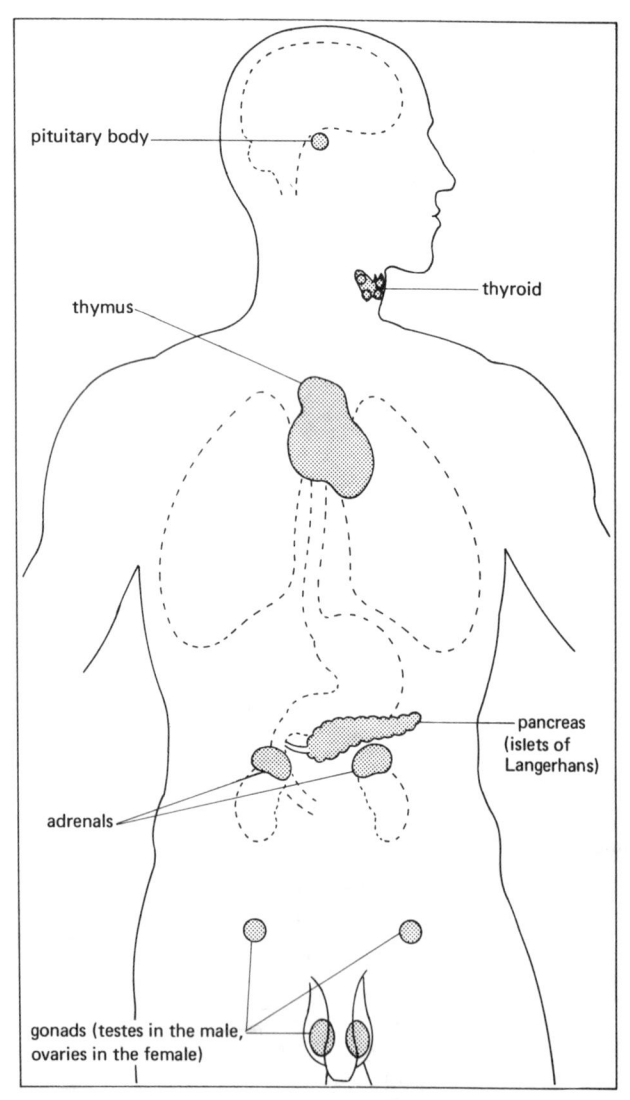

Fig. 8.42 Location of endocrine glands

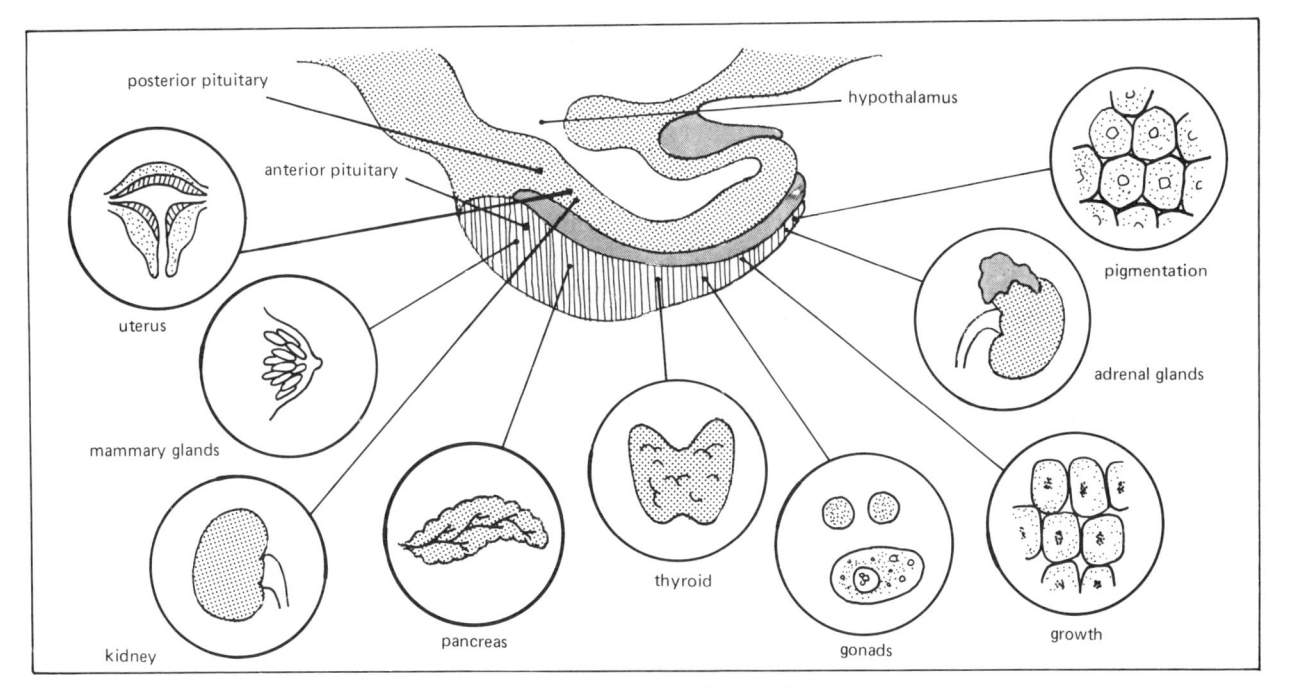

Fig. 8.43 Target organs of pituitary hormones

Thyroid Gland

This is the largest endocrine gland. It is a double structure situated in the neck in front of the trachea and is richly supplied with blood. *Thyroxine,* its hormone, controls the rate of growth in the young: deficiency of this in children can cause dwarfism and mental retardation (known as cretinism). If the condition is diagnosed quickly, it can be cured by administering thyroxine.

In adults thyroxine affects the rate of metabolism: excess amounts raise the metabolic rate, causing thinness and hyperactivity, while a deficiency lowers the basal metabolic rate causing obesity, lethargy and mental confusion.

Any disturbance in the activity of the gland causes it to enlarge, resulting in *goitre*. The hormone thyroxine contains iodine, and if there is a deficiency of iodine in the diet, goitre will result.

Parathyroid Glands

Imbedded in the tissues at the back of the thyroid are four small glands knows as parathyroids. They release a hormone *parathormone* (PTH), which controls the level of calcium in the blood. Excess PTH causes brittle, badly formed bones and may cause kidney stones, while deficiency can be fatal unless calcium is administered.

Adrenal Glands

These are two small structures which lie one on each kidney.

Adrenal cortex. This is the outer layer of the glands. It produces at least three types of hormones:
1. The *glucocorticoids* – steroids including *cortisone* which is involved with metabolism of carbohydrate, stimulating the laying down of glycogen in the liver. They are also involved with general body metabolism. The glucocorticoids reduce inflammation; this is why cortisone is important in the treatment of many diseases. They also take over from adrenaline after its initial stimulation has worn off, helping the body to adapt to stress.
2. *Aldestrone,* which regulates salt/water balance in the blood.
3. *Sex hormones,* particularly male androgens. Excess of these hormones causes male characteristics in women.

A general deficiency of the hormones of the adrenal cortex causes *Addison's Disease* which results in muscular weakness, apathy and eventually death. Administration of cortical hormones is usually successful, however.

Adrenal medulla. This is the interior of the adrenal gland. It secretes two hormones, *adrenaline* and *noradrenaline.* The first is secreted in large quantities when the body suffers stress, anger, fright or injury; it causes the heartbeat to quicken, diverts blood from less important organs to those concerned with 'fright or flight', speeds up the rate of breathing, raises the blood pressure and blood sugar and increases the rate of metabolism. All of these changes prepare the body for violent physical action. The second hormone, noradrenaline, is secreted soon after adrenaline in order to sustain these reactions by causing a further increase in blood pressure.

Pancreas

Apart from its function in the digestion of food, the pancreas contains groups of endocrine cells called *Islets of*

Langerhans. These secrete *insulin,* a hormone which controls carbohydrate metabolism and hence the level of sugar in the blood. A rise in blood sugar releases insulin into the blood. This passes through the portal vein to the liver where it hastens the conversion of glucose to fat and glycogen, quickly returning the blood sugar level to normal.

Insufficient insulin causes *diabetes mellitus.* As the body is unable to convert extra glucose to glycogen, large amounts circulate in the blood and pass into the urine. Protein and fat are converted into glucose and through osmosis, large amounts of water remain in the tubules to be excreted as urine. A low carbohydrate diet is often effective in coping with mild cases of diabetes, and injections of animal insulin control the disease in more serious cases so that diabetics can lead a normal life. Excessive insulin can reduce the level of glucose in the blood to such an extent that the brain cells are deprived of food, and shock and even coma may result.

Gonads

These are the male and female reproductive organs. As well as producing the sperm and ova, they are endocrine glands which secrete important hormones controlling reproduction and sex drive.

Testes

These produce the male sex hormone testosterone which influences the development of secondary male characteristics such as the deepening of the voice and growth of hair on the face and body. It is stimulated by the LH from the pituitary gland.

Ovaries

These secrete two hormones in response to stimulation by the pituitary gland:

Oestrogen which causes various changes to take place at puberty, e.g. breast development. This hormone also initiates and controls menstruation.

Progesterone controls the changes which occur during pregnancy and lactation. Deficiency causes miscarriage.

Other Organs with Endocrine Functions

The stomach secretes a hormone *gastrin* into the bloodstream when we eat and when this returns to the stomach it stimulates the production of gastric juices.

The duodenum produces *secretin* which, when it reaches the pancreas in the bloodstream, activates the production of pancreatic enzymes.

The kidneys produce a hormone concerned with blood pressure.

Feedback

Many hormones operate a self-regulating feed-back system whereby the release of one hormone into the bloodstream triggers off a response in another hormone, causing it to suppress production. This ensures that excessive amounts of any one hormone are not released. For example, presence of thyroxine inhibits TSH production; TSH stimulates the formation of thyroxine.

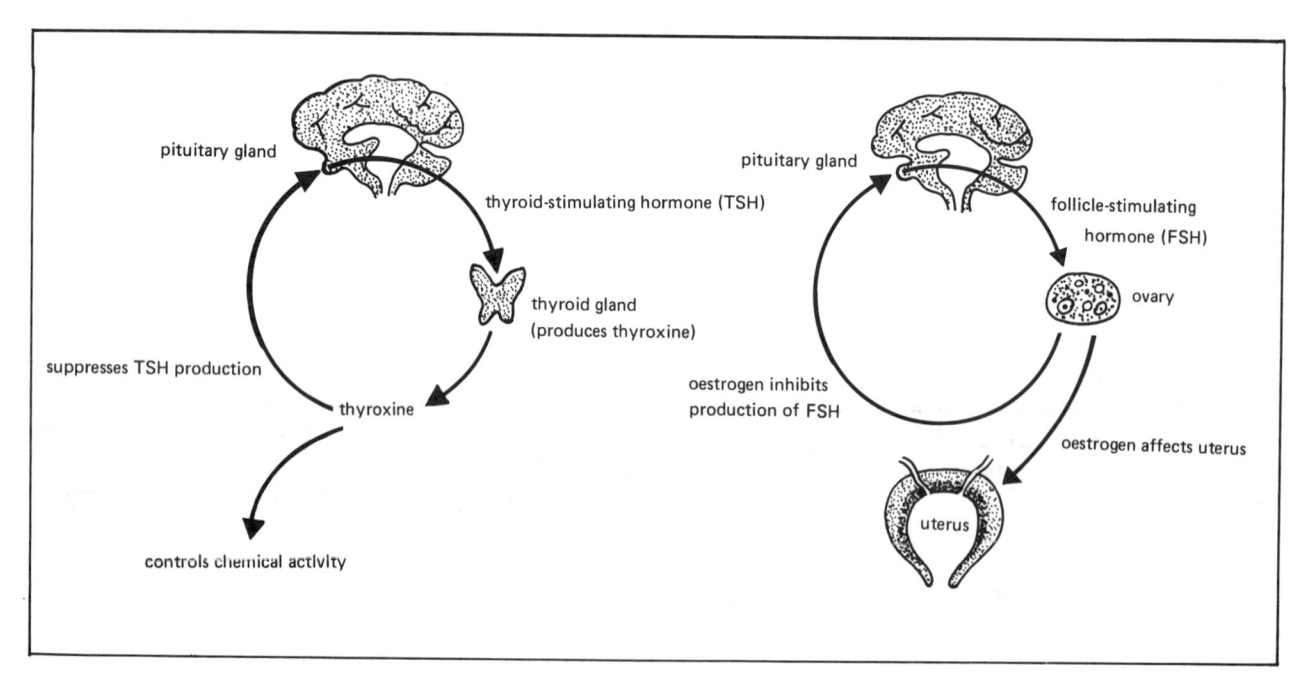

Fig. 8.44 Hormone 'feedback'

HORMONES

Gland	Position	Hormone	Effect	Excess (Hyper-)	Deficiency (Hypo-)
Pituitary (Anterior Lobe)	At base of brain	1. Human growth hormone (HGH)	Controls growth	Gigantism	Dwarfism
		2. Lactogenic hormone (LTH)	Stimulates milk production in females		
		3. Thyroid stimulating hormone (TSH)	Stimulates secretion of thyroxine		Depression of thyroid functions
		4. Adrenocorticotropic hormone (ACTH)	Stimulates secretions of adrenal cortex	Increased adrenal activity	Depression of adrenal functions
		5. Follicle stimulating hormone (FSH)	Stimulates production of male and female sex hormones		Depression of sex functions
		6. Luteinising hormone (LH)	Stimulates secretion of male and female sex hormones		
		7. Melanocyte stimulating hormone (MSH)	Controls laying down of pigment in skin		
Pituitary (Posterior Lobe)	At base of brain	Oxytocin	Initiates labour in women		
		Antidiuretic hormone (ADH) (also called vasopressin)	a. Controls water absorption in kidneys b. Causes changes in blood pressure		*Diabetes insipidus* (symptoms – excessive urination and dehydration)
Thyroid Gland	In neck	Thyroxine	a. Controls growth of the young b. Affects metabolism of adults	Hyperactivity Weight loss Overheating	Cretinism (infants) Lethargy Obesity
Parathyroid Gland	Embedded in thyroid	Parathormone (PTH)	Controls level of calcium	Brittle bones Kidney stones	Muscular spasm. Death unless treated
Adrenal Medulla	On top of each kidney	Adrenaline Noradrenaline	Prepares body for stress situations Sustains reactions of adrenaline	Increased heartbeat and blood pressure	Rare – can cause a tumour which increases blood pressure
Adrenal Cortex	On top of each kidney	Cortisone Aldosterone	Metabolism, storage of glycogen Regulates salt balance in blood	Cushings Disease (fat body, thin limbs) Excessive maleness in women, female characteristics in males	Addisons Disease Death
		Sex hormones (e.g. androgens)	Produce male and female characteristics		
Pancreas	Under stomach	Insulin	Controls glucose level in blood	Shock, coma, death	Diabetes
Gonads	Testes Ovaries	Testosterone a. Oestrogen b. Progesterone	Controls male sexual development a. Controls female development, menstruation b. Controls uterus during pregnancy		Lessened development a. Interferes with menstrual cycle b. Miscarriage
Digestive Glands	Stomach	Gastrin Secretin	Activates secretion of gastric juice Activates secretion of pancreatic juice		Interference with digestion

REPRODUCTION

Reproduction is the means whereby a species is perpetuated. In humans the reproductive function also satisfies emotional needs and drives such as love, protection and security.

Male Reproductive Organs

The reproductive organs of the male lie outside the body because the cooler temperature is more conducive to the production of sperm. They consist of the *testes*, two oval organs which lie in a loose covering of skin called the *scrotum*. Each testis contains a large mass of seminiferous tubules, the walls of which contain cells which manufacture sperm. The mature sperm pass into a duct which leads from inside the testes to another coiled tube, this time outside each testis (the *epididymis*). From this a

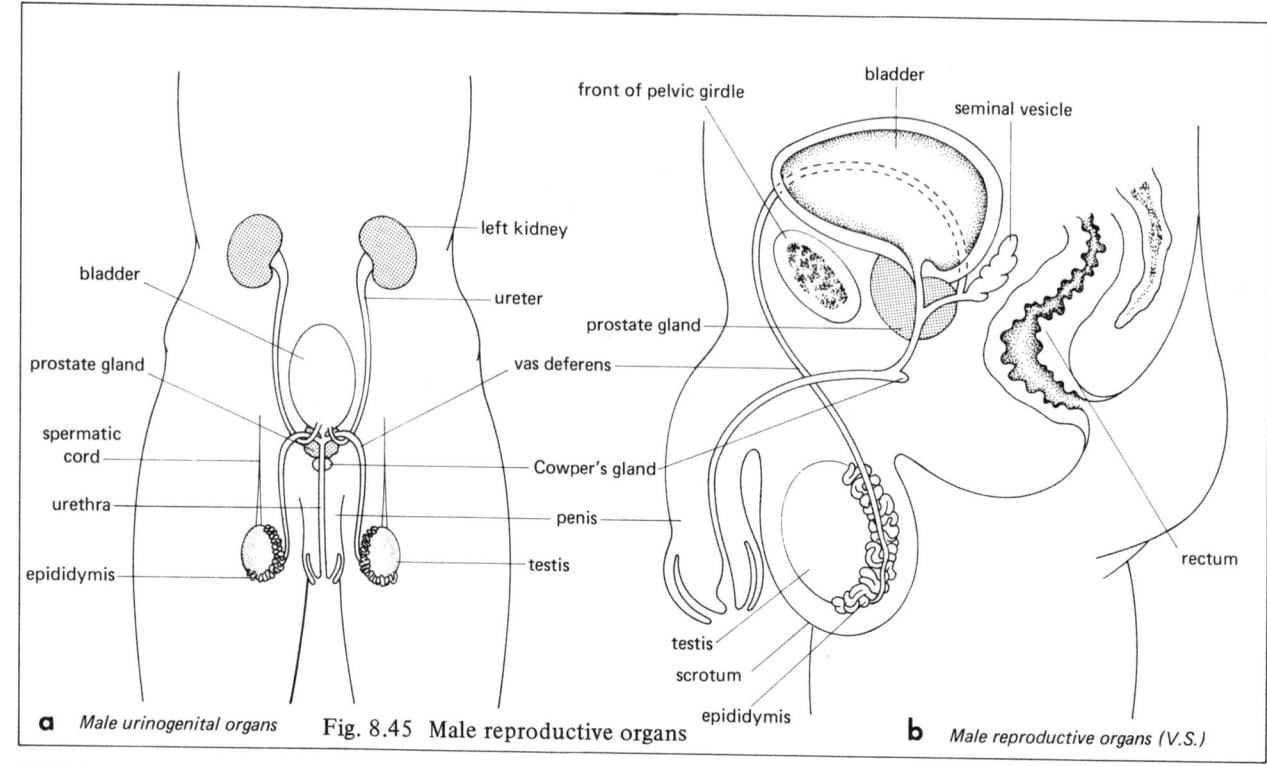

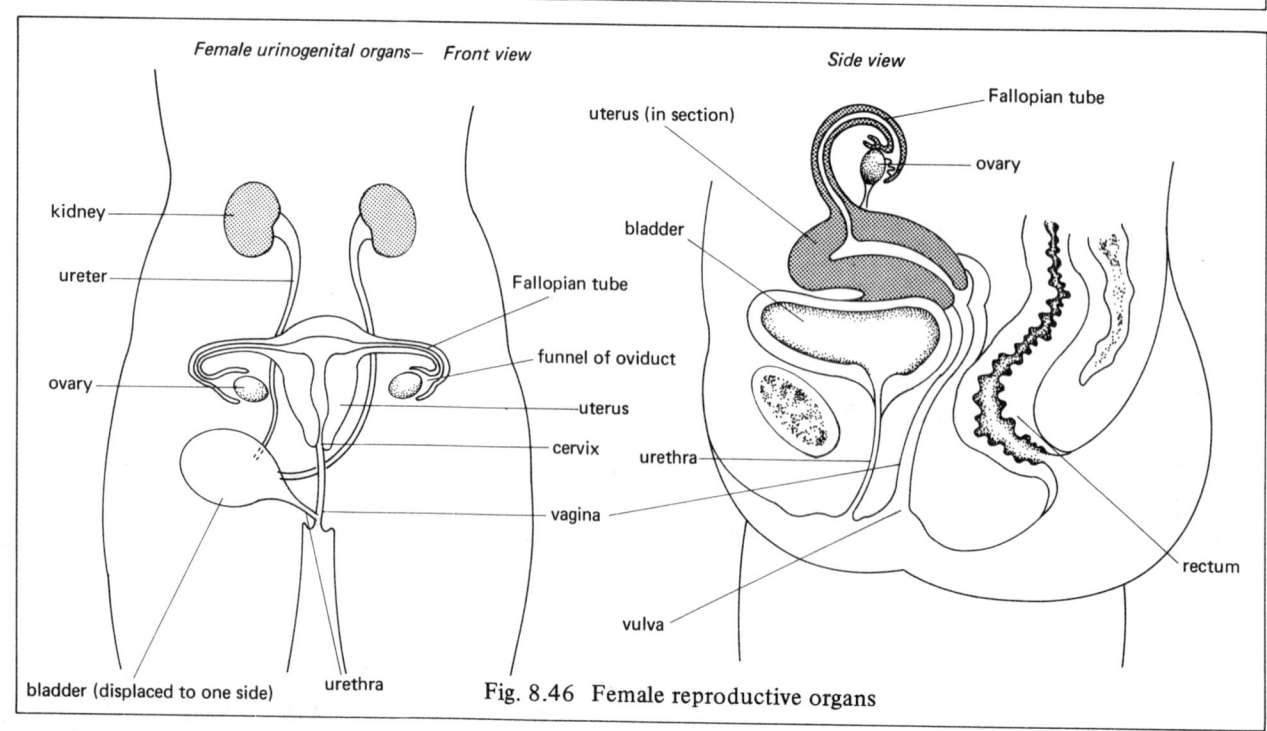

Fig. 8.45 Male reproductive organs

a *Male urinogenital organs*

b *Male reproductive organs (V.S.)*

Fig. 8.46 Female reproductive organs

larger tube, called the *vas deferens* or sperm duct, passes out of the scrotum. Near the bladder it branches into two, one branch leading to the coiled seminal vesicle, the other opening into the urethra at the base of the bladder. Near this point are two glands, the *prostate gland* and *Cowpers gland*. Together with the seminal vesicles, these produce the seminal fluid in which the *spermatozoa* (sperm) are suspended. The urethra travels downward into the penis, which is composed of connective tissue and in which there are large numbers of spaces resembling the holes in a sponge. These fill with blood during intercourse.

At puberty the pituitary hormone LH stimulates production of testosterone which brings about secondary male characteristics (e.g. hair growth, deepening of the voice) and also initiates sperm production.

Female Reproductive Organs

The two ovaries lie within the pelvis, towards the back and underneath the kidneys. They are small, oval organs, each about the size of a bean, composed of connective tissue and containing many thousands of immature ova or eggs. The *oviduct* or *Fallopian tube* with its funnel-shaped opening commences beside each ovary and passes into the uterus or womb, a pear-shaped muscular organ. The uterus is about 8 cm long and opens at the base where there is a strong band of muscle called the *cervix*. From this a narrow muscular tube, the *vagina*, passes outwards and opens between the labia into the groin. The bladder, which lies to the front of the uterus, releases its urine through the urethra which opens close to the vagina.

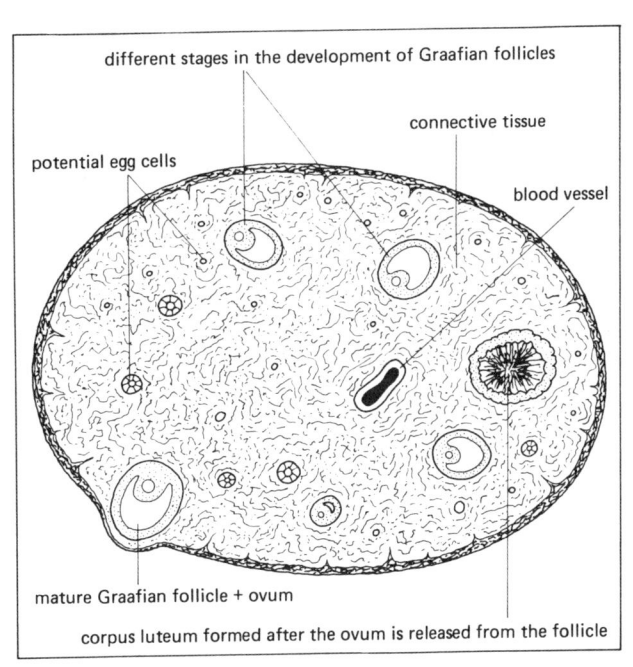

Fig. 8.47 Structure of ovary

Ovulation

Thousands of potential ova (egg cells) are present in the ovaries of a girl even before birth. These lie dormant until puberty when, upon stimulation of the ovaries by the pituitary hormone FSH, the ova begin to mature one at a time about once a month. One by one, each ovum grows larger and a sheath of cells containing fluid rich in nutrients develops around it. This *Graafian* or *ovarian follice*, as it is called, grows larger as it ripens and pushes

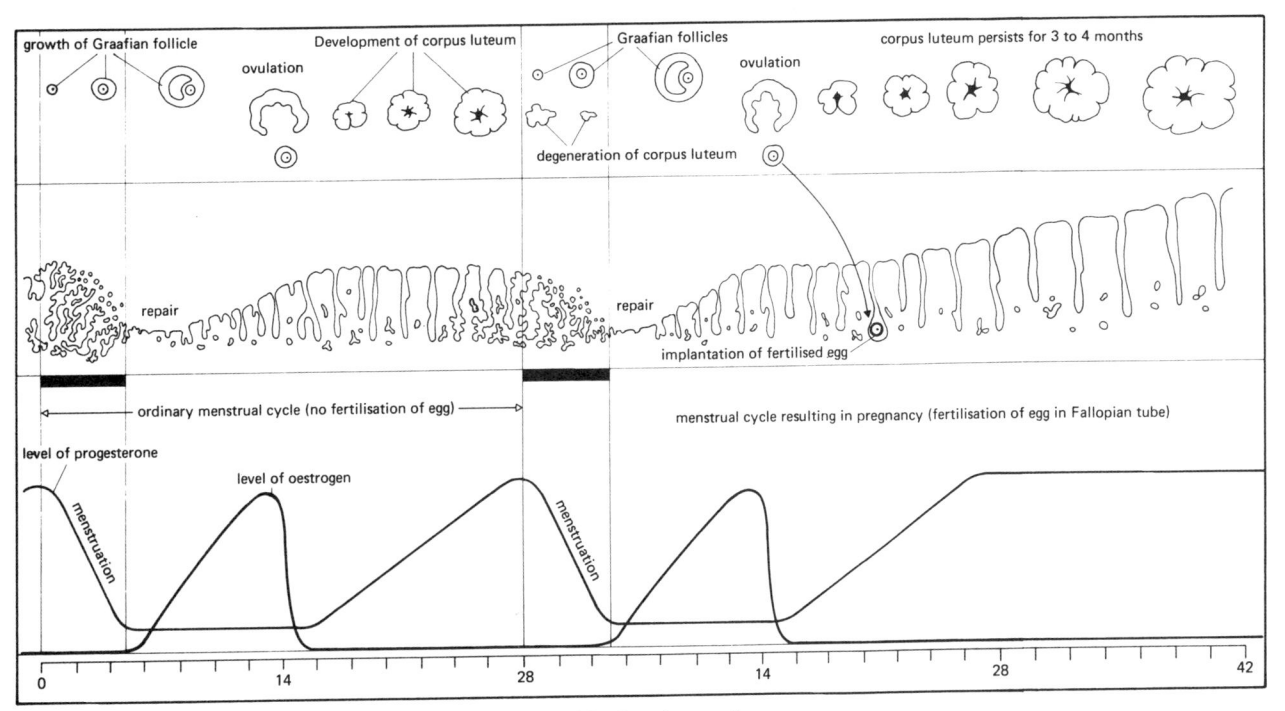

Fig. 8.48 Ovarian cycle

the wall of the ovary outwards until it eventually bursts, discharging the ovum into the funnel-shaped entrance to the Fallopian tube. The process just described is *ovulation*. It occurs in the second week after a period and is the time during which fertilisation is most likely to occur.

As the ovum travels along the Fallopian tube to the uterus, the follicle in the ovary continues to grow, eventually forming the *corpus luteum*. This secretes the hormone progesterone which begins to prepare the body for pregnancy by thickening the uterine lining and enlarging the breasts. If fertilisation does not occur the corpus luteum shrivels up and the progesterone level falls rapidly. If the ovum is fertilised the corpus luteum enlarges, secreting even greater amounts of progesterone. This increases the blood supply to the uterus, stimulates further thickening of the uterine wall and halts the development of follicles in the ovaries.

Menstruation

The menstrual cycle is directly related to the hormonal influence on the uterine lining. This lining is thin following menstruation. As the follicle ripens it releases oestrogen which causes thickening of the lining and an increase in the blood flow to the uterus. After ovulation, the corpus luteum secretes progesterone which causes further thickening of the lining. When fertilisation does not take place, production of progesterone slows down and stops, the corpus luteum shrinks, the uterine lining disintegrates and as a result of contractions of the uterus, its contents are expelled through the vagina.

Menstruation — or a period as it is usually called — occurs about fourteen days after ovulation. The cramps often experienced on the first or second day of a period are due to uterine contraction; some people also experience pre-menstrual tension, which is thought to be related to the withdrawal of progesterone.

Menstruation begins at puberty, usually between the ages of eleven and fifteen, and continues every month (except during pregnancy) until the menopause, when it becomes more erratic and finally ceases altogether. During the menopause, which generally occurs between the mid-forties and mid-fifties, some women experience unpleasant symptoms (e.g. depression, hot flushes) because of the withdrawal of the female hormones. In some cases these symptoms can be alleviated by the use of hormone replacement therapy (HRT), a treatment which is still quite new.

Fertilisation

During sexual intercourse the blood vessels in the penis dilate and an extra supply of blood fills the spaces, causing the penis to become erect. This facilitates its entry into the female vagina and upon further sexual stimulation the urethra contracts, causing ejaculation of the semen. In the female the walls of the vagina relax during intercourse and secrete lubricating mucus.

Each sperm consists of a 'head' containing a nucleus and a long 'tail'. The strong, lashing movements of the tail enable the hundreds of millions of sperm released to swim upward through the cervix into the Fallopian tubes. Many millions die before they reach the tubes, but others survive and if an ovum is present it is likely that fertilisation will take place. Several sperm surround the ovum but only one penetrates its protective covering. Soon afterwards the nucleus of the sperm enters the ovum and unites with the nucleus of the ovum. This is the moment of fertilisation.

Pregnancy

The fertilised ovum, now called a zygote, divides in two. Each cell then divides again and thus the growth of the embryo begins. The cells multiply rapidly as the embryo moves along the Fallopian tube until within a week it reaches the uterus. Here it sinks into the thick uterine lining *(endometrium)* and sends small projections (villi) into the lining to hold it in position and facilitate absorption of nutrients from its walls. The area around the villi develops into the *placenta* which is attached to the abdomen of the foetus by the umbilical cord.

The Placenta

This is concerned with providing the embryo with nutrients and oxygen from its mother's blood. It also

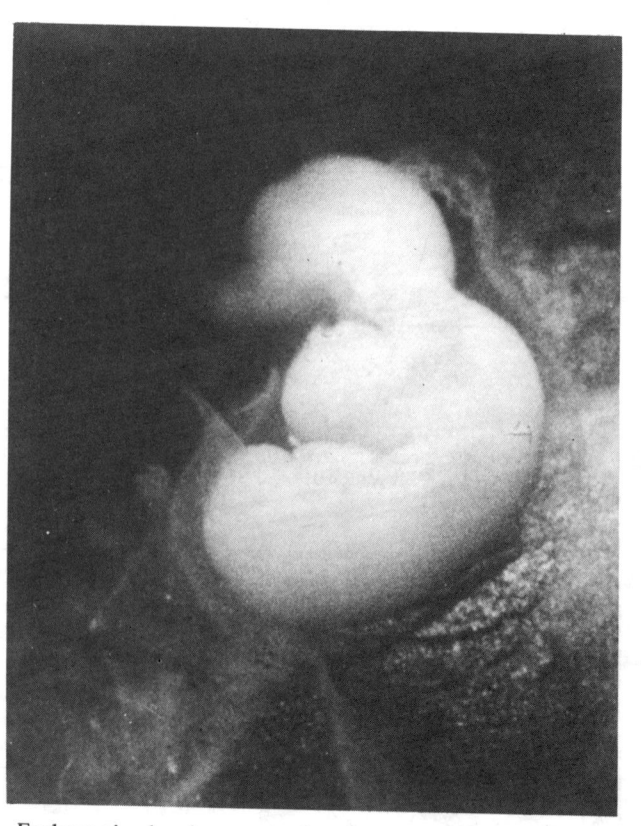

Embryonic development after four weeks (N. Guigoz, C. Edelman and J. M. Baufle)

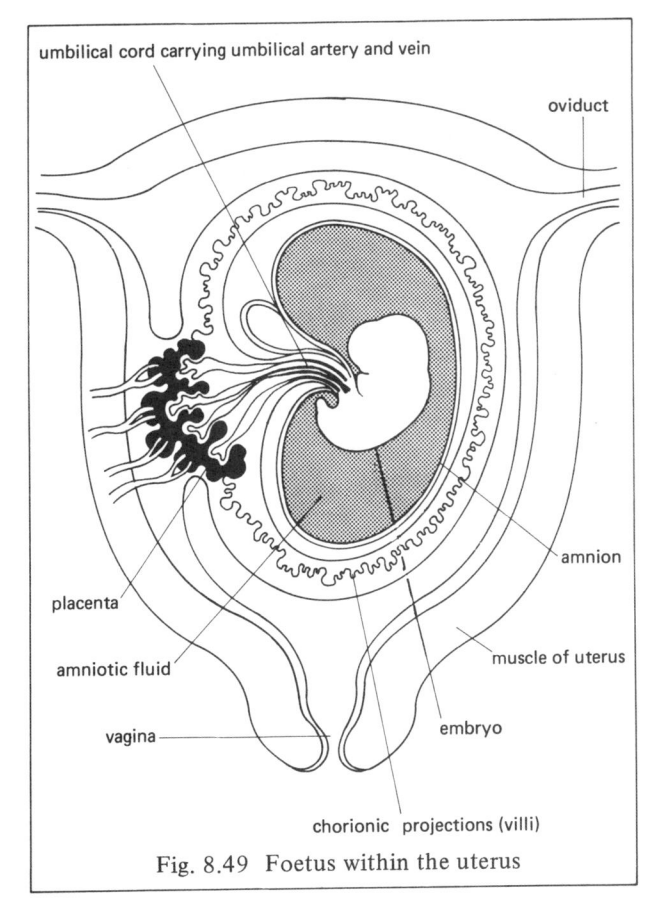

Fig. 8.49 Foetus within the uterus

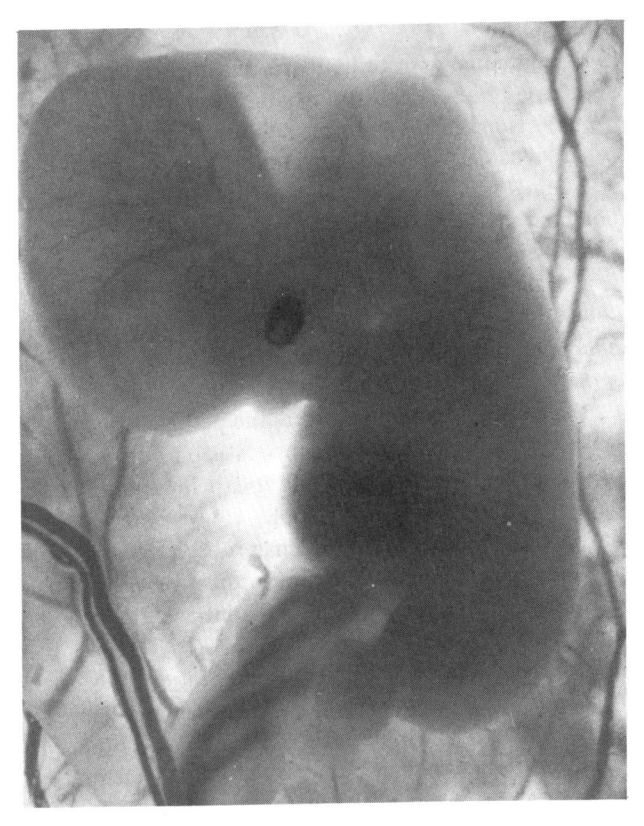

The embryo after seven weeks (N. Guigoz, C. Edelman and J. M. Baufle)

receives waste products such as carbon dioxide and urea from the blood of the embryo. Although the blood of mother and embryo are closely connected by the capillaries in the placenta, they do not mix; rather the placenta acts as a filter through which certain substances are allowed to pass and by which many harmful products are held back. It does not prevent the passage of all dangerous substances, however: some drugs and viruses such as German measles can pass through and may have unfortunate consequences for the foetus. For this reason it is inadvisable to take any form of drugs, even aspirin, or to smoke during pregnancy.

The placenta stimulates increased production of oestrogen and progesterone and eventually begins to produce its own progesterone. These hormones are responsible for the many changes which occur during pregnancy, preparing the body for birth and lactation:

1. Enlargement and thickening of the uterus.
2. Accumulation of fat and fluid.
3. Inhibition of contraction of the uterus.
4. Enlargement of the mammary glands.

The uterus grows larger to accommodate the growing embryo and the cells in the embryo begin to form tissues. Within eight weeks the embryo, now called a foetus, has a recognisable form with a head, trunk and limbs. The foetus is surrounded by a water sac or *amnion* (containing amniotic fluid) which protects it from injury. It continues to develop and by about six months all the organs are fully formed, but it remains in the uterus until it is strong enough to survive independently. The whole period of pregnancy, known as the gestation period, is about forty weeks.

Birth

By the time the baby is ready to be born it lies with its head down, facing the cervix. Secretion of progesterone diminishes and, stimulated by the hormone oxytocin, labour begins. The uterus begins to contract rhythmically and the contractions gradually increase in strength and frequency. After some time the cervix dilates, the membrane of the amniotic sac ruptures and the amniotic fluid flows out. In the final stages of labour, vigorous contractions of the uterus and abdomen expel the baby out through the cervix and vagina. The baby now begins to breathe on its own. The umbilical cord is tied and severed and soon afterwards, as a result of further contractions of the uterus, the placenta, now called the afterbirth, is expelled.

Within a couple of days the mammary glands, stimulated by the lactogenic hormone from the pituitary gland, begin to secrete milk with which to suckle the baby.

Birth Control

When people wish to limit the size of their family, it is necessary to use some form of birth control or contraception. None of the known methods of contraception is totally reliable for every individual. A couple should seek a doctor's advice as to the merits, disadvantages and potential side-effects of any particular method they are considering.

Certain forms of birth control involve restricting intercourse to the times of the month when fertilisation is unlikely to occur. A woman using the *rhythm method* or *safe period* keeps records of the lengths of her cycle so that she can gauge approximately the period during which it is inadvisable to have intercourse. Keeping daily records of *basal body temperature* can also help determine the days when abstinence is advisable. However few women have either monthly cycles which are absolutely regular at all times, or basal body temperatures with invariable patterns. Thus these methods of contraception are far from reliable in every case.

Another natural method of contraception, the *ovulation, or Billings, method* is based on observing changes in cervical mucus which occur at ovulation. With proper instruction regarding such changes, the Billings method can offer a more reliable indication of safe period.

The Billings, basal body temperature and rhythm methods of contraception all demand a high degree of communication and co-operation between the couple. There must be mutual agreement to abstain completely from intercourse except during the safe period, as well as a sophisticated understanding of the method (or methods) being followed if its use is not to be a failure.

The *contraceptive pill* contains the hormones oestrogen and/or progesterone which restrict the production of pituitary sex hormones, suppressing ovulation and therefore conception. If taken correctly contraceptive pills are almost 100 per cent effective, but they have many side-effects and cannot be taken by women who suffer from certain conditions such as high blood pressure.

Venereal Disease

Venereal diseases are highly infectious diseases which are passed from one person to another during sexual intercourse. They are generally the result of promiscuous sexual behaviour. Gonorrhoea and syphilis are the two most common types of venereal disease and, owing to increasing permissiveness, both are becoming a major social problem.

Gonorrhoea is caused by a bacterium. It first manifests itself through itching and a burning sensation when urinating. Without treatment it can affect the heart and cause blindness and even death. Mothers infected with this disease can pass it on to the foetus and the baby may be born blind.

Syphilis is also caused by a bacterium. It first appears as a sore or rash in the genital area and is followed a few weeks later by a general body rash and sore throat. If untreated it may lie dormant and become active only years later, when it may cause heart and brain damage, insanity and death.

QUESTIONS

Cells and tissues

1. Sketch and describe in detail the structure of an animal cell.
2. Classify the various types and human tissue. Write a brief note on
   a. connective tissue
   b. epithelial tissue
   c. skeletal muscle tissue
   using diagrams where appropriate.

Heart and blood

3. Describe with the aid of diagrams the blood flow through the heart.
   Write a note on
   a. coronary circulation
   b. systemic circulation
4. Write a note on the factors which control heartbeat.
   What is blood pressure?
   Explain the importance of haemoglobin and fibrinogen in the blood.
   What nutritional factors affect the formation of haemoglobin?
5. Differentiate between an artery, a vein and a capillary. How does the structure of each affect its function?
   Give the composition of blood and enumerate the functions of its constituent parts.
6. Write a detailed description of the structure of the human heart, using diagrams where necessary. Refer to the structure and function of heart valves.
   Explain the terms
   a. sino-atrial node
   b. systole
   c. blood group
   d. Rhesus factor

Lymphatic system

7. Outline the functions of lymph in the body.
   Describe how lymph is formed in the tissues.
   What is the role of lymph in the digestion/absorption process?
8. Explain the main differences between the lymph system and the circulatory system of the blood.
   Write a brief note on the following
   a. lymphocytes
   b. lymph nodes
   c. lacteals

Breathing

9. Discuss the breathing (respiratory) system of the body under the following headings
   a. mechanical action
   b. gaseous exchange
   c. nervous control

   What other factors influence the rate of breathing?

10. Describe the structure of the human lung, using diagrams.

    Explain the terms
    a. tidal volume
    b. vital capacity
    c. emphysema

    State the average composition of inspired and expired air.

Digestive system

11. Sketch the alimentary canal, labelling all organs clearly.

    Define the following terms
    a. digestion
    b. peristalsis
    c. hypothalamus

12. Write a note on enzymes, stating their function and how they operate.

    Describe the digestion of a protein food, listing the enzymes which act upon protein.

13. Describe in detail the structure and secretions of the stomach, using diagrams where necessary.

    List some of the factors which affect the digestion of food and state how these factors may be controlled.

14. Describe the role played by the small intestine in the digestion and absorption of food.

    Show with the aid of a diagram the structure of the small intestine.

    What are the functions of the large intestine?

15. Describe the process by which glucose is oxidised in the body.

    How does the body cope with excess proteins, fats and carbohydrates in the diet?

16. Write a detailed description of the position, shape and structure of the liver.

    What are the functions of the liver?

    Describe briefly the hepatic circulation, using diagrams where appropriate.

Excretion

17. Draw a labelled diagram of the structure of the skin.

    Describe *in detail* the functions of the skin.

    How does the skin act as a sensory organ?

18. How do the kidneys assist the homeostatic regulation of the body?

    Describe with the aid of diagrams the position, shape and structure of the kidneys.

    What is the composition of urine?

19. Describe in detail, with the aid of diagrams, the renal circulation.

    Explain clearly the filtering action of the kidneys and the chemical and nervous factors which control renal filtration.

Nervous system

20. Name two systems which bring about co-ordination in the human body.

    Describe the position, shape and structure of *(a)* the brain and *(b)* a nerve cell.

    In the case of *(b)* state how impulses are passed along nerves and across the synapse.

21. Differentiate between the central nervous system and the peripheral nervous system.

    Describe in detail how the autonomic nervous system operates.

Endocrine system

22. Describe the principles of the endocrine system.

    Draw a simple outline of a man, showing the most important endocrine glands.

    Select two hormones and describe the effect of *(a)* under-secretion and *(b)* over-secretion in each case.

23. Describe the position, shape and structure of *(a)* the pituitary gland and *(b)* another endocrine gland of your choice.

    List the hormones produced by each gland and their effects on the body.

Reproduction

24. Draw a labelled diagram of *(a)* the male *and (b)* the female reproductive organs.

    In the case of *one* describe how they function.

25. Describe the menstrual cycle, using a diagram to show the varying levels of female hormones.

    Describe briefly the stages which occur if the ovum is fertilised.

APPENDIX A
Basic Science

THE ATOM

The atom is the smallest part of an element (see below) which can take part in a chemical change. At one time it was thought that atoms were the smallest or simplest particle of matter — about one hundred millionth of a centimetre — but it is now known that sub-atomic particles exist. The three most important of these are protons, neutrons and electrons.

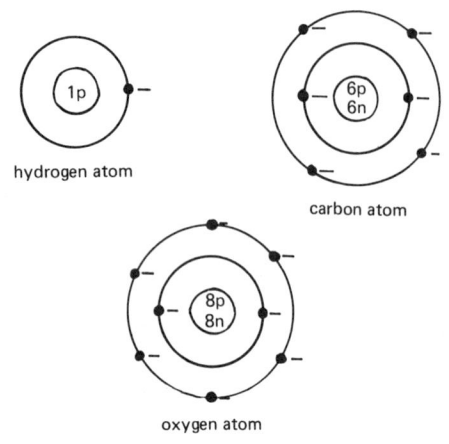

Fig. A.1 Atoms

Protons and neutrons make up the central core or nucleus of the atom. Protons have a positive (+) charge and neutrons, as their name implies, are neutral, i.e. they have no electric charge.

Electrons have a negative charge (−) and they are almost two thousand times lighter than either protons or neutrons. They are in constant motion spinning around the nucleus in much the same way as planets orbit the sun. Because the number of positively charged protons and negatively charged electrons is equal, the atom is electrically neutral. The spinning electrons are kept in orbit by the force of attraction between the positive protons and the negative electrons.

Atomic number. All the atoms of any one element have an identical number of protons in each nucleus. The number of protons within a nucleus is called the atomic number of the atom. The hydrogen atom is the simplest, containing just one proton and one electron. Carbon contains six protons and oxygen eight.

Energy shells. The electrons spinning around the nucleus rotate in a series of orbits or shells. The shell nearest the nucleus is called the K shell. Outside that is found the L shell and so on. Each shell has a limit to the number of electrons it may contain. The K shell may take up to two, the L shell up to eight, the M shell up to eighteen electrons and so on. If an atom contains the maximum of electrons in its outer shell, it has a complete shell and is incapable of reacting with other atoms. Such an atom is said to be *inert*. This occurs in gases such as helium with two electrons in its K shell and neon with eight electrons in its L shell.

The chemical properties of an atom or element (i.e. their ability to link up with each other) are determined by the number of electrons in the outer shell of the atom. Atoms are most stable when the outer shell contains its maximum number of electrons, which is often eight, or two in the case of helium. During chemical reactions atoms exchange or transfer electrons in an attempt to attain the stable number of electrons.

Elements

An element is a substance made up of atoms with the same atomic number. It cannot be broken down into a simpler substance by chemical means. Over a hundred elements are known to man and are usually classified in order of increasing atomic number. They are set out in horizontal rows (periods) with a new period beginning when a new energy shell starts to fill. This classification is known as the *periodic table* and can be found in logarithm table books. Another important property of an element is its relative *atomic mass;* this is the average mass of an atom taking the mass of carbon as 12.

Here are some of the more common elements found in foodstuffs:

Element	Symbol*	Atomic Number	Atomic Mass
Hydrogen	H	1	1
Carbon	C	6	12
Oxygen	O	8	16
Fluorine	F	9	19
Sodium	Na	11	23
Magnesium	Mg	12	24
Phosphorus	P	15	30
Sulphur	S	16	32
Chlorine	Cl	17	35.5

Potassium	K	19	39
Calcium	Ca	20	40
Iron	Fe	26	55
Copper	Cu	29	63
Iodine	I	53	126

Symbols of elements. In order to simplify the writing of chemical formulae, the first letter, or in some cases, two letters from the Latin name for the element are used to identify each element (e.g. C – carbon; Ca – calcium).

Molecules

A molecule is the smallest group of atoms capable of a separate existence. It is a minute particle consisting of two or more atoms which are chemically combined. When the atoms of a molecule are all identical, molecules of an *element* are formed. When different atoms combine molecules of a *compound* are formed.

Compounds

When atoms of two or more different elements are chemically joined together they form a compound. This may be broken down into its component elements only by chemical changes. The atoms of each element present in a compound must combine in a fixed ratio, for example common salt (NaCl) is composed of one atom of sodium for every one atom of chlorine. A water molecule contains one atom of oxygen and two atoms of hydrogen. When a compound is formed the elements making up the compound may lose their individual properties such as state and colour and become totally unrecognisable – for example the gases hydrogen and oxygen when combined in the correct ratio will produce a liquid, H_2O, which is water.

Compounds may be classified as organic or inorganic.

Organic compounds always contain carbon and these are usually joined by covalent bonds (see below). They include many nutrients such as proteins, lipids and carbohydrates.

Inorganic compounds are mostly compounds of elements other than carbon and include the mineral elements and water. Their atoms may be joined by ionic bonds or covalent bonds.

Formulae. The formula for a molecule of a compound can be compared with the recipe for a cake. It shows the components or ingredients (atoms) present, together with the number of atoms of each element. *Examples*

H_2 (H + H) – a hydrogen molecule
$C_{12}H_{22}O_{11}$ – a sucrose molecule

Just as symbols are used to represent elements, so chemical formulae are used to represent molecules.

When *chemical changes* take place the reactions of the atoms and molecules involved are shown by equations. The formulae of the reactants are placed on the left, the change is represented by an arrow or equality sign and the formulae of the new products are listed on the right. *Example:* ethyl alcohol, when exposed to air, changes to acetic acid:

$$C_2H_5OH + O_2 \rightarrow CH_3COOH + H_2O$$

A Mixture

This occurs when two or more substances are blended together without any of the components losing their individual properties. A mixture is usually easy to separate as the components are not chemically bonded together and therefore no chemical reaction is necessary to separate them. The ratio between the components of a mixture is not a fixed ratio as in compounds. Air, for example, is a mixture containing oxygen, nitrogen and carbon dioxide, but the amount of oxygen and carbon dioxide present may vary considerably.

Alloys are mixtures of two or more metals such as brass and bronze, both compounds of copper. Other mixtures include solutions and colloids.

PHYSICAL AND CHEMICAL CHANGES

A *physical change* does not produce a new substance. It usually involves a change of state and appearance but the change is often only temporary. *Example:* ice changing to water.

A *chemical change* involves the breaking up and rearranging of molecules. It produces a new substance which has different properties from the original substances. Chemical changes are usually difficult to reverse.

Heat can produce a physical or chemical change: the boiling of water is a physical change (liquid into gas) while the burning of coal is a chemical change (carbon into carbon dioxide – $C + O_2 \rightarrow CO_2$)

Chemical Changes	*Physical Changes*
Burning of coal	Freezing of food
Combustion of petrol	Dissolving of sugar or salt in water
Rusting of metals	
Digestion of protein	Boiling of water
Action of baking powder in bread making	Condensation
	Melting of butter or lard
Many changes which take place in cooking, e.g. browning	

Bonding

There are several ways by which the atoms in molecules may be joined or bonded together but all involve sharing or transfer of the *valence* electrons in the outer shell.

The valency of an atom is a measure of its ability to link up chemically to form molecules.

The most usual types of bond are:
a. ionic bonds
b. covalent bonds
c. hydrogen bonds

a. Ionic bonds (electrovalent bonds). These involve the transfer of electrons: one atom loses one or more electrons to another atom. The method by which it occurs can be explained by looking at the formation of sodium chloride (NaCl). Sodium has one electron in its outer shell and chlorine has seven. The sodium electron passes to the chlorine atom to give it the necessary eight electrons in its outer shell. Although the sodium and chlorine atoms were neutral, the loss of an electron has given the sodium atom a positive charge (11 protons – 10 electrons) and the chlorine atoms becomes negatively charged by gaining an electron (17 protons – 18 electrons). These charged atoms are called *ions*. The attraction of opposite electrical charges holds the ions together in an ionic or electrovalent bond.

The compound formed consists of a three-dimensional lattice of ions forming a crystal. It is not a true molecule. The bonds holding the ions in position can be readily broken – for example when NaCl is mixed with water, the ions are released and the salt dissolves in the water. Ionic compounds usually contain metals which lose their electrons easily. This makes them good conductors of electricity.

Charged particles either attract or repel one another.
Like charges repel. ←(+) (+)→
Unlike charges attract. (+)→ ←(−)

b. Covalent bonds. In a covalent bond two atoms share a pair of electrons. *Example 1.* A hydrogen molecule consists of two atoms of hydrogen each contributing one electron to the bond.

$$H^\bullet + H^\bullet \rightarrow H:H \ [H-H]$$

Example 2. Water: oxygen has six electrons in the outer shell. Two of these electrons join up with the single electrons from two hydrogen atoms producing a stable eight electrons.

$$H^\bullet + \cdot\ddot{O}\cdot + \cdot H \rightarrow H\!:\!\ddot{O}\!:\!H \text{ or } H_2O.$$

When representing covalent molecules a dash is used to indicate a shared pair of electrons.

Example

H–N(–H)–H [NH$_3$] *Ammonia*

Double or treble covalent bonds can be formed when two or three pairs of electrons are shared. They are stronger than single bonds but not necessarily twice or three times as strong.

The carbon dioxide molecule, for example, is formed when two oxygen atoms unite with one carbon atom.

O :: C :: O   [CO$_2$]

Double or treble bonds are represented by using two or three dashes to indicate the paired electrons. *Example.* O=C=O. Covalent compounds are generally:

1. Gases, volatile (easily evaporated) liquids or solids with a low melting point.
2. Insoluble in water.
3. Bad conductors of electricity.
4. Reactions are slow and may not go to completion.

c. Polar bonds and hydrogen bonds. When a covalent bond is formed between elements, the electrons may not be shared equally between the atoms involved. One atom may have a greater attraction for the shared electrons and as a result of its greater *electronegativity* this atom will acquire a small negative charge and the other atom a small positive charge. This bond is termed a polar covalent bond and occurs in the water molecule.

By means of these small charges, separate water molecules will become attracted to one another with the positively charged atoms of one water molecule bonding to the negatively charged atom of another molecule of water. The hydrogen atoms act as a bridge, as it were, holding the water molecules together (as in the liquid and solid states of water). For this reason these bonds are termed hydrogen bonds.

The hydrogen bond is much weaker than normal bonds and is much easier to break. A slight increase in temperature, for example, would cause molecules to vibrate sufficiently to break the bonds. This occurs in boiling, when the vibrations of the molecules are strong enough to bounce them off the surface, causing the liquid to evaporate.

Hydrogen bonds are quite common in foodstuffs: carbohydrates and lipids contain O–H bonds; proteins (amino acids) are made up of N–H bonds. Although individually weak, these bonds are capable of holding large molecules in place. Protein, for example, and many polymers (see below) may contain hundreds of hydrogen bonds.

ORGANIC COMPOUNDS

Organic compounds contain the element carbon and are also known as *carbon compounds*. Food containing carbon compounds include proteins, fats (lipids) and carbohydrates.

As already mentioned, the carbon atom has six protons and six neutrons in its nucleus with six electrons in orbit around it. Two electrons complete the K shell but the L shell contains only four electrons. It therefore requires four electrons to make up the eight required to produce a stable atom.

Example. The methane molecule (CH_4) has the following structure:

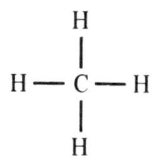

$$\begin{array}{c} H \\ | \\ H - C - H \\ | \\ H \end{array}$$

Four hydrogen atoms are bonded covalently to the carbon atom.

It is important to note that many molecules are not two-dimensional as shown on paper but are three-dimensional structures. An example of this is the ammonia molecule which is pyramidal. Carbon atoms link together readily to form a chain. Complicated branched chains and rings can be formed with units which repeat themselves several times, forming molecules of high molecular mass. These repeating units are called *monomers*. A large molecule made up of many monomers is called a *polymer*. Protein and starch molecules are natural polymers.

Plastics and synthetic fabrics are often made up of polymers, e.g. polyethylene, polypropelene, polyester, nylon. Although on paper the structural formula of one of these polymers appears to be a long straight chain, in practice they are arranged in three-dimensional patterns including zig-zags, circles or spirals with the atoms tightly packed together. If the carbon atoms in the molecule are joined by single covalent bonds, it is known as a *saturated* molecule. Carbon atoms joined by double or triple covalent bonds form molecules of *unsaturated* compounds. These are more reactive than saturated compounds.

Organic compounds are divided into families or groups. The compounds within each family have similar structures and reactions. This similarity in properties occurs because members of the same family have the same group of reacting atoms (the functional group).

1. **Hydrocarbons: CH**
   These compounds contain hydrogen and carbon only. They are not found in food and include many powerful fuels.

   a. Alkanes or paraffins are joined by single bonds only and include the gases methane, propane and butane (bottled gas).

   b. Alkenes or olefins contain at least one double bond; ethylene is an important member of this group. It is used in the manufacture of plastics.

   c. Alkynes or acetylenes. The carbon atoms in this molecule are joined by a triple bond. Acetylene torches are used for welding. As alkenes and alkynes are unsaturated hydrocarbons they are very reactive compounds.

2. **Oxygen Compounds**
   a. Alcohols. The compounds in this group are derived from hydrocarbons and include methanol CH_3OH which is found in methylated spirits and commonly used as a solvent. Ethanol C_2H_5OH is found in wines and spirits and is used in the manufacture of perfume, paint and flavouring essences such as vanilla. Commercially it may be obtained from ethylene but the alcohol present in alcoholic drinks is produced by fermentation, i.e. yeast acting on the sugar present in the various fruits and cereals to produce

Group	Functional Group	Examples
1. *Hydrocarbons*	CH	
a. Alkanes	single bond	Butane, paraffin etc.
b. Alkenes	double bond	Ethlene
c. Alkynes	triple bond	Acetylene
d. Aromatic hydrocarbons	rings of carbon atoms	Benzene
2. *Oxygen Compounds*		
a. Alcohols	OH	Ethanol – C_2H_5OH
b. Aldehydes	CHO	Formaldehyde – HCHO
c. Esters	COO	Fats
d. Organic acids	COOH	Acetic acid (vinegar) – CH_3COOH
3. *Nitrogen compounds*		
a. Amides	$CONH_2$	Urea (a component of urine) – $CO(NH_2)_2$
b. Amines	NH_2	Analine (used in dyes) – $C_6H_5NH_2$
		Proteins

alcohol and carbon dioxide. Glycerol (glycerine) containing 3–OH groups is called a trihydric alcohol and is an essential part of the structure of lipids.

b. Aldehydes. These are often derived from alcohols and are used to produce many synthetic flavourings such as acetaldehyde (apple flavour) and benzaldehyde (almond flavour).

c. Esters. These are substances produced when an organic acid (see below) reacts with alcohol. They have an attractive odour and taste and are found naturally in strongly-flavoured fruits such as bananas and pineapple. Many artificial flavourings are manufactured from chemically produced esters, e.g. benzyl acetate – strawberry flavour produced from benzoyl alcohol and acetic acid.

d. Organic acids (carboxylic acids). Organic acids contain a *carboxyl group* COOH. They are weak acids and are found in many fruits. All contain a double bond. Vinegar is a weak solution of acetic acid CH_3COOH – a carboxylic acid with the following structural formula.

$$H-\overset{\overset{\displaystyle H}{|}}{\underset{\underset{\displaystyle H}{|}}{C}}-\overset{\displaystyle O}{\overset{\displaystyle \diagup\!\diagup}{C}}-OH$$

Other foods containing the carboxyl group include rhubarb, which contains oxalic acid; lemons, which contain citric acid, and baking powder, which contains tartaric acid (cream of tartar). Lactic acid (hydroxypropanoic acid) in sour milk contains both a carboxyl and hydroxyl group, i.e. it reacts both as an acid and an alcohol. Lactic acid is produced in milk when lactic acid bacilli convert the lactose (milk sugar) into lactic acid by enzyme action.

$$C_{12}H_{22}O_{11} + H_2O \xrightarrow[\text{bacilli}]{\text{lactic acid}} 4CH_3CHOHCOOH$$
lactose   lactic acid.

Fatty acids (see p. 13) are complicated organic acids of high molecular mass which combine with glycerol to form fats. They include *stearic* $C_{17}H_{35}COOH$ and *palmitic* acids $C_{15}H_{31}COOH$

ACIDS, BASES AND SALTS

Many foods and chemicals used in the home contain acids, bases or salts.

Acids

An acid can be described as a substance which releases hydrogen ions (protons) in a solution of water, for example organic acids contain the carboxyl group – COOH – and give H^+ ions in solution. Most acids have the following properties:
1. They have a sour taste.
2. They are corrosive.
3. They turn litmus paper red.
4. They neutralise bases producing salt and water.

Organic acids used in the home include: ascorbic acid (vitamin C); acetic acid (in vinegar); citric acid (in lemons); lactic acid (sour milk).

Acids react with carbonates, e.g. sodium bicarbonate, producing salt, water and carbon dioxide. They act as preservatives because strong concentrations of acid are toxic to most micro-organisms.

Inorganic acids include hydrochloric acid (HCl) secreted in the stomach and sulphuric acid (H_2SO_4) used in the manufacture of synthetic fibres. They are much stronger than organic acids and should not be handled.

Bases

A base is defined as a substance which accepts a hydrogen ion (proton) and neutralises an acid. A base which is soluble in water and contains a hydroxyl (–OH) ion is known as an *alkali*.

Properties of bases and alkalis:
1. They turn litmus paper blue.
2. They have an alkaline (soapy) feel.
3. They neutralise acids.
4. Strong alkalis damage fabrics.
5. They combine with fats or oils to form soap (saponification).

Bases and alkalis used in the home are found in many cleaning agents and washing powders. These include sodium hydroxide (caustic soda) which is used in the manufacture of soap and powerful oven cleaners. Strong alkalis should not be allowed come in contact with the skin as they react with the oils and fats of the tissue.

Salts

When an acid and a base neutralise each other, a salt and water are produced. Metal atoms usually take the place of the hydrogen atoms in the acid and the metal forms a bond with the remainder of the acid molecule.

Example:

$$H_2SO_4 \rightarrow Na_2SO_4$$
sulphuric acid   sodium sulphate

The most widely used salt is probably sodium chloride or common salt. Other salts include sodium bicarbonate, $NaHCO_3$ (bread soda); sodium carbonate, Na_2CO_3 (washing soda).

The pH Scale

This is a measurement of acidity and alkalinity. It indicates the concentration of hydrogen ions in a substance.

Pure water is neutral and has a pH value of 7. Acids have a pH value below 7; alkaline substance have a pH over 7.

Each unit on the scale indicates a tenfold increase or decrease in the concentration of hydrogen ions.

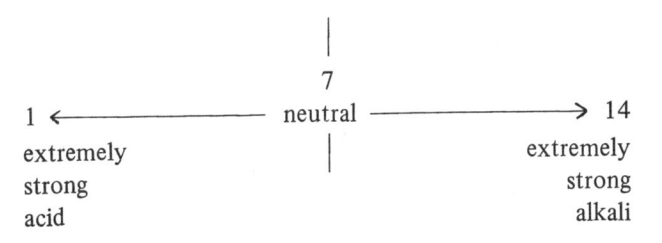

```
                    7
1  ←———————— neutral ————————→  14
extremely                      extremely
strong                         strong
acid                           alkali
```

The pH scale

acid	2	hydrochloric acid
	2.7	vinegar
	3	lemons
	5	bread
	6	milk, eggs, rain water
neutral	7	water
	7.4	plasma
alkaline	8	egg white
	9	soda bread
	14	caustic soda

Most foods are slightly acid. A low pH is necessary for the setting of jam. Soaps and many cleaning agents have a high pH value. The pH value of a solution can be measured by special testing papers (universal indicators) which show a range of colours, each indicating a degree of acidity or alkalinity.

OXIDATION AND REDUCTION

Oxidation

Oxygen is one of the main constituents of air and is essential to life. It is necessary for the burning of fuel and is therefore essential for metabolism of nutrients in the body. When an element joins with oxygen, loss of electrons often takes place. This process is called oxidation.

$$\text{element} + \text{oxygen} \xrightarrow{\text{oxidation}} \text{element oxides}$$

Example. Combustion: $C + O_2 \rightarrow CO_2$

Examples of oxidation

1. Combustion: carbon unites with oxygen in the air producing energy, e.g. the burning of coal or petrol. Rapid combustion occurs when coal gas or other explosives ignite, causing an explosion.
2. Cellular respiration: a form of controlled combustion whereby carbon in food reacts with oxygen producing heat/energy and CO_2.
3. Discolouration: many fruits and vegetables will brown if cut surfaces are left exposed to the air. Enzymes within plants hasten this process. Steeping in water or an acid substance prevents discolouration. Sodium sulphite is commonly used to inhibit browning.
4. Bleaching of clothes: oxidising agents such as sodium perborate unite with the oxygen in the air, removing colour or stains.
5. Bleaching of flour: oxidising agents are added to speed up the bleaching and maturing process.
6. Food production: to prevent or delay oxidation of foods many are packed in a vacuum or in carbon dioxide, e.g. coffee, milk powder. Antioxidants (see below) are added to foods with a high fat content to delay rancidity. Such foods include margarine, lard and many dehydrated foods.

Reduction

This involves the gain of electrons and takes place when oxygen is removed from a substance. It is, in fact, the opposite to oxidation and occurs simultaneously with it. Obviously if one substance loses electrons another must gain electrons. The process of adding hydrogen to a compound is also known as reduction. Hydrogen and some other elements have an affinity for oxygen so they tend to remove oxygen from compounds 'reducing' them in mass. These are known as reducing agents.

Examples of reduction

1. Antioxidants (see p. 117): these are reducing agents which prevent fats from becoming rancid. Natural antioxidants include vitamins C and E.
2. Hydrogenation: the chemical process of using hydrogen to convert soft fats and oils into solid fats.

Hydrogenation

During the manufacture of margarine and other solid cooking fats hydrogen is added to the mixture of oils used. As these are unsaturated oils, their molecules contain at least one double bond. The hydrogen attaches itself readily to the oil molecule; two atoms of hydrogen are absorbed by each double bond, thereby 'saturating' the molecule and solidifying the fat. To speed up the reaction a catalyst (nickel) is used and filtered out at the end of the process. Hydrogenation delays rancidity, makes fats easier to spread and enables them to cream easily with sugar.

Hydrolysis

This occurs when a compound reacts with water: the compound is broken down into simpler substances by reacting with water molecules. Hydrolysis occurs at each stage of the digestion of protein, fats and carbohydrates, breaking down the molecules so that they react more readily with digestive enzymes.

MATTER

Anything which occupies space and has mass is called matter. The properties of matter vary considerably. It may be hard or soft, heavy or light, cold or hot. It exists in three forms – solid, liquid and gas. Water, for example, is solid below 0°C, liquid between 0°C and 100°C and gas (vapour) over this temperature. Gas, although often not visible, takes up space and has weight just like a solid or liquid. When liquid changes into a gas the process is known as *evaporation*. When a gas changes to a liquid, it is said to *condense*.

$$\text{SOLID} \underset{\text{freezing}}{\overset{\text{melting}}{\rightleftarrows}} \text{LIQUID} \underset{\text{condensing}}{\overset{\text{evaporating}}{\rightleftarrows}} \text{GAS}$$

Most substances can be changed from one state to another by adding or taking away energy, i.e. applying heat or cooling. In a refrigerator, for example, the refrigerant is heated and changes from a liquid to a gas. Some substances change directly from a solid to a gas by the process of *sublimation*. This occurs in accelerated freeze drying (see p. 139) when ice crystals in frozen food are converted directly into steam without going through the intermediate liquid state.

Almost all matter expands on heating and contracts when cooled.

Solids

The molecules in a solid are tightly packed together in a fixed manner, giving them little freedom to move. This means that they have a definite shape. The more closely the molecules are packed, the greater the density of the solid – for example the molecules in metals are very tightly packed, but those in a sponge or polystyrene packing are less tightly packed and consequently less dense.

When heated, the molecules vibrate and the vibration increases with further heating until the particles break loose and begin to flow (move). The temperature at which this occurs is called the *melting point*.

Liquids

The molecules in a liquid are a little further apart and can move about more freely. This enables the liquid to flow and take up the space of any container into which it is poured. In liquids, the molecules on the surface escape, causing evaporation to take place, and when a liquid boils, gas evaporates from all parts of it as bubbles. Liquids vary in density; oils are less dense than water and are therefore lighter so that they float on water (e.g. cream rises to the top of the milk). Some liquids flow more quickly than others; syrup and oil flow slowly due to the force of attraction between the molecules in these substances. These are known as viscous liquids.

Gases

The molecules of gases are very far apart and they can expand to fill any space. They move rapidly about at random, colliding frequently with each other and with the walls of their container. This freedom of movement occurs because there is little force of attraction between the molecules. Different gases will mingle with one another readily; this process is known as *diffusion*. When gas is heated it expands, becomes less dense and rises. This principle is utilised, for example, in bread-making. Ventilation of many rooms and buildings is also based on the same principle.

Summary. Almost all materials can exist in the form of a solid, liquid or gas. Substances which are normally considered solid (iron), liquid (water) or gas (oxygen) are so at normal temperatures, but as temperatures rise so too does the motion of the molecules in the substance, eventually causing a change in state. All three states are capable of exerting pressure. Consider the pressure of ice in a water pipe causing it to crack; water gushing from a tap or spring and steam forcing the lid off a kettle.

Evaporation

The molecules in liquids move about freely and some of the more active ones escape from the surface, thereby reducing the volume of the liquid. As already mentioned, this is known as evaporation. The presence of wind speeds up evaporation by sweeping off the surface molecules more readily. As the temperature rises the movement of the molecules and rate of evaporation increase until boiling point, when large amounts of molecules are lost from all parts of the liquid.

Energy is lost upon evaporation, lowering the temperature on the surface. This occurs during evaporation of sweat from the skin and around the evaporator of a refrigerator. Volatile liquids such as alcohol evaporate more quickly than water.

Evaporation may be increased by reducing the pressure above a liquid; this occurs during freeze drying in order to speed up the drying (evaporation) process. Liquid boils at a lower temperature at high altitudes where atmospheric pressure is low, e.g. on Mount Everest liquid boils at 45°C. When pressure is increased – for example by sealing the container – the steam cannot readily escape and so it exerts pressure on the surface of the liquid which increases the boiling point of water. This occurs in pressure cooking when water boils at 120°C instead of 100°C. Because of the increased temperature, the food cooks more quickly (see p. 104).

Humidity

This is used to describe the amount of water vapour in the air. Breathing, evaporation and steam add to the humidity of a room. Many kitchens and bathrooms have high humidity levels. When the air finally becomes saturated with water vapour, the vapour condenses on cold walls and windows.

Surface Tension

Molecules in a liquid are equally attracted in all directions by the other molecules in the liquid. On the surface of a liquid the molecules are only attracted by forces beneath and beside them and there is no upward pull. This results in a skin-like surface on the liquid called the surface tension.

Surface tension explains why small insects can walk on water and why water will form droplets on the surface of a greasy object.

Experiment

Place a needle or razor blade carefully on the surface of a container of water. It will float. Add a few drops of liquid detergent and the object will sink, proving that detergent lowers the surface tension of water.

When clothes or fabrics are waterproofed, substances which repel water, such as oils or silicones, are applied to the surface of the fabric to repel water molecules.

Diffusion

The molecules in fluids (liquids and gases) move from areas where there is a high concentration of molecules to areas where the concentration is less dense, until both are equal. This happens because the molecules which are moving about continually collide with one another and scatter in all directions so that they mingle easily with other molecules. Ammonia and other strong smelling gases diffuse readily. Diffusion can be speeded up by air movements (by stirring or wind).

Osmosis

All living things consist of cells with semi-permeable walls — i.e. the membraneous walls allow certain substances (small molecules) through but larger molecules cannot penetrate. Many solids are impermeable, so nothing can pass through them. Others are porous or permeable, and most things can pass through.

When molecules of cell fluids collide, they throw each other outwards so that they often hit the walls of the cell. As these walls are semi-permeable, smaller molecules such as water or salt can pass through the membraneous wall until the concentration on either side of the wall is equal. This process is known as osmosis and is, in fact, a form of diffusion.

Osmosis could be defined as the diffusion of water molecules through a semi-permeable membrane from the weaker solution to the stronger one until their concentrations are equal.

Examples

1. Salting of fish as a method of preserving draws the liquid out of the cells, drying the fish and so preventing bacterial decomposition.
2. Meat should not be sprinkled with salt before or during cooking as this will have the effect of drawing the juices from the meat.
3. Osmosis occurs in the body:
    a. in the intestine as digested food is absorbed
    b. in the nephron of the kidney

SOLUTIONS

A solution is a homogeneous mixture of two or more substances. The most common solutions involve a solid dissolving in a liquid, which is usually water. In a solution the molecules of solute are evenly dispersed throughout the liquid, giving the solution a uniform concentration throughout.

Solute. The solid which dissolves, e.g. sugar, salt.

Solvent. The liquid in which it dissolves. Water is capable of dissolving many substances, both solid and gas. Other solvents include turpentine, alcohol and many cleaning agents.

A substance which dissolves easily in a liquid is said to be soluble in that solvent. One which does not dissolve is insoluble.

When a solution contains as much solute as the solvent can hold, it is said to be a *saturated* solution. The solubility of most substances increases with temperature and when the liquid evaporates from the solution or the solution is cooled, crystals of solute come out of it. This process is termed *crystallisation*.

When chemicals are in solution they are broken down into smaller particles such as molecules or ions, and this makes it easier for them to react together. Most chemical reactions take place in solution (e.g. enzyme action in digestion).

Note. Most solutions are of solid in liquid. Solutions of gas in liquid (C_2O in water) and gas in solids are also

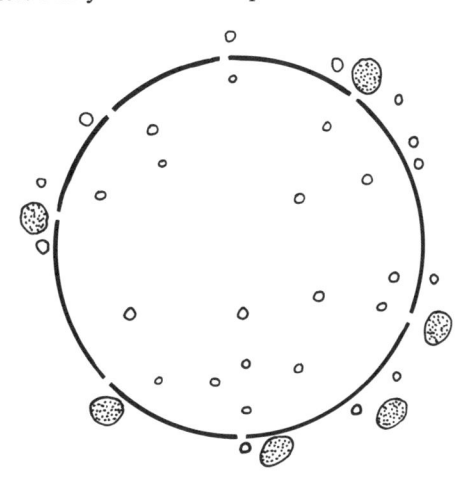

Fig. A.2 Semi-permeable cell membrane

possible, e.g. gases absorbed by charcoal filters in cooker hoods.

Colloids

A colloid is another type of mixture where large molecules or clumps of smaller molecules are suspended in a liquid. Many food molecules form colloidal suspensions. Most colloids appear slightly milky or opaque and they absorb water.
Examples: egg white, milk.

Emulsion

This is a colloid of two liquids which would not normally mix together. It can be *temporary*, as in the case of oil and water when shaken, or a *permanent* emulsion where the emulsion is stabilised by the addition of an emulsifying agent (emulsifier).
Examples of emulsifiers: lecithin in eggs; gluten in flour; casein in milk; starch in soups and sauces.
Examples of emulsions: cream – fat in water; mayonnaise – vinegar in oil.

APPENDIX B
Cookery Terms

Though similar cooking techniques are used throughout the world, many of the terms which we use to describe them are French, because since the eighteenth century France has produced many of the world's greatest chefs. However, as the following list shows, modern domestic cooking draws on the varied cuisines of many other nations, too.

Agar-Agar: A setting agent prepared from seaweed and used to set cold sweets instead of gelatine. It is useful in vegetarian diets where gelatine may not be eaten.

Aerate: To entrap air in a mixture (e.g. by sieving or beating).

Aigrette: A savoury dish made with choux pastry or batter and deep fried (e.g. cheese aigrettes).

Anchovy: A small Mediterranean fish with a strong, salty taste, which is usually preserved in oil. It is used on pizza.

Arrowroot: A starchy powder obtained from the root of a tropical plant. It is used to thicken sauces in much the same way as cornflour.

Aspic: A clarified jelly made from meat stock. It is used to coat or garnish salad dishes and savouries.

Baba: A spongy yeast pudding, usually soaked in syrup (e.g. baba au rhum).

Bainmarie: A tin or large flat pan containing simmering water in which one or several saucepans may be standing. It is used to cook foods very gently or to keep food, perhaps a sauce, warm. A double saucepan is based on the same principle.

Ballotine: Poultry which has been boned and stuffed. Served hot or cold.

Barbecue: To cook over an open fire. Charcoal is the most suitable fuel. A barbecue is also the name of the cooking fire.

Barding: Covering lean cuts of meat or poultry with fat meat (e.g. covering a chicken with bacon before roasting).

Baste: To spoon hot fat over lean meat during roasting or grilling. This helps to keep it moist and juicy.

Batter: A mixture of eggs, flour and milk used for pancakes, fritters, etc.

Bavarois: A rich custard made from milk, cream and egg yolks, and set with gelatine. Vanilla, chocolate and other flavourings may be used.

Béarnaise: A rich brown herb-flavoured sauce served with beef.

Béchamel: A basic white sauce made from milk which has been infused with onion and herbs.

Beignet: A piece of choux pastry which has been deep-fried and is served sweet or savoury.

Bercy: Food poached in a mixture of rich stock and wine and well flavoured with herbs.

Beurre manié: Equal quantities of flour and butter used as a liaison for soups and sauces.

Bigarade: A sauce made from bitter oranges, served with duck or game.

Bisque: A shellfish soup.

Blanch: To put in cold water and bring to the boil. Used to whiten or cleanse food (e.g. onions, leeks, meat), remove skin from almonds, and destroy enzymes in vegetables before freezing.

Blanquette: A white stew made from white meat such as chicken or veal, thickened with egg yolk and cream.

Bombe: A dome or sphere-shaped mould or an ice cream dessert frozen in such a mould.

Bonne femme: Food cooked in a rich sauce containing mushrooms.

Bordelaise: Bordeaux style – cooked in red wine.

Bouchée: Literally a mouthful. A tiny puff pastry case filled with mushrooms, chicken or other savoury foods in a white sauce.

Bouillabaisse: A rich Mediterranean fish stew containing several types of fish.

Bouillon: Stock.

Bouquet garni: A bunch of herbs used to flavour stews, stock etc. This may be tied with string or wrapped in muslin for easy removal. The herbs are usually parsley, thyme, bayleaf.

Bourguignonne: Burgundy style — cooked in red wine with mushrooms and onions.

Brioche: A plain yeast cake.

Brochette: A skewer or food cooked on a skewer (e.g. kebabs).

Canapé: Small pieces of savoury food served on a small biscuit, pastry or toast. They are sometimes called *savouries*.

Capiscum: The family of plants which includes chillies, red and green peppers and pimentos.

Caramel: The brown colour and flavour obtained from prolonged boiling of sugar.

Carbonnade: Meat stewed in beer.

Chantilly: Vanilla flavoured whipped cream mixed with whipped egg white, or a dessert containing such cream.

Charlotte: A hot or cold sweet with fingers of biscuit, cake or bread surrounding it.

Chasseur: A sauce or dish containing wine, tomatoes and mushrooms.

Chateaubriand: A double steak cut from the centre of the fillet.

Chaudfroid: A white or brown sauce to which aspic jelly is normally added. It is used to coat meat, fowl or fish.

Choux: A pastry made by heating water and fat and then beating in flour and eggs. Used for éclairs, profiteroles and beignets.

Chowder: A thick soup made from fish — usually shellfish.

Clarify: To remove impurities (a) from fat by boiling with water, so that the sediment falls to the bottom of the fat; (b) from liquid (e.g. clearing soup and wine by whisking with egg white and straining).

Compôte: Fruit poached in syrup.

Condiments: Seasonings and spices which are used to improve the flavour of a dish. Served at table they include pepper, salt, mustard and vinegar.

Crécy: A dish containing carrots.

Creole: Food cooked with tomatoes and spices and served with rice.

Crêpe: Pancake; *Crêpes Suzette* are pancakes flamed with orange and liqueur.

Croissant: A crescent-shaped breakfast roll made from a rich yeast dough.

Croquette: A savoury mixture of meat, fish or vegetables, shaped into rolls, coated in egg and breadcrumbs and deep-fried.

Croûte: Pieces of toast or fried bread used as a base for savoury dishes or as a garnish.

Croûtons: Small dice of fried bread served with soup.

Crudités: A dish of raw vegetables such as carrot, celery, onion, radish, served as an hors d'oeuvre, usually with mayonnaise.

Crumpet: A thick yeast pancake served with butter.

Daube: A rich stew containing red wine.

Dégorger: To remove the strong flavour and/or excess liquid from a food (e.g. cucumber is sprinkled with salt, allowed to stand and the liquid is drained off before a cucumber salad is made).

Dessert: Strictly speaking the last course of a meal, such as fresh fruit or nuts, served with port. In modern usage it has come to mean the sweet course.

Devilled: Food made 'hot' by the use of curry, mustard or spicy sauces.

Dice: Cut into small cubes.

Duchesse: Purée of potatoes containing egg yolk piped into shapes and baked.

Entrée: A course in a meal which is complete in itself, with a sauce and vegetable garnish. It is served before the main course.

Entremets: A savoury or sweet course served after the main course.

Escalope: A very thin slice of meat.

Escargot: Edible snail.

Espagnole: A rich brown sauce containing tomato purée and sherry.

Farce/Forcemeat: Stuffing.

Fines herbes: A mixture of chopped fresh herbs as in *omelette aux fines herbes*.

Flambé: A dish which has alcohol poured over it and is then set alight.

Flan: A shallow pastry case which may be filled with sweet or savoury filling.

Florentine: A dish containing spinach.

Fondant: A cooled sugar syrup, worked on a cold surface until thick and white. It is used in sweet-making and icings.

Fondue: A communal dish kept hot over a spirit lamp, into which diners dip cubes of bread or meat before eating.

Fool: A sweet dish consisting of fruit purée, custard and/or cream.

Frappé: Iced desserts or drinks.

Fricadelle: Meat ball.

Fricassée: A white stew.

Fritter: A small piece of savoury or sweet food dipped in batter and deep fried.

Galantine: A cold dish of minced meat, breadcrumbs, herbs and flavourings usually used to stuff a boned joint or fowl. This is boiled or roasted, pressed and glazed.

Gâteau: A rich cake served as a dessert.

Glacé: Glazed or iced.

Glaze: Strained purée of jam used to brush over sweet flans.

Gratin, au gratin: A dish which is put in a shallow container, covered with a sauce, sprinkled with breadcrumbs and sometimes cheese and baked in the oven. Vegetables and fish are suitable gratin foods.

Hors d'oeuvre: Light first course of small pieces of meat, fish, eggs and/or vegetables.

Isinglass: A form of gelatine made from fish.

Jardinère: Served with diced fresh garden vegetables.

Julienne: Thin strips of vegetables or dishes garnished with these vegetables.

Kebabs: Cubes of meat and vegetables threaded on skewers and grilled.

Knead: To manipulate dough, usually by hand, in order to develop the gluten and blend raising agents thoroughly.

Kromeski: Savouries of minced meat, rolled in bacon, coated with batter and fried.

Larding: Threading pieces of fat through lean meat in order to moisten it.

Liaison: Thickening agent for soups and sauces.

Lyonnaise: A dish containing onions and sometimes potatoes.

Macaroon: A biscuit made from sugar, egg white and ground almonds.

Macédoine: A mixture of assorted fruits (e.g. fruit salad) or vegetables.

Maître d'hôtel butter: Butter flavoured with parsley and lemon juice.

Marinade: A mixture of oil and wine, lemon juice or vinegar containing sliced vegetables and seasoning, in which meat or fish is soaked (marinated) to tenderise it and improve its flavour.

Marrons: Edible chestnuts.

Meringue: A mixture of egg white and sugar used to make cakes and puddings.

Meunière: Fried fish served with flavoured melted butter.

Mirepoix: A mixture of diced vegetables used as a base on which to braise meat.

Mocha: With a coffee flavour.

Mornay: A dish coated with a rich cheese sauce.

Mousse: A light, aerated dish usually containing egg yolks and cream.

Muesli: Rolled oats, raisins, nuts and sometimes fresh fruit, usually served with milk as a breakfast dish.

Muffin: A yeast cake similar to a crumpet.

Navarin: A brown mutton or lamb stew with root vegetables.

Panada (panard): A thick white sauce.

Papillote: A buttered paper or foil envelope in which small pieces of meat or fish, vegetables and seasonings are cooked.

Parboil: To half-cook.

Parmentier: A dish containing potato.

Pâté: A savoury paste made from minced meat, usually liver, fat and seasoning and served with toast.

Pâté de Foie Gras: A rich pâté made from the liver of force-fed geese. It is very expensive.

Petits fours: Small iced cakes.

Piccalilli: A mustard flavoured pickle.

Pilau or *Pilaff:* A savoury rice, cooked in stock.

Pimiento: Sweet canned red peppers.

Poach: To cook very gently in water.

Portuguaise: A tomato flavoured dish.

Pot-au-feu: Fresh beef and vegetables cooked in a casserole. The liquid is served as soup.

Praline: Almonds and caramelised sugar, ground and served with sweet dishes.

Provençale: A dish containing tomatoes, garlic and possibly green peppers and olives.

Purée: A mixture of meat, vegetables or fruit made smooth by sieving or liquidising. Also a thick soup.

Quenelle: An oval-shaped mixture of meat or fish poached in liquid.

Quiche: A savoury egg custard flan.

Ragoût: A stew, usually brown.

Ramekin: A small individual soufflé dish used for baked eggs or soufflés.

Ratatouille: A vegetable casserole consisting of tomatoes, courgettes, peppers and aubergines simmered in garlic flavoured oil.

Réchauffé: A reheated dish (e.g. shepherd's pie).

Reform: A rich brown sauce.

Reduce: To boil a liquid until it evaporates and becomes thick.

Remove or *Relève:* Obsolete term for the main course of a meal.

Rissole: A dish made from cooked minced meat, which is rolled in breadcrumbs and deep-fried.

Rollmop: A raw herring pickled in vinegar.

Roux: A liaison made from flour and fat.

Royale: A dish garnished with shapes of cold custard.

Sabayon: A dessert made from whisked eggs, sugar and wine.

Saignant: Underdone meat.

Salmis: A game stew.

Saltpetre: A preservative used in curing or pickling meat.

Sauté: To toss lightly in fat.

Savarin: Rich yeast pudding baked in a ring-shaped tin.

Scald: To dip into boiling water. (This makes it easier to remove the skin of tomatoes etc.)

Scalloped: Cooked and served in a scallop shell.

Shortening: Any fat used in baking which produces a 'short', brittle texture. Examples: butter, lard, margarine.

Skillet: Frying pan (American)

Smorrebrod: Savoury open sandwiches (Danish).

Sorbet: A water ice served as a sweet, or between courses to cleanse the palate.

Soubise: A purée of onions.

Soufflé: A light sweet or savoury egg dish which can be hot or cold.

Soya: A high protein bean, used to make synthetic meat, soy sauce and soya flour.

Strudel: A very thin pastry in which a savoury or sweet filling is rolled. It is then baked.

Sundae or *coupe:* A stemmed glass in which ice cream is served.

Terrine: A dish like pâté, consisting of minced meat, served

cold in a straight-sided dish (also called a terrine).
Tournedos: A thick slice from the centre of a fillet of beef.
Truffle: A dark fungus which grows underground and is considered a great delicacy. It is used for garnishes and in *foie gras*.
Vacherin: A sweet consisting of a meringue base, filled with fruit and/or cream.
Velouté: A rich white sauce, made from stock, thickened with egg yolks and cream.
Vichyssoise: A rich leek and potato soup, served cold.
Vinaigrette: salad dressing made from oil and vinegar.
Vol-au-vent: A hollow puff pastry case filled with a savoury or sweet mixture.
Waffle: A batter like mixture baked in a special frying 'iron' with a lattice design.
Whitebait: The young of herrings, usually fried whole in deep fat.
Zabaglione: A sweet made from whipped egg yolk, sugar and Marsala or sherry.
Zest: The outer rind of citrus fruit.

SOME FOREIGN DISHES

Dish	Country of Origin	Description
Bamboo shoots	China	Vegetable, sold in cans
Blini	Russia	Yeast pancake
Bortsch	Russia/Poland	Beetroot soup
Cannelloni	Italy	Tubes of pasta filled with meat or cheese
Chapatti	India	Unleavened bread served with curry
Cous cous	North Africa	A type of semolina made from millet, served with lamb stew
Dhal	India	Purée of lentils served with curry
Dolmades	Greece	Stuffed vine or cabbage leaves filled with lamb and rice, served with tomato or lemon sauce
Frankfurter	Germany	A red-brown sausage (used for hot-dogs)
Gazpacho	Spain	Chilled soup made from tomatoes, onions, cucumber, garlic
Gnocchi	Italy	Savoury cakes made from maize meal and cheese
Lasagne	Italy	A wide pasta ribbon or a meat dish containing this pasta
Lychee	China	A fruit
Minestrone	Italy	A thick tomato and vegetable soup thickened with pasta
Moussaka	Turkey/Greece	Layers of lamb, tomato, aubergine and cheese sauce baked in casserole
Osso buco	Italy	A veal stew containing marrow bones and tomato
Paella	Spain	Rice, shellfish, chicken and vegetables cooked together
Pizza	Italy	A base of bread dough topped with tomato, cheese, olives etc.
Pretzel	Germany	A knot-shaped, salty biscuit
Puchero	Mexico	Meat, sausage and bean stew
Pumpernickel	Germany	Black rye bread
Ravioli	Italy	Little envelopes of pasta filled with minced meat or cheese
Risotto	Italy	Savoury rice dish using short-grained rice
Sauerkraut	Germany	Pickled cabbage
Sweet and sour	China	Dishes using a sauce made from vinegar and pineapple
Wiener Schnitzel	Austria	Thin slices of veal coated in bread crumbs and fried

FLAVOURINGS

Good cooking involves the use of flavourings, seasonings and essences. Flavourings can do much to improve the taste of dull and insipid foods and can give a lift to ordinary everyday dishes, but they should be used carefully as the aim is to produce a subtle flavour, not to overpower natural flavours in the food. Any flavouring used should blend with the ingredients in the dish.

The list below suggest the flavourings most suitable for many foods.

During cooking food should be tasted at each stage to see if the required flavour is developing. Do not use the cooking spoon to taste dishes: test them on a teaspoon which is rinsed after each tasting.

Herbs

Herbs are strongly flavoured plants which are grown for their aromatic properties and used to introduce flavour into foods and/or bring out their natural flavours. Most are used in savoury dishes. They are best used fresh but can also be used dried, both for medicinal and for culinary purposes. Here are some popular herbs:

Angelica: Crystallised green stems, used in sweet dishes mainly as a decoration.

Balm: A lemon-flavoured herb used in tea and other beverages, salads and perfumes.

Basil: A pungent herb with a slightly clove-like flavour. It is used in savoury dishes, stuffings, soups, salads and mixed herbs.

Bay: The leaf from a type of laurel tree or bush, it is used in stock, soups and meat dishes and is also included in a bouquet garni.

Chive: A member of the onion family, it resembles fine spring onion tops. It is used in salads, omelettes and cheese dishes.

Cress: A small green plant, easy to grow, it is used in salads, sandwiches and garnishes.

Dill: Small green leaves used with fish, cucumber and pickles.

Fennel: A tall, mild flavoured herb which is easily cultivated. It enhances fish dishes and sauces served with fish.

Garlic: A very pungent, onion-like bulb, divided into sections called cloves; it is widely used in all the Mediterranean countries. Use sparingly in soups, stews, meat dishes and salads.

Horseradish: A white root which is grated in a sauce served with roast beef or beef salads.

Marjoram: A pleasant savoury flavoured herb used with lamb, in salads, stuffings and savoury dishes, especially those of Italian origin.

Mint: There are many varieties of this large leaved aromatic herb, which is easily grown. It is used with lamb, in mint sauce, jelly, salads and also when cooking vegetables such as new potatoes and peas.

Nasturtium: A fast growing plant with orange flowers. Both flowers and leaves are eaten in salads; they taste like watercress.

Oregano: A wild variety of marjoram with the same uses.

Parsley: Probably the best-known herb. It has bright green leaves and is used in salads, stuffings, sauces, soups, stews and sandwiches and extensively as a garnish. Rich in vitamin C and iron.

Rosemary: An attractive shrub with thin leaves. It is easily grown and has a strong flavour. Especially good with lamb.

Sage: A strong flavoured herb used in stuffing for pork, duck and goose.

Shallot: A member of the onion family, used in stuffings, stews and salads.

Tarragon: An aniseed flavoured herb with long spikey leaves used with chicken, fish, eggs and tomatoes. It is also used to flavour tarragon vinegar.

Thyme: There are two types – black thyme and lemon thyme. It is used in mixed herbs, stuffings, soups, stews, salads and egg dishes.

Watercress: A large-leaved herb which needs to be thoroughly washed. It is used mainly as a garnish, especially for game dishes.

Spices

This term includes most of the condiments and seasonings used in cookery. Many are obtained from the seeds, root, bark or fruit of plants. They are dried and used in cookery for their strong flavours and slightly preservative effect.

Allspice: A berry so named because it tastes like a mixture of cloves, nutmeg and cinnamon (not mixed spice). Use whole or ground in meat dishes.

Caraway: Small black seeds with a strong pungent flavour. They are used in cakes such as seed cake and on top of some breads.

Cayenne: A red powder made from dried chillies, it is very hot and spicy and should be used sparingly in savoury dishes.

Chilli: A powder ground from a special variety of chilli, it is very hot. Used in pickles and some hot stews, e.g. chilli con carne.

Cinnamon: A pleasant flavoured brown spice available in sticks or ground. It is used in cakes, biscuits, sweet dishes and pickles.

Cloves: Dried flower buds with a distinctively strong flavour. Used whole or ground in apple dishes, sauces, cakes and biscuits.

Curry powder: A mixture of some or all of the following spices: cayenne, coriander, cumin, cardamon, cloves, ginger, fenugreek, pepper and turmeric. The 'hotness' of a curry depends on the amount of chilli used. Many traditional Indian dishes contain no chilli and are only mildly spiced.

Ginger: A dried root which is used whole or ground, in cakes, puddings, chutneys and preserves. Crystallised ginger is used in cakes and puddings.

Mace: The outer covering of a nutmeg. It is sold ground and in sticks and is used in savoury sauces, soups and pickles.

Mustard: Powdered seeds of the mustard plant, available in powder or paste form. Used as a condiment and in sauces and preserves. It complements cheese dishes.

Nutmeg: The pungent kernel of a fruit from the Moluccan Islands (the original Spice Islands) and Jamaica. Used ground in sauces and sweet and savoury dishes.

Paprika: A sweet pepper made from dried capiscum pods, it is native to Central America, but Hungarian paprika and the slightly sweeter Spanish paprika are now more familiar to us. Use in meat dishes and as a colouring or garnish, e.g. on egg mayonnaise.

Pepper: A dried berry from Malabar, used whole or ground. *Black pepper* is dried with the skin on, while *white pepper* is dried after the skin is removed; it is used as a condiment and is stronger than black pepper, which is widely used in cooking.

Saffron: An expensive spice obtained from a species of wild crocus. Mildly flavoured, it is used for colouring cakes and rice dishes.

Turmeric: A ground dried tuber, it is brownish yellow in colour. Used in curries, pickles and as a cheap alternative to saffron.

Vanilla: The dried pods of orchids, vanilla can be used whole to flavour the sugar in cake and pudding recipes. Pods and seeds are used to make vanilla essence.

Essences

These contain alcohol in which the volatile oils from strongly flavoured plants such as vanilla or peppermint are dissolved. Most essences are now made from coal tar, a cheap source of many flavourings and colourings. Synthetic flavours have been developed to resemble vanilla, lemon, pineapple, banana and so on. (See also *Additives* p. 116).

APPENDIX C
Composition of Foods

Composition per 100 g (raw edible weight except where stated)

Food	In-edible waste %	Energy kcal	kj	Protein g	Fat g	Carbo-hydrate (as mono-saccharide) g	Water g	Calcium mg	Iron mg	Vitamin A (retinol equivalent) μg	Thia-min mg	Ribo-flavin mg	Nicotinic acid equivalent mg	Vitamin C mg	Vitamin D μg
Milk															
Cream, double	0	449	1,848	1.8	48.0	2.6	47	65	0	420	0.02	0.08	0.4	0	0.28
Cream, single	0	189	781	2.8	18.0	4.2	75	100	0.1	155[1]	0.03	0.13	0.8	0	0.10[1]
Milk, liquid, whole	0	65	274	3.3	3.8	4.8	88	120	0.1	44[1]	0.04	0.15	0.9	1	0.05[1]
										37[2]					0.01[2]
Milk, condensed, whole, sweetened	0	322	1,361	8.2	9.2	55.1	28	290	0.2	112	0.10	0.37	2.0	3	0.09
Milk, whole, evaporated	0	165	690	8.5	9.2	12.8	69	290	0.2	112	0.06	0.37	2.0	2	2.91
National dried milk	0	490	2,053	27.6	26.8	37.0	3	917	6.7[3]	341	0.28	1.31	7.1	60[3]	13.20[3]
Milk, dried, skimmed	0	352	1,498	36.0	0.9	53.3	5	1,260	0.5	4	0.30	1.73	9.7	10	0
Yogurt, low-fat, natural	0	53	224	5.0	1.0	6.4	86	180	0.1	10	0.05	0.26	1.3	1	0.02
Yogurt, low-fat, fruit	0	96	410	4.8	1.0	18.2	75	160	0.2	10	0.05	0.23	1.2	1	0.02
Cheese															
Cheese, Cheddar	0	412	1,708	25.4	34.5	0	37	810	0.6	420	0.04	0.50	5.2	0	0.35
Cheese, cottage	0	114	480	15.3	4.0	4.5	75	80	0.4	27	0.03	0.27	3.2	0	0.02
Meat															
Bacon, rashers, raw	13	422	1,743	14.4	40.5	0	41	7	1.0	0	0.36	0.14	5.8	0	0
Bacon, rashers, cooked	0	447	1,852	24.5	38.8	0	32	12	1.4	0	0.40	0.19	9.2	0	0
Beef, average	17	226	940	18.1	17.1	0	64	7	1.9	0	0.06	0.19	8.1	0	0
Beef, corned	0	216	905	26.9	12.1	0	58	14	2.9	0	0.01	0.23	9.0	0	0
Beef, stewing steak, raw	0	176	736	20.2	10.6	0	69	8	2.1	0	0.06	0.23	8.5	0	0
Beef, stewing steak, cooked	0	223	932	30.9	11.0	0	57	15	3.0	0	0.03	0.33	10.2	0	0
Black pudding	0	305	1,270	12.9	21.9	15.0	44	35	20.0	0	0.9	0.07	3.8	0	0
Chicken, raw	31	144	602	20.8	6.7	0	73	11	1.5	0	0.04	0.17	9.5	0	0
Chicken, roast	0	148	621	24.8	5.4	0	68	9	0.8	0	0.08	0.19	12.8	0	0
Ham, cooked	0	269	1,119	24.7	18.9	0	54	9	1.3	0	0.44	0.15	8.0	0	0
Kidney, average	11	89	375	16.2	2.7	0	79	9	6.0	300	0.39	1.90	10.7	12	0
Lamb, average, raw	17	335	1,388	15.9	30.2	0	53	7	1.3	0	0.09	0.19	7.4	0	0
Lamb, roast	0	291	1,209	23.0	22.1	0	54	9	2.1	0	0.10	0.25	9.2	0	0
Liver, average, raw	0	163	683	20.7	8.0	2.2	69	6	11.4	6,000	0.26	3.10	18.1	30	0.75
Liver, fried	0	244	1,020	24.9	13.7	5.6	56	14	8.8	6,000	0.27	4.30	20.7	20	0.75
Luncheon meat	0	313	1,298	12.6	26.9	5.5	52	15	1.0	0	0.07	0.12	4.5	0	0
Pork, average	15	330	1,364	15.8	29.6	0	54	8	0.8	0	0.58	0.16	6.9	0	0
Pork chop, grilled	40	332	1,380	28.5	24.2	0	46	8	1.2	0	0.66	0.20	11.0	0	0
Sausage, pork	0	367	1,520	10.6	32.1	9.5	45	41	1.1	0	0.04	0.21	5.7	0	0
Sausage, beef	0	299	1,242	9.6	24.1	11.7	50	48	1.4	0	0.03	0.13	7.1	0	0
Steak and kidney pie, cooked	0	304	1,266	13.3	21.1	14.6	51	37	5.1	126	0.11	0.47	6.0	0	0.55
Tripe	0	60	252	9.4	2.5	0	88	75	0.5	10	0	0.01	2.1	0	0

Food	In-edible waste %	Energy kcal	kj	Protein g	Fat g	Carbo-hydrate (as mono-saccharide) g	Water g	Calcium mg	Iron mg	Vitamin A (retinol equivalent) μg	Thiamin mg	Ribo-flavin mg	Nicotinic acid equivalent mg	Vitamin C mg	Vitamin D μg
Fish															
Cod; haddock; white fish	40	76	321	17.4	0.7	0	82	16	0.3	0	0.08	0.07	4.8	0	0
Cod, fried in batter	0	199	834	19.6	10.3	7.5	61	80	0.5	0	0.04	0.10	6.7	0	0
Fish fingers	0	178	749	12.6	7.5	16.1	64	43	0.7	0	0.09	0.06	3.1	0	0
Herring	37	234	970	16.8	18.5	0	64	33	0.8	45	0	0.18	7.1	0	22.20
Kipper	40	184	770	19.8	11.7	0	68	60	1.2	45	0.02	0.30	6.9	0	22.20
Salmon, canned	2	155	648	20.3	8.2	0	70	93	1.4	90	0.04	0.18	10.7	0	12.50
Sardines, canned in oil	0	217	906	23.7	13.6	0	58	550	2.9	30	0.04	0.36	12.4	0	7.50
Eggs															
Eggs, fresh	12	147	612	12.3	10.9	0	75	54	2.1	140	0.09	0.47	3.7	0	1.50
Fats															
Butter	0	731	3,006	0.5	81.0	0	16	15	0.2	995	0	0	0.1	0	1.25
Lard; cooking fat; dripping	0	894	3,674	0	99.3	0	1	0	0	0	0	0	0	0	0
Low-fat spread	0	365	1,500	0	40.5	0	57	0	0	900[4]	0	0	0	0	8.00[4]
Margarine	0	734	3,019	0.2	81.5	0	15	4	0.3	900[4]	0	0	0.1	0	8.00[4]
Oils, cooking and salad	0	899	3,696	0	99.9	0	0	0	0	0	0	0	0	0	0
Preserves, etc.															
Chocolate, milk	0	578	2,411	8.7	37.6	54.5	0	246	1.7	6.6	0.03	0.35	2.5	0	0
Honey	0	288	1,229	0.4	0	76.4	23	5	0.4	0	0	0.05	0.2	0	0
Jam	0	262	1,116	0.5	0	69.2	30	18	1.2	2	0	0	0	10	0
Marmalade	0	261	1,114	0.1	0	69.5	28	35	0.6	8	0	0	0	10	0
Sugar, white	0	394	1,680	0	0	105.0	0	1	0	0	0	0	0	0	0
Syrup	0	298	1,269	0.3	0	79.0	20	26	1.4	0	0	0	0	0	0
Vegetables															
Beans, canned in tomato sauce	0	63	266	5.1	0.4	10.3	74	45	1.4	50	0.07	0.05	1.4	3	0
Beans, broad	75	69	293	7.2	0.5	9.5	77	30	1.1	22	0.28	0.05	5.0	30	0
Beans, haricot	0	256	1,092	21.4	0	45.5	11	180	6.7	0	0.45	0.13	6.1	0	0
Beans, runner	14	23	100	2.2	0	3.9	89	27	0.8	50	0.05	0.10	1.4	20	0
Beetroot, boiled	20	44	189	1.8	0	9.9	83	30	0.7	0	0.02	0.04	0.4	5	0
Brussels sprouts, raw	25	26	111	4.0	0	2.7	88	32	0.7	67	0.10	0.16	1.5	87	0
Brussels sprouts, boiled	0	17	75	2.8	0	1.7	92	25	0.5	67	0.06	0.10	1.0	41	0
Cabbage, green, raw	30	22	92	2.8	0	2.8	88	57	0.6	50	0.06	0.05	0.7	53	0
Cabbage, green, boiled	0	15	66	1.7	0	2.3	93	38	0.4	50	0.03	0.03	0.5	23	0
Carrots, old	4	23	98	0.7	0	5.4	90	48	0.6	2,000	0.06	0.05	0.7	6	0
Cauliflower	30	13	56	1.9	0	1.5	93	21	0.5	5	0.10	0.10	1.0	64	0
Celery	27	8	36	0.9	0	1.3	94	52	0.6	0	0.03	0.03	0.5	7	0
Crisps, potato	0	533	2,222	6.2	35.9	49.3	3	37	2.1	0	0.19	0.07	6.3	20	0
Cucumber	23	9	39	0.6	0	1.8	96	23	0.3	0	0.04	0.04	0.3	17	0
Lentils, dry	0	295	1,256	23.8	0	53.2	12	39	7.6	6	0.50	0.25	6.3	0	0

Food	In-edible waste %	Energy kcal	kj	Protein g	Fat g	Carbo-hydrate (as mono-saccharide) g	Water g	Calcium mg	Iron mg	Vitamin A (retinol equivalent) μg	Thiamin mg	Ribo-flavin mg	Nicotinic acid equivalent mg	Vitamin C mg	Vitamin D μg
Lettuce	20	8	36	1.0	0	1.2	96	23	0.9	167	0.07	0.08	0.4	15	0
Mushrooms	25	7	31	1.8	0	0	92	3	1.0	0	0.10	0.40	4.5	3	0
Onions	3	23	98	0.9	0	5.2	93	31	0.3	0	0.03	0.05	0.4	10	0
Parsnips	26	49	210	1.7	0	11.3	83	55	0.6	0	0.10	0.09	1.3	15	0
Peas, fresh or frozen, raw	63	62	273	5.8	0	10.6	84	15	1.9	50	0.32	0.15	3.5	25	0
Peas, fresh or frozen, boiled	0	49	208	5.0	0	7.7	80	13	1.2	50	0.25	0.11	2.3	15	0
Peas, canned, processed	0	76	325	6.2	0	13.7	72	27	1.5	67	0.10	0.04	1.4	0	0
Peppers, green	16.5	14	59	0.9	0.2	2.2	94	9	0.4	42	0.08	0.03	0.9	91	0
Potatoes, raw	27.5 / 14.6	76	324	2.1	0	18.0	78	8	0.7	0	0.11	0.04	1.8	8–30[7]	0
Potatoes, boiled	0	80	339	1.4	0	19.7	81	4	0.5	0	0.08	0.03	1.2	4–15[7]	0
Potato chips, fried	0	236	1,028	3.8	9.0	37.3	48	14	1.4	0	0.10	0.04	2.2	6–20[7]	0
Potatoes, roast	0	111	474	2.8	1.0	27.3	64	10	1.0	0	0.10	0.04	2.0	6–23[7]	0
Spinach	25	21	91	2.7	0	2.8	91	70	3.2	1,000	0.12	0.20	1.3	60	0
Sweet corn, canned	0	79	336	2.9	0.8	16.1	73	3	0.1	35	0.05	0.08	0.3	4	0
Tomatoes, fresh	0	12	52	0.8	0	2.4	93	13	0.4	117	0.06	0.04	0.7	*	0
Turnips	16	18	74	0.8	0	3.8	93	59	0.4	0	0.04	0.05	0.8	25	0
Watercress	23	14	60	2.9	0	0.7	91	222	1.6	500	0.10	0.16	2.0	60	0
Fruit															
Apples	20	46	197	0.3	0	12.0	84	4	0.3	5	0.04	0.02	0.1	5	0
Apricots, canned (including syrup)	0	106	452	0.5	0	27.7	68	12	0.7	166	0.02	0.01	0.3	5	0
Apricots, dried	0	182	776	4.8	0	43.4	15	92	4.1	600	0	0.20	3.4	0	0
Bananas	40	76	326	1.1	0	19.2	71	7	0.4	33	0.04	0.07	0.8	10	0
Blackcurrants	2	28	121	0.9	0	6.6	77	60	1.3	33	0.03	0.06	0.3	200	0
Cherries	15	47	199	0.6	0	11.8	81	18	0.4	20	0.05	0.06	0.4	5	0
Dates, dried	14	248	1,056	2.0	0	63.9	15	68	1.6	10	0.07	0.04	2.3	0	0
Figs, dried	0	213	908	3.6	0	52.9	17	284	4.2	8	0.10	0.13	2.2	0	0
Gooseberries	1	27	116	0.9	0	6.3	87	22	0.4	30	0.04	0.03	0.4	40	0
Grapefruit	50	22	95	0.6	0	5.3	91	17	0.3	0	0.05	0.02	0.3	40	0
Lemons	60	7	31	0.3	0	1.6	91	8	0.1	0	0.02	0	0.1	50	0
Melon	40	23	97	0.8	0	5.2	94	16	0.4	160	0.05	0.03	0.5	25	0
Oranges	30	35	150	0.8	0	8.5	86	41	0.3	8	0.10	0.03	0.3	50	0
Orange juice, canned unconcentrated	0	47	201	0.8	0	11.7	87	10	0.4	8	0.07	0.02	0.2	40	0
Peaches, fresh	13	36	156	0.6	0	9.1	86	5	0.4	83	0.02	0.05	1.1	8	0
Peaches, canned (including syrup)	0	88	373	0.4	0	22.9	74	4	1.9	41	0.01	0.02	0.6	4	0
Pears, fresh	25	41	175	0.3	0	10.6	83	8	0.2	2	0.03	0.03	0.3	3	0
Pineapple canned (including syrup)	0	76	325	0.3	0	20.0	77	13	1.7	7	0.05	0.02	0.3	8	0
Plums	8	32	137	0.6	0	7.9	85	12	0.3	37	0.05	0.03	0.6	3	0
Prunes, dried	17	161	686	2.4	0	40.3	23	38	2.9	160	0.10	0.20	1.7	0	0
Raspberries	0	25	105	0.9	0	5.6	83	41	1.2	13	0.02	0.03	0.5	25	0

Food	In-edible waste %	Energy kcal	kj	Protein g	Fat g	Carbo-hydrate (as mono-saccharide) g	Water g	Calcium mg	Iron mg	Vitamin A (retinol equivalent) µg	Thiamin mg	Ribo-flavin mg	Nicotinic acid equivalent mg	Vitamin C mg	Vitamin D µg
Rhubarb	33	6	26	0.6	0	1.0	94	103	0.4	10	0.01	0.07	0.3	10	0
Strawberries	3	26	109	0.6	0	6.2	89	22	0.7	5	0.02	0.03	0.5	60	0
Sultanas	0	249	1,064	1.7	0	64.7	18	52	1.8	0	0.10	0.30	0.6	0	0
Nuts															
Almonds	63	580	2,397	20.5	53.5	4.3	5	247	4.2	0	0.32	0.25	4.9	0	0
Coconut, desiccated	0	608	2,509	6.6	62.0	6.4	3	22	3.6	0	0.06	0.04	1.8	0	0
Peanuts, roasted	0	586	2,428	28.1	49.0	8.6	5	61	2.0	0	0.23	0.10	20.8	0	0
Cereals															
Barley, pearl, dry	0	360	1,531	7.7	1.7	83.6	11	10	0.7	0	0.12	0.08	2.2	0	0
Biscuits, chocolate	0	497	2,087	7.1	24.9	65.3	3	131	1.5	0	0.11	0.04	1.9	0	0
Biscuits, cream crackers	0	471	1,985	8.1	16.2	78.0	4	145	2.2	0	0.22	0.05	2.3	0	0
Biscuits, plain, semi-sweet	0	431	1,819	7.4	13.2	75.3	3	126	1.8	0	0.17	0.06	2.0	0	0
Biscuits, rich, sweet	0	496	2,084	5.6	22.3	72.7	3	92	1.3	0	0.12	0.04	1.5	0	0
Bread, brown	0	230	981	9.2	1.4	48.3	39	88	2.5	0	0.28	0.07	2.7	0	0
Bread, starch reduced	0	234	996	10.5	1.5	47.6	36	100	1.3	0	0.18	0.03	2.7	0	0
Bread, white	0	251	1,068	8.0	1.7	54.3	39	100	1.7	0	0.18	0.03	2.6	0	0
Bread, wholemeal	0	241	1,025	9.6	3.1	46.7	38	28	3.0	0	0.24	0.09	1.9	0	0
Cornflakes	0	354	1,507	7.4	0.4	85.4	2	5	0.3	0	1.13[3] 0.04[8]	1.41[3] 0.10[8]	10.6[3] 0.8[8]	0	0
Custard powder; instant pudding; cornflour	0	353	1,506	0.5	0.7	92.0	12	15	1.4	0	0	0	0.1	0	0
Crispbread, Ryvita	0	318	1,352	10.0	2.1	69.0	6	86	3.3	0	0.37	0.24	1.3	0	0
Flour, white	0	348	1,483	10.0	0.9	80.0	13	138	2.1	0	0.30	0.03	2.7	0	0
Oatmeal	0	400	1,692	12.1	8.7	72.8	9	55	4.1	0	0.50	0.10	2.8	0	0
Rice	0	359	1,531	6.2	1.0	86.8	12	4	0.4	0	0.08	0.03	1.5	0	0
Spaghetti	0	364	1,549	9.9	1.0	84.0	12	23	1.2	0	0.09	0.06	1.8	0	0
Beverages															
Chocolate, drinking	0	397	1,683	5.5	6.3	84.8	3	5	2.8	2	0.03	0.09	1.4	0	0
Cocoa powder	0	443	1,860	19.0	21.9	45.4	14	13	7.0	7	0.08	0.30[9]	4.9	0	0
Coffee, ground	0	0	0	0	0	0	4	0	0	0	0	0.20[9]	10.0[9]	0	0
Coffee, instant	0	155	662	4.0	0.7	35.5	4	140	4.0	0	0	0.10	45.7	0	0
Cola drink	0	46	195	0	0	12.2	2	0	0	0	0	0	0	0	0
Tea, dry	0	0	0	0	0	0	90	0	0	0	0	0.90[9]	6.0[9]	0	0
Squash, fruit, undiluted	0	122	521	0.1	0.1	32.2	63	16	0.2	0	0	0.01	0	1	0
Alcoholic beverages per 100 ml															
Beer, bitter, draught	0	30	127	0	0	2.3	—	11	0	0	0	0.05	0.7	0	0
Spirits, 70° proof	0	221	914	0	0	0	—	0	0	0	0	0	0	0	0
Wine, red	0	67	277	0	0	0.3	—	6	0.8	0	0.01	0.02	0.2	0	0
Puddings and cakes etc.															
Apple pie	0	281	1,179	3.2	14.4	40.4	42	42	0.8	2	0.08	0.02	0.9	2	0

Food	In-edible waste %	Energy kcal	kj	Protein g	Fat g	Carbo-hydrate (as mono-saccharide) g	Water g	Calcium mg	Iron mg	Vitamin A (retinol equivalent) µg	Thiamin mg	Ribo-flavin mg	Nicotinic acid equivalent mg	Vitamin C mg	Vitamin D µg
Bread and butter pudding	0	154	648	5.6	7.0	18.3	67	112	0.7	81	0.06	0.21	1.6	0	0.48
Buns, currant	0	328	1,385	7.8	8.5	58.6	25	88	1.6	24	0.15	0.10	2.0	0	0.27
Custard	0	92	387	3.0	3.5	12.9	81	110	0.2	37	0.04	0.14	0.8	0	0.03
Fruit cake, rich	0	368	1,546	4.6	15.9	55.0	28	71	1.6	57	0.07	0.07	1.2	0	0.80
Jam tarts	0	391	1,648	3.2	13.8	67.7	18	50	1.3	0	0.05	0.01	0.8	0	0
Plain cake, Madeira	0	426	1,785	6.0	24.0	49.7	20	67	1.4	82	0.08	0.11	1.7	0	1.20
Rice pudding	0	142	594	3.6	7.6	15.7	75	116	0.1	96	0.05	0.14	1.0	1	0.08
Soup, tomato, canned	0	55	230	0.8	3.3	5.9	84	17	0.4	35	0.03	0.02	0.2	6	0
Trifle	0	162	684	3.1	5.6	26.5	65	75	0.6	73	0.04	0.10	1.1	2	0.30
Yeast extract	0	6	25	1.4	0	0	25	123	6.9	0	3.00	5.90	67.7	0	0
Ice cream, vanilla	0	192	805	4.1	11.3	19.8	62	137	0.3	1	0.05	0.20	1.1	1	0

[1] Summer value
[2] Winter value
[3] fortified
[4] some margarines contain carotene
[5] old potatoes
[6] new potatoes
[7] vitamin C falls during storage
[8] unfortified
[9] 90 to 100 per cent is extracted into an infusion

Index

absorption,
  in gut, 6, 14, 174–175
  of iron, 25
accelerated freeze drying (AFD), 22, 54, 114, 139, 202
acetylcholine, 185–186
acid,
  in jam making, 142
  in stomach, 200
acne, 38, 46
acrolein, 15, *15*
Addison's disease, 187
additives, 115, **116–118**, 131–132
adipose tissue, 9, 10, 14, 15, 31, 38, 39, 43, 53, 82, 154, 175, **179**
adolescents, 38, 43, 46, 53, 74, 78, 95
ADP (adenosine diphosphate), 25, 176
adrenal glands, 20, 180, 187, 189
  cortex, 186, 187, 189
  medulla, 187, 189
adrenalin, 168, 187, **189**
AFD (see accelerated freeze drying)
agar agar, 61, 204
air pollution, 167
alanine, 3
albumen, 4, 51, 72, 73, 93
  a globular protein, 4
alcohol, 35, 46, 92, 117, 121, 124, 157, **199**, 202, 203
  as a preservative, 142
aleurone layer, 86
algae, 119
alginates, 15, 117
alkalis, 22, 117–118, 200
alkanoic acids (see also fatty acids), 12
alloys, 197
alveoli, 154, 165–168
amino acids, 3–7, 74, 79, 122, **175–176**
  essential and non-essential, 4
  naturally occurring, 3
  utilisation and storage, 175–176
amino group (NH$_2$), 3
  deamination, 5, 178
amylase, 10
anabolism, 3, 29, 168, 176
anaemia, 2, 20, 26, 35, 39, 43, **46, 161**
anaerobic,
  bacterial action, 133
androgens, 186
angina, 157
animal fats (see also lipids), 82
animal proteins (see also proteins), 86
animal starch (see glycogen)
anorexia nervosa, 45
antibiotics, 75, 115, 121, 127, 134, 141
antibodies, 5, 133–134, 160, 161, 163, 178
antidiuretic hormone, ADH, 182
anti-oxidants, 16, 19, **117**, 142, 201
antispattering agents, 15, 118
aorta, 156
appetite, 31
arachidonic fatty acid, *13*, 14, 20
arteries, 159
arteriosclerosis, 15, 40, 157
arthritis, 32, 39
artificial sweeteners, 44, 46
ascorbic acid (see also vitamin C), 20
  antioxidant, 20
  fermentation, 111
  heat on, 21–22
  oxidising agent, 22
  sources, 20–21
aspic jelly, 61
atom, **196–204**
  atomic number, 196
  energy shells, 196
  protons and neutrons, 196
atomic number, 196
ATP (adenosine triphosphate), 25, 153, 155, 176
  co-enzyme, 170
autolytic enzymes, 4
autonomic nervous system, **184**
axon, 185

babies (and children), **36–38**, 43, 53, 73, 74, 78, 95, 107, 172
  dietary needs, 36–38
bacon cuts, 59
bacteria, 74, 100, 117, 119, **124–127**, 137, 152, 165, 172
  classification, 119, 125
  disease, 124–125
  gut, 175
  nutrition, 124–127
  practical work, 126–128
  reproduction and structure, 124–127
  source of infection in food, 129–130
baking, 69, 100, **108–113**
balanced diet, 1, **33–35**
bases, **200**
batch cooking, 48
beef cuts, 56
beer making, 124
beri-beri, 2, **23**, 90
bile, 15, 40, 46, 160, **173, 177**
binary fission, 126
birth, **193**
  control, 194
Biuret test, 6, 7
bladder, 181
blanching, 22, 139, 169
blast freezing, 141
blindness, 18
blood, 25, 154, 155, **160–163**
  cell structure, 152
  clots and clotting mechanism, 20, 24, 40, 157, **161**
  functions, 6, 15, 162
  groups, 162–163
  hepatic circulation, 177
  renal circulation, 181–182
blood plasma, 155, **160**, 201
blood pressure, 159
blood vessels, 156–160
boiling, 11, 102–103
  pressure cooking, 202
bonding, **197–198**
  covalent bonds, 198, 199
  hydrogen bonds, 3, 14, 198
bone, 19, 24–25, 154
  bone marrow, 25, 160, 161
Bord Báinne, 73
Bord Iascaigh Mhara, 66
bottlefeeding, 37
bottling, 95, 137–138, 145–148
botulism, 127, **133**
bowel disorder, 9, 35, 42
Bowman's capsule, 180, 181
brain, 154, **182–183**
braising, 103–104
bran, 9, 86, 88, 91
bread, 5, 74, **109–112**, 119, 197, 202
  dough elasticity, 6
  freezing, 151
  heat, 9
  pH, 201
bread soda, 22, 76, 96, 109, 110, 201
breastfeeding, 36–37, 39
breathing, **164–168**
  exchange of gases, 166
  mechanism, 164–165
brewing, 119
brine, 54, 71, 80
British Thermal Unit (BTU), 29
bronchioles, 165–168
brucellosis (undulent fever), 74
'budding' (in yeast), 123
budgeting food bills, 50–51
buffets, 50
butter, 77, 82
buttermilk, 76
butyric acid, 13, 74

cake making, 16, 113
  freezing, 151
calciferols (see vitamin D)
calcium, 19, **24–25**, 34, 65, 95, 197
  daily requirements, 34
calcium caseinate, 74, 172
calorie, 28–32
cancer (of the bowel), 9, 42, 43, 93
candling (of eggs), 71
canning and canned foods, 22, 54, 67, 96, 99, 113, 114, 137–138
  'blown' cans, 133
capillaries, 159
caramelise, 10
carbohydrates, 1, **7–11**, 51, 65, 74, 91, 93, 94
  digestion in ileum, 174–175
  effect of heat, 10
  energy value, 7, **29**
  functions, 9
  overprocessing, 9
  requirements in diet, 9, 35
  tests for, 11
  utilisation, 175
carbon, 3, 196–200

atomic structure, 196
  carboxyl group, 3, 13, 200
  inorganic compounds, 197
  organic compounds, 197, 198–200
  rings, 14
carbon dioxide, 7, 90, 92, 99, 124, 160
  as a raising agent, 109–111, 198
carbonise, 10
carcase meat, 51, 82
carcinogens, 35, 93
carotene, **18**, 71, 74, 94, 97, 116
cartilage, 154, 165
casein, 4, 15, 172, 204
caseinogen, 16, 73, 74, 78, 172
casserole, 48, 50, 97, 151
catabolism, 3, 168, 176
catalyst, 16, 84, 169, 201
cells,
  cell types, 152–155
  nerve cells, 152, 154, 183–185
  semi-permeable membrane, 203
  tissues, 154–155
  typical animal cell structure, 153
  yeast cell, 123–124
cellulose, 8–10, 38, 41–42, 46, 88, 91, 94, 98, 99, 153, 168, 175
central nervous system, CNS, **182–184**
cereals, 41, 42, 43, 50, 61, **86–91**, 121
cerebellum, 182, 183
cerebrum, 182, 183
charcoal, 16, 92, 204
cheese, 74, **78–81**, 127
chicken, **62–63**
chlorophyll, 7, 116, 119, 153
chloroplast, 153
cholcaliferol (see vitamin D)
cholesterol, 14, 15, 39, 40, 74, 79, 82, 83, 84, 157
  controversy, 85
  low-cholesterol diet, 40–41, 71, 74, 79
chopping, 22
Chorley Wood Process, 111
chyme, 6, 15, 172
cilia, 154, 165
citric acid cycle, 176
classification of living matter, 119
Clostridium, 138
clotting, 74
  of blood, 20, 24, 40, 157, **161**
coagulation,
  milk, 6
  proteins, 4, 6, 68, 80
  test for a protein, 7
coconut oil, 13, 14
co-enzymes, 170
cod liver oil, 13, 41
coeliac disease, 40, **42**, 89
collagen, 4, 51, 53, 60, 66, 154
colloids (see also emulsions), 11, 15, **204**

215

colon, 42, **175**
coma, 46
complex molecules, 9
composition of food (tables), **210–214**
compounds, **197–200**
   inorganic, 197
   organic, 197, 198–200
condensation, 3, 8, 12, 197, 202
condensed milk, 76
conjugated proteins
   classification and examples, 4, 5
connective tissue, 4, 6, 51, 53, 66, 103, **154**, 155
constipation, 1, 9, 35, 38, 42, 94, 175
contaminants, **115–116**
continuous phase (see emulsions)
contraceptive pill, 193
convalescence, 40, 107
convection, 29
convenience foods, 33, 40, 42, 50, **113–116**, 117, 118
cookery terms, **204–207**
cooking methods,
   dry heat, **100–102**
   haybox cooking, 106
   in fat or oil, **102**
   moist heat, **102–103**
   pressure cooking, **104–106**
cooking oils, 41
   decomposition temperature, 15
cordon bleu, 49
coronary arteries, 14, **156–157**
coronary thrombosis (see also heart attacks), **40–41**, 161
corpus luteum, 186, 189, 191–192
cortex, 180, 181
cortisone, 187
cottage cheese, 80–81
covalent bonds, 198, 199
cream, 76, 78, 204
cream cheese, 80
creaming,
   of fats, plasticity, 16
crosslinks, 3, 4
crustaceans, 63
crystallisation, 203
curd cheese, 81
curdling, 4, 6, 73, 74, 107
curing, of meat, 54
curry and curries, 49, 50, 61, 103
cuts (of protein foods), **56–60**
cyanocobalamin (B_{12}), see vitamins, 24
cyclamate, 93, 116, 118
cytoplasm, 123, 153
cyrogenic freezing, 141

DC PIP (see test for vitamin C), 22
DDT (insecticide), 118
deamination, 5, 35, 175, **178**
decomposition,
   in fats and oils, 15
deficiency diseases, 2, 17, 18, 19, 20, 23–24, 25, 26, 27, 42, 186–189
dehydrated foods, 114, 117
dehydration, 15, 28, 37, 113, 114, 175
   in food preservation, 138–140
dementia, 24
denaturation,
   enzymes, 169
   proteins, 4, 6
dendrite, 185
dendron, 184–185
deoxyribonucleic acid (see DNA), 4, 153
dermatitis, 24
dextrin, 9, 10
dextrose (see glucose), 8
diabetes, 32, 40, 46
   diabetic food, 40, 46, 93, 116
   diabetes insipidus, 182
   diabetes mellitus, 188
diarrhoea, 24, 28, 46, 131, 175, 182
diastase, 110
diet, 1, **33–51**
   adolescents, 38
   adults, 38–39
   babies and children, 36–38
   balanced, 33–35
   carbohydrates, 35
   effect of poverty, 42, 44
   essential amino acids, 4
   essential fatty acids, 14
   fats, 35
   invalids, 39–40, 71
   minerals, 35
   old people, 39
   proteins, 33–35
   variety and flavour, 14
   vegans, 5, 43
   vitamins, 35
diets, 40–47
   bland, 40
   European, 41
   gluten-free, 40
   high fibre, 40, **41–42**, 43
   high or low protein, 40
   low calorie (kilocalorie), 40, **43–45**, 71, 74, 81, 86, 95
   low cholesterol, **40–41**, 71, 74, 79
   low fat, 40
   low residue, 40, 95
   low salt, 40
   macrobiotic, 43
   sugar free (diabetes), 40, 46, 116
   vegetarian, 40, 41, **42–43**, 53, 61, 98, 111
diffusion, 202, 203
digestion, 3, 6, 9–10, 15, **168–176**
   absorption, 174–175
diphtheria, 74, 119, 131
disaccharides (see also carbohydrates), 8, 74, 91, 99, 110
disease,
   immunity, **133–134**
   infectious diseases, 131
   methods of transmission, 128–131
   resistance, 131
disperse phase (see emulsions)
disulphide links, 3, **4**
diverticular disease, 9
diverticulosis, 42, 43

DNA (deoxyribonucleic acid), 4, 153
double bonds, 13, 14, 198
dough, 109
drugs,
   in pregnancy, 193
dry heat, 10
drying (see also AFD), 22, 54, 96, 99, 136, 137, 139–140
ductless glands (see endocrine system)
duodenum, 6
   endocrine function, 188
economic, value of foods, 79, 95, 99
   oven economy, 106
   seasonality, 47, 68, 136
eggs (chicken), **70–73**, 109, 152, 204
   cooking uses, 72–73
   test for freshness, 71
   pH, 201
egg (human), see ova, 152
   follicle stimulating hormone FSH, 186
elasticity,
   of proteins, 6
elastin, 4, 51, 60, 154
electric slow cookers, 103
elements, **196–197**
elimination, 178
embryo,
   human, 192–193
   plant, 86
emulsions and emulsifiers, 14, 15, 71, 72, 73, 84, 93, 108, 109, 117, 173, **204**
   action of bile, 173
   naturally occurring, 15
   stabilisers, 16, 117
   temporary and permanent, 15, 204
endocrine system, 182, **186–189**
endoplasmic reticulum, 153
endosperm, 86, 88, 90
energy,
   calculation of energy value, 29
   daily requirements, 30, 31, 33–35
   food tables, 210–214
   from sugar, 91
   glycolysis, 176
   muscle tissue, 155
   photosynthesis, 7
   release of, 10, 15
   storage, 9, 10, 176
   value from food, 3, 5, 7, 9, 11, 14, **28–32**, 82, 91
energy shells, 196
entertaining, 49–50
enzymes, 5, 110, 113, 117, 136, 160, **168–171**, 173, 201
   cause of decay, 137
   co-enzymes, 170
   effect of heat on activity, 169
   in digestion, 6, 9–10, **168–176**, 203
   Kreb's cycle, 176
   'lock and key' mechanism, 169–170

epithelial tissue, 154
erepsin, 6, 74
Escherichia coli, 125
essences, **209**
essential amino acids, **4**, 43, 61, 71, 73
   non-essential, **4**
essential fatty acids, 14, **19**
ester, 12
evaporated milk, 76
evaporation, 28, 29, **202**
excretion, 178–182
   kidneys, 180–182
   skin, 178–180
extracellular fluid (ECF), 28

factory farming, 115
faeces, 168, 175, 182
FAO (Food and Agriculture Organisation), 32, 118
fats (see lipids)
fats in cooking, 14
   plasticity, 16
fat soluble vitamins, 14
fats and oils, **82–86**
fatty acids, **12–17**, 74, 83, 174, 200
   absorption in ileum, 174–175
   chemical composition, 12, 200
   essential, 14
   melting point, 14
   proportion in some lipids, 13
   saturated, 13, 41, 71, 82, 83, 84
   unsaturated, 13, 16, 35, 40, 83, 84, 199, 201
   utilisation, 175
feedback mechanisms, 188
Fehlings test, 10, 11
fermentation, **110–111**, 121, 124, 199
   in jam making, 143
fertility and fertilisation, 19, 192
fever, 40, 131, 180
fibrin, 161
fibrinogen, 160, 161
fibrous protein, 4
fish, **63–70**
   commercial processing, 67–68, 203
   composition and structure, 64, 65–66
   cuts of, 69
   fishing methods, 66–67
   frozen, 151
   oily fish, 63–65, 68
   shellfish, 63–65
   stocks (soups), 106
flagellae, 125
flash point, 15
flavourings, 207
flo-freezing, 141
flour, types, of, 88–90
fluorine, 65, 196
foetus, 192–193
folic acid, **24**, 39, 46, 160, 161
follicle stimulating hormone, FSH, 186, 191
food,
   composition tables, 210–214
   cooking methods, 100–106

diets, 36–47
energy foods, 82–100
protein foods, 51–81
science, a definition, 1
food composition tables, **210–214**
food poisoning, 66, 124, 125, 127–134, 175
   bacterial, 132–133
   chemical, 131–132
   prevention, 134
foreign dishes, **207**
formal table setting, 49
freeze drying, 140, 202
freezer, 47, 48
   freezer accessories, 149–150
   chart, 150
freezing and frozen foods (see also AFD), 22, 54, 67, 95, 96, 99, 110, 113, 114, 136, 137, 140–141, 148–151
fructose, 8, 10, 91
fruits, **98–99**
   bottling, 146
   frozen, 151
   jamming, 142–143
frying, 69, 85, **102**
FSH (see follicle stimulating hormone)
Fullers earth, 16
fungi, 110, 115, 119–124
   classification, 119–121
   conditions for growth, 122
   disease, 121
   poisonous fungi, 123
   reproduction, 120–124

galactose, 8
gallstones, 14, 40, 46, 47
game, 51
ganglion, 184–185
gastric juice, 6, 168
gastrin, 172, 188
gastroenteritis, 28, 36, 46, 74, 127
gel, 11, 142
gelatine, 4, 6, **60–61**, 66, 68, 103, 106, 117
   stabiliser, 16
   supplementary protein, 5
   types of, 61
germ, 86, 88, 90
German measles, 193
gestation period (in woman), 193
globulin, 4, 51
glomerulus, 180, 181
glucose, **8–11,** 74, 83
   test for, 10, 12, 31
   utilisation and storage, 175
gluten, 3, 4, 15, 40, 42, 86, 90, 93, 109, 111, 204
   disulphide links, 3
   gluten free diet, **42,** 89
glycerine (see also glycerol), 12
glycerol (glycerine), 12, 15, 74, 174, 200
glyceryl monosterate (GMS), 15, 16, 117
glycine, 3
glycogen, 8, 10, 31, 51, 53, 123, 155, 175, 176, 178, 187

glycolysis, 53, 176
GMS (see glycerol monosterate)
goitre, 2, 27, **187**
golgi apparatus, 153
gonads, 188
   ovaries, 188, 191
   testes, 188
gonorrhoea, 131, 193
Graafian follicle, 191, **191–192**
gram stain bacteria, 125
grape sugar (see glucose), 8
grilling, 69, 101–102
growth spurt, 31
gums, 15, 117

haemoglobin, 3, 4, 25, 39, 60, **160,** 161, 178
haemophilia, 161
haemorrhoids, 39, 42
hair follicles, 179
hanging of meats, 53, 54
hard cheese, 78–80
haybox cooking, 106
H bonds, 3
heart, **155–158,**
   circulation, 157–158
   blood pressure, 159
   blood vessels, 156–157, 159
   heart beat, 157–158
heart attacks, 14, 85, 157
heat, 5, 9, 10, 15, **28–29,** 55, 68, 69, 155, 197
   on cereals, 91
   on cheese, 80
   on eggs, 72
   on enzymes, 169
   on fruit, 99
   on milk, 76
   on vegetables, 95
   pasteurisation, 75
Henle, loop of, 181
herbs and spices, 117, **208**
hereditary,
   tendency to diabetes, 46
high blood pressure, 32, 193
homeostasis, 162, 164
   kidney, 180–182
   liver, 178
homogenisation, 75, 76, 78
hormones, 5, 115, 160, 172, **186–189**
   duodenal hormone, 173
   endocrine system, 186–189
   feedback mechanism, 188
   osmoregulation, 182
hormone replacement therapy (HRT), 192
humidity, **202–203**
hydrocarbon chain, 13, *13*
hydrochloric acid,
   in digestion, 172, 200
hydrogen,
   atomic structure, 196
   bonds, 3, 14, 198
hydrogenation, 13, 15, 16, 41, 83, 84, **201**
hydrolysis, 3, 8, 12, 110, 168, **201**
   chemical digestion, 168
   definition, 10
   saponification, 16
hydrophilic group, 15, 173

hydrophobic, 11, 173
   group, 15
hydroxyl group, 12, *12*
hygroscopic, 10
hypercalcaemia, 24
hypertension, 159
hypervitaminosis, 18, 19
hypha, 120, 122
hypothalamus, 31, 32, 44, 168, 182, **183**

ice cream, 76
immunity, 37, 119, **133–134**
immunisation, 134
improver, 20, 88, 117
impurities,
   in fats and oils, 15
insecticides and insects, 115, 118
insulin, 3, 46, 172, 188
intestinal glands, 10, 174
intestinal juice, 174
invalids, 39, 40, 107
invertase (see also sucrase), 10
'invert sugar' (see also glucose and fructose), 10
iodine, 2, 11, **27,** 65, 197
   test for unsaturated fat, *17*
   test for iodine, 27
   thyroid gland, 187
Irish Sea Fisheries Association, 66
iron, 1, 20, **25,** 34, 39, 44, 46, 65, 95, 178, 197
   anaemia, 2, 20, 26, 35, 39, 43, **46, 161**
   daily requirements, 34
   in haemoglobin, 160, 161
   test for iron, 27
islets of Langerhans, 172, 187–188

jam setting, 9, 10
   diabetic jam, 46
   hydrolysis in jam making, 10
   making, 142–145
   mould growth, 144
   pH, 201
jelly making, 143
Joule, **29**

kidneys, 14, 44, **180–182**
   homeostasis, 182
   in diabetes, 46
   offal, **60**
   organ of excretion, 5, 180–182
   structure, 180–181
   urine composition, 182
kilocalorie, **28–32**
   diet, **43–45**
knock knees, 38
kwashiorkor, 2, 6
Krebs cycle, 176

lactalbumin, 73
lactase, 10, 74, 174
lactation, 38, 186, 189, 193
lacteals, 15, 163, 164, 173, 174, 175
lactoglobulin, 73
lacto-vegetarians (see also vegetarians), 43
lactose (milk sugar), 8, 10, 74,

174, 200
lamb cuts, 57
lard, 13, 14, 201
larynx, 165
lecithin, 14, 15, 16, 71, 72, 74, 117, 204
liaisons (thickening agents), 107
linking bonds,
   effect of heat, 4
linolenic fatty acid, *13,* 14, 20
lipase, 12, 15, 74
lipids (or fats), 1, **11–17,** 51, 65, 74, 78, 79, 82–86, 94
   action of bile on, **173**
   composition, 12, 82, 174
   digestion, 15, 173–175
   effect of heat, 15
   emulsions (see also emulsions), 15
   energy value, 11, 12, **29,** 82
   from glucose, 10
   functions, 14
   heart disease, 40
   in meat, 51, 53
   plant and animal sources, 14, 82
   refining, 16
   requirements in
   requirements in diet, 14, 35
   tests for, 16–17
   utilisation, 175
lipoproteins, 4
liver, 14, 25, 39, **176–178**
   deamination, 5, 6, 175, 178
   detoxication, 178
   functions, 177–178
   homeostasis, 178
   offal, **60**
   structure, 177
lungs, 154, **164–168**
   composition of air, 168
   structure, 166
lymph system, 15, 28, 131, 161, **163–164**
   functions, 163–164
   lymphocytes, 161
   lymph nodes, 161, 163, 164
lysosomes, 153

Maillard reaction, 6
malnutrition, 2, 6
malt, 3
maltase, 10, 168
maltose, 8, 9, 10, 168
marasmus, 2, 6
margarine, 13, 16, **83–85,** 114, 201
marinating, 102
marine (fish) oils, 14, 16, 41, 83
matter,
   properties of, 202
   solids, liquids, gases, 202
mayonnaise, 16, 72, 204
meal planning, **47–50,** 106
   appearance of food, 47, 50
   availability, 47
   economy and seasonality, 47, 48, 50, 68, 136
   flavourings, 207
   texture, 47–48
meat, **51–62**
   composition, 52

cuts, 56–60
dietetic value, 53
frozen, 151
processing, 54
production in Ireland, 53–54
products, 60
smoked meats, 54
medulla,
of adrenal gland, 187, 189
of kidney, 180, 181
medulla oblongata, 182, 183
melting point, 202
menopause, 192
menstruation, 25, 38, 45, 46, **192**
menu planning, 35–50
metabolism, 3, 19, 20, 24, 25, 168, **176**
basal metabolic rate (BMR), **29**, 187
metabolic rate, 27, 30, 39, 51
methionine, 3, 61
methylene blue test, 75
methyl group, 13
microbiology, **119–136**
micro-organisms, 110, 113, 117, **119–136**, 137, 200
rancidity in lipids, 16
temperature, 75
milk, 40, 73–**81**, 200, 204
breast and bottlefeeding, 36–37, 73
casein (conjugated protein), 4
coagulation, 6
cow and human (composition), 73
digestion, 74
globular protein, 4
in diet, 38, 41, 74
lactogenic hormone (LTH), 186, 193
pH, 201
processing, 75–76
milk sugar (see lactose), 8, 174, 200
Millons test, 6
mineral oils, 11
minerals, 2, **24–28**, 33, 51, 65, 74, 97, 99, 114, 187
deficiency diseases, 2, 24–27, 187, 189
photosynthesis, 7
tests for, 27–28
trace elements, 24
mitochondria, 153, 176
mitosis, 154
mixed feeding in infants, 37
mixtures, **197**
molasses, 92
molecules, **197**
molluscs, 63–65
monomers, 199
monosaccharides, 8, 91
monosodium glutamate, 108, **116–117**
monounsaturated fatty acids, 14
mould, 117, 119, 121–122, 137
growth on jam, 142
Mucor, 120–121
mucus, 154, 172, 192
multiplesclerosis, 42
muscle cell, 152, 154
action of stomach muscles, 171–172
types of muscle tissue, 155
mushrooms, 121–123
spore print, 122
mutton cuts, 57
mycelium, 120
myosin, 51, 53
myristic acid, 13

NAD (nicotinamide adenine dinucleotide), 170
nerve cells, 152, 154, **183–185**
in skin, 179
motor, 183–185
sensory, 183–185
nerve fibres (see nerve cells)
nervous system, **182–186**
autonomic nervous system, 184
brain, 182–183
CNS, 182–184
nerve cells, 152, 154, 184–185
nerve impulses, 185
periferal nervous system, 184
reflex, 184
spinal cord, 183–184
synapse, 184, 185
neurons (see nerve cells)
niacin, *23–24*
nickel, 16
nicotinic acid, 23
night blindness, 18
nitrates, 3
nitrogen, 3
conversion to urea and heat, 6
noradrenalin, 187, **189**
nucleic acid, 4
nucleoproteins, 4
nucleus, 123, 153
nutrients, 1–2, 33–35
nuts, 98

obesity, 2, 6, 9, 15, 31–32, 35, 37, 41, 43, 175, 178, 187
general causes, **44**
in children, 38
thyroid deficiency, 187
oesophagus, 9, 165, **171**
oestrogen, 115, 186, 188, 193
offal, 33, 38, 41, 44, 46, 51, 54, **60**
off flavours and odours, 16, 55, 60, 61, 67, 72, 80
taste and smell, 170
oils, 11, 12, 41, 61, 82, 83–85
unsaturated oils, 201
old people, 39
oleic acid, 13, 74, 83
omelettes, 73
osmoregulation, **182**
osmosis, 92, 141, **203**
osteomalacia, **19**, 25
ova, 152, 186, 191
ovaries, 188, 191
ovarian cycle in woman, **191**
ovulation, **191**
oxidase, 22
oxidation, 86, 102, 117, 128, 136, 153, **201**
oxygen, 2, 3
atomic structure, 196
carbohydrates, 7
composition in air, 168
exchange in lungs, 165
photosynthesis, 7
rancidity, 14, 16
transport, 162
oxyhaemoglobin, 160
oxyns, 16

palmitic acid, 13, 82, 200
pancreas, 6, 46, 60, **172–173**
endocrine function, 187–188
pancreatic hormone (see insulin), 3
pancreatic juice, 6, 10, 15, 74, **172–173**
parasites, 55, 60, 68, 100, 120, 126
pasta, 90, 103
Pasteur, Louis, 75, 119
pasteurisation, 74, 75, 76, 80, 119
pastry, 112–113
pastry making, 16
pâté, 60, 63, 206
pathogenic organisms, 53, 55, 60, 74, 75, 76, 100, 119–136
pectin, 8, 9, 11, 15, 16, 99, 117, 142
pellagra, **24**
pelvis,
of kidney, 180, 181
Penicillium, 80, 121
pepsin, 6, 74, 172
test effect on protein, 7
pepsinogen, 172
peptides, 6
peptide link, 3
peptones, 6, 172
periferal nervous system, **184**
peristalsis, peristaltic movement, 9, 94, 155, 168, 171
pernicious anaemia, 24, 160
perspiration (see also sweat), 28, 29
phosphate group, 4
radical, 14
phospholipids, 14, 15
phosphoproteins (see also conjugated proteins), 4
phosphorus, 3, 19, 24, **25**, 51, 65, 196
photosynthesis, 7, 92, 119
pH scale, **200–201**
effect of pH on enzymes, 169
physical change (see also matter), 197
Phytophtora infestans, 121
pickling, 95
pigments,
chlorophyll, 7, 116, 153
haemoglobin, 3, 4, 25, 39, 60, 160
rhodopsin, 18
pituitary gland, **186–187**
anterior lobe, 186
posterior lobe, 186
placenta, **192–193**
plasticity of lipids, 16
platelets, 161
pleura, 166
poaching, 103
poached foods, 39, 40

pollution (see air pollution)
polymers, 198, 199
polypeptide chains, 3
polysaccharides (see also carbohydrates), 8, 99, 122
polyunsaturated fatty acids, 14, 40–41
pork cuts, 58
portal vein, 6, 10
potassium, 27, 65
potatoes, 8
blight, 121
cooking, 10
protein source, 4, 5
structure, **9**
vitamin C content, 22
potential energy, 43
poultry, 51, 55, **62**
frozen, 151
poverty,
effect on diet, 42
pregnancy, **192–194**
balanced diet, 1, 44
drug danger, 193
energy requirements, 30, **31**, 38–39
German measles, 193
milk, 74
mineral needs, 1, 24–28, 53
protein needs, 5, 53, 78
vitamin needs, 17–24, 53
preservation methods, **136–151**
chemical, 141–142
dehydration, 138–140
freezing, 140–141, 148–151
heat treatment, 137–138
home preserving, 142–151
preservatives, 117
smoking meats, 54
vegetables, 95–96
pressure cookers and pressure cooking, 23, 103, **104–106**,
boiling point, 202
in jam making, 145
progesterone, 14, 186, 188, 192–193
proteins, 1, **3–7**, 51, 65, 74, 78, 94
amino acids, 3–7
conjugated, 4, 5
deamination, 5, 35, 175, 178
digestion, 6, 174–175
effect of heat, 4, 6
energy value, 3, 4–5, **29**
fibrous, 4, 5, 6, 154
functions, 5
globular, 4, 5
high or low protein diets, 40
manufacture of TVP, 62
plant and animal sources, 4, 5, 33, 86
requirements in diet, 5, 33–35
tests for, 6–7
utilisation, 175–176
proteolytic enzymes, 53, 172
prothrombin, 20
protons and neutrons, 196
provitamin, 18
ptyalin (see saliva)
puberty, 31, 45
sexual changes, 165, 188, 190, 192

pulse, 159
pulse vegetables, 33, 38, 43, 50, 94, 97
pyridoxine (B₆) (see also vitamins), 24

R (hydrocarbon groups) in proteins and amino acids, 3
radiation, 142
raising agent, 76, **109–111**
rancidity, 14, **16**, 82, 86, 117, 118, 128, 201
   hydrolytic, 16
   off flavours, 16, 118
   oxidative, 16, 19, 86
   oxyns, 16
   recontamination, 136
red blood cells, 160
   destruction in liver, 178
reducing sugars, 10
reduction, **201**
refining of oils and fats, 16
reflex, **184**
refuse disposal, **134–136**
regularity of eating, 37, 40, 46, 48
rendering, 16, 82
rennet, 80
rennin, 6, 73, 172
reproduction,
   in humans, 190–194
   in *Mucor*, 120
   in *Rhizopus*, 120–121
residual air, 165
respiration, 164, **201**
retina, 18
retinol (see vitamin A)
rhesus factor, 162
rhodopsin, 18
riboflavin (B₂) (see also vitamins), **23**, 34, 71
ribosomes, 124, 153
rice, 42, 90, 103
   structure, **9**
rickets, **19**, 25, 35
roasting, 100–101
roller drying, 140
roughage, 1, 9, 35, 38, 39, 40, **41–42**, 90, 95, 97, 99
roux, 107, 108
rubbing in,
   of fats, plasticity, 16, 85
rusts, 121

Saccharomyces cerevisiae (see under yeast), 119
salads, **97–98**
saliva, 9, 168, **170–171**
   test to show action, 171
salivary glands, 170–171
salmonella poisoning, 55, 129, 132
salt (see also sodium), 54
   as a preservative, 141
   loss via skin, 179
salt links, 3
salts,
   chemical, **200**
saponification, 16
saprophytes, 120, 121, 126
saturated fatty acids, 13, 16, 17, 41, 71, 82, 83, 84
   test for, 17

sauces, 15, **108**
sausages, 54, 102, 130, 132
Scientific Committee for Food of the EEC, 118
scurvy, **20**, 35, 99
seasonality of foods, 47, 68, 136
sebaceous glands, 179
secretin, 188
semen, 192
semi-permeable membrane, 203
serum, 161
sewage, 121, 135
sex hormones, 14, 186, 187, 189
sexual intercourse, 192
   veneral diseases, 193
shellfish, 41, 63–65, 70, 132
skimmed milk, 74
skin, **178–180**, 202
   as an organ of excretion, 178
   structure, 179
small intestine, 6, 15
smoke point, 15, 102
smoking, 41, 46, 85, 117, 142
   effect on lungs, 166
   fish, 67–68, 70
   meats, 54
   pregnancy, 193
smuts, 121
soapmaking, 16
sodium, **26**, 65, 196
soft cheese, 78–81
solute, 203
solutions, **203–204**
solvents, 11, 116, 199, 203
soufflés, 73
soups, and stocks, **106–108**
souring bacteria, 76
soya bean (see also synthetic meat), 6, 41, 43, 61, 96
   biological value, **5**
sperm, 188, 191
spices and herbs, 117, **208**
spinal cord, 154, **183–184**
spleen, 25, 160, 161, **164**
spoilage of food, 16, 119–134
   by micro-organisms, **128–131**
sporangium, 120
spores, 120
   endospore formation in bacteria, 126
spray drying, 140
stabilisers, 16
starch (see also carbohydrates), 8–11, 15, 16, 42, 86, 168–176
steamed foods and steaming, 39, 40, 69, 104
stearic acid, 13, *13,* 82, 200
sterilisation (of milk), 76
steroids (or sterols), 14, 187
stewed foods and stewing, 39, 40, 53, 69, 102, **103,** 151
stomach, **171–172**
   endocrine function, 188
storage effect,
   on vitamin C, 22
stress, 41, 45, 46, 85
subcutaneous fat (see also adipose tissue), 179
sublimation, 139, 202
sucrase, 10
sucrose, 8, 10, 197

suet, 13, 14, 17, 82
sugars, 8–11, **91–93**
   as a preservative, 141
   hydrolysis, 10
   test for reducing sugars, 10, 11, 12
sulphur, 3, 196
sunlight,
   effect on vitamins, 23
   photosynthesis, 7
   vitamin D formation, 19
superglycerinated fats, 85
surface tension, 203
sweat, 179, 180, 182, 202
   gland, 179
   pore, 179
sweetbreads, 60
sweeteners, 93, 116, 118
synapse, 184, **185**
synthetic meat (see textured vegetable protein)
syphilis, 131, 193
syrup, 10, 11, **93**
   in bottling, 146

tannin, 116
taste buds, 170
teeth, 19, 24–25
temperature, 4, 11
   control by blood, 162
   flash point, 15
   for pasteurisation, 75
   for roasting, 100
   maintainance in body, 29, 178, 179
   of caramelisation, 93
   on carbohydrates, 10
   on cheese, 80
   on enzymes, 169
   on lipids, 13, 15
   on milk, 76
   on vitamin C, 21–22
   smoke point, 15
   solubility, 203
   to destroy pathogens, 130
   to keep meat, 55
   to store fish, 68
tenderising, 53, 102
tendons, 154
testes, 188, 191
testosterone, 14, 188, 191
textured vegetable protein (TVP), 6, 43, 61, 62, 114, 115
textured soya protein (TSP), 61
thawing, 151
thiamine (vitamin B₁), 2, **23,** 34, 71, 91
thickeners, 117
thrombin, 161
thyroid gland, 27, 186, **187**
thyroxine, 27, 186, 187
tidal volume, 165
tissues,
   types of animal tissue, 154–155
tocopherols (see vitamin E)
tongue and taste, **170**
toxaemia, 39
toxins, 126, 127, 128, 131, 162
   toxic food poisoning, 132–133
trace elements, 24
trachea, 165

triglyceride, 12, 16
trihydric alcohol, 12
tripe, 51
trypsin, 6
tryptophan, 23, 61, 86
TSP (see textured soya protein)
tuberculosis, 74, 119, 125, 131, 168
TVP (see textured vegetable protein)
typhoid, 74, 132

ulcers, 40, **46,** 47, 172
ultra-violet light, 23
umbilical cord, 192–193
UNESCO (United Nations Educational, Scientific and Cultural Organisation), 32
UNICEF (United Nations Children's Fund), 32
United States Food and Drugs Administration, 118
unsaturated fatty acids, 13, 16, 35, 40, 83, 84
   test for, 17
urea, 5, 160, 178, 199
ureter, 180, 181
urine,
   composition, 182
UTH (Ultra Heat Treatment), 76
UHT (ultra-high temperature), 138

vaccines, 133
vacuoles, 123, 154
vacuum packing, 54
varicose veins, 39
veal cuts, 56
Vegans, 5, 43
vegetables, **93–96,** 97, 103
   bottling, 147–148
   frozen, 151
vegetable fats, 83
vegetable oils, 16
vegetarians, 24, 40, 41, **42–43,** 53, 61, 98, 111
veins, 159–160
   structure, 159
venae cavae, 156, 177
venereal diseases, 193
villi, 6, 74, 173, **174–175**
vinegar, as a preservative, 141
viruses, 119, **127–128**
   in pregnancy, 193
vital capacity, 165
vitamins, **17–24,** 51, 65, 74, 78, 84, 85, 86, 88, 93–94, 97, 114, 127, 170
   A, 14, **18–19,** 34, 65, 84
   avoiding loss, 21, 22, **96**
   B₁ (see thiamine)
   B complex, 6, **22–24,** 34, 39, 43, 170, 88, 127, 175
   B₁₂, 43, 160, 161, 175
   C (ascorbic acid), **20–22,** 34, 99, 111
   co-enzyme, 170
   daily requirements, 34–40
   D (cholcalciferol), 14, **19,** 24, 34, 65, 84, **179**
   E, 14, **19**
   F, **19–20**

219

fat soluble, 14, **18–20**, 65, 84
in cheese, 78
in manufacture of haemoglobin, 160, 161
in milk, 74
in protein metabolism, 6
in vegetables, 93–94
K, 14, **20**, 175
tests for, in foods, 22
vitamin deficiency diseases, 18–24

water, 2, 11, 12, **28**, 51, 65, 74, 95, 99, 201
balance, 26
borne diseases, 125–134
formation of sugars, 8
loss in man, 182
photosynthesis, 7
smoke point, 15
surface tension in emulsions, 15
waterglass, 72
waxes, 11

weight reducing or low kilocalorie diet, **43–45** (see also diets)
wheat, 9, 42, **86–88**, 109
whey, 74, 80, 81
white (of egg), 70, 72, 201, 204
white blood cells, **161**
white sauce, 10
WHO, 1, 18, 32, 118
wine making, 124
World Health Organisation (see WHO)

yeast, 88, 93, 99, 110–112, 119, **123–124**, 137, 199
'budding', 123
yoghurt, 76–77
yolk (of egg), 70, 71, 72, 117

Zen, 43
zerophthalmia, 2, **18**
zygospore, 120
zygote, 192
zymase, 110, 124